SUGAR

AND

YOUR HEALTH

For the first time, a book about sugar that presents the beliefs of the medical establishment as well as the viewpoints of others who care for persons with sugar-related disorders of health.

SUGAR
AND
YOUR HEALTH

Nutritional Problems, Diabetes, and Low Blood Sugar

by

Ray C. Wunderlich Jr., M.D.

Physician specializing in Preventive Medicine, health promotion, and degenerative disorders

Good Health Publications
Johnny Reads, Inc.
St. Petersburg, Florida

Hardcover ISBN 0-910812-21-7
Paperback ISBN 0-910812-22-5

Library of Congress Catalog Card Number 78-50566

Illustrated by Susan P. Prescott

Edited by Edwin R. Rodgers

Designed by Lillian Harris and Edwin R. Rodgers

Front cover designed by Sharon Manley

Indexed by Sandi Schroeder

Manufactured in the United States of America

Published by Good Health Publications, Johnny Reads, Inc.
Box 12834, St. Petersburg, Florida 33733

CONTENTS

This book is dedicated to
my firstborn,
Mary Palliser,
who grew up and away too fast.
She was and is as sweet as sugar —
but much more wholesome!

ACKNOWLEDGEMENTS

The following individuals have provided background information for this book, or have otherwise interacted with the author to foster the progression of this work. Mention in this section does not imply agreement with the author's viewpoints or the material presented in the book.

Dr. Ross Cameron (M.D.), retired, Pinellas County Health Department, St. Petersburg, Florida.

Dr. Robert Davis (D.Sc.), Bay Pines Veterans Hospital, Bay Pines, Florida.

Dr. John McCamy (M.D.), Preventive Medicine, St. Petersburg, Florida.

Dr. Robert Kilmark (M.D.), Preventive Medicine, St. Petersburg, Florida.

Dr. J. D. Phillips (M.D.), Pathology Department, Bayfront Medical Center, St. Petersburg, Florida.

Dr. Allen Root (M.D.), All Children's Hospital, St. Petersburg, Florida.

Dr. Arlan Rosenbloom (M.D.), J. Hillis Miller Health Center, Gainesville, Florida.

Dr. William Crook (M.D.), Child Health Centers of America, Inc., Jackson, Tennessee.

Mrs. Mae McBath, Nutritionist, St. Petersburg, Florida.

Mr. Warren Ankerberg, former Director of Laboratory, Suncoast Medical Clinic, St. Petersburg, Florida.

Mr. W. R. Bell (R.Ph.), Chief Pharmacist, All Children's Hospital, St. Petersburg, Florida.

Miss Rebecca Joplin, Medical Technician, All Children's Hospital, St. Petersburg, Florida.

Reverend John Apolus, Naturalism, Inc., Los Angeles, California.

Mrs. Susan Prescott, Artist, Pittsburgh, Pennsylvania.

Dr. Bruce Pacetti (D.D.S.), formerly of the Page Foundation, St. Petersburg Beach, Florida.

Mr. Louis Moore, Biobehaviorist, Spartanburg, South Carolina.

Mrs. Karen Buel, former Medical Assistant, St. Petersburg, Florida.

Mrs. Kierstine Del Sol, L.P.N., Medical Assistant, St. Petersburg, Florida.

Mrs. Luanne Topp, former Medical Assistant, St. Petersburg, Florida.

Mrs. Peggy Dougherty, former Medical Assistant, St. Petersburg, Florida.

Mr. Steven Goodman (R.Ph.), Pharmacist, Medical Arts Pharmacy, St. Petersburg, Florida.

Mr. Peter Reuter, Natural Hygiene Society of America, Inc., Clearwater, Florida.

Dr. Marion Cole (D.O.), Audiologist, Metropolitan General Hospital, Pinellas Park, Florida.

Mrs. Ruth Gyland, Tampa, Florida.

Dr. E. H. Culbertson (D.C.), Kinesiologist, St.Petersburg, Florida.

I am indebted to Dr. Ross Cameron and Mr. Louis Moore for time and effort expended in reviewing an early phase of the manuscript, and for their suggestions. Mrs. Janet Snipes, (deceased), former Medical Librarian, Bayfront Medical Center, St. Petersburg, Florida, assisted in locating important reference material.

WHAT'S IN THIS BOOK

This book has been written so it will be of use to many persons. Physicians, technologists, nurses, and other professionals, as well as interested lay persons and students, will find down-to-earth explanations, tables, figures, and diagrams about sugar and its relationship to disease and health.

Many readers will want to start at page 1 and read straight through this book to the very end. Some readers will want to read only about certain aspects of sugar and health that appeal to them. Other readers will want to read the entire book and thereafter review certain chapters which particularly interest them, or which provide information that they need. Students may use this book as a source of information and study.

A rather extensive glossary is available in the rear of the book. It should be consulted to clarify the meaning of any words or phrases that are not defined in the text.

The first five chapters do not contain any consistently technical material. In order to explain carbohydrate metabolism in some detail, however, Chapters 6 and 8 have been written for readers with some scientific background. Chapter 6 is oriented toward basic physiology; Chapter 8 deals extensively with clinical material as well as physiology. The remaining chapters are less technical. Although Chapters 6 and 8 are more technical than the rest of the book, the person who wishes to learn and who seeks challenge will find considerable "meat" therein.

Explanatory illustrations throughout the book assist in interpreting concepts put forth in the text.

In order to assist all readers, this section has been prepared as a description guide to the contents of this book. A brief abstract of each chapter follows:

Chapter 1, "Sickness from Sugar," introduces the reader to some of the clinical problems that are associated with eating sugar. Some eating habits of Adolf Hitler are discussed. The amount of sugar consumed by the average person is given, and hidden sources of sugar in the diet are pointed out. Sugar craving is explained. The content of sugar in smoking tobacco is given. The various forms of sugar are discussed. The desirability of using honey and molasses in the diet is examined. The relationship of sugar to health disorders and dental problems is pursued. Allergy to sugar and natural versus refined sugar are discussed. The origin of nearsightedness (myopia) is examined and related to nutritional factors. The very important effect of sugar in the pregnant mother's diet is reviewed. Finally, the disrupting effect of sugar on the body's chemical equilibrium is pointed out.

Chapter 2, "Some Definitions," tells the reader what sugar is. The types of carbohydrates are defined. Chemical and structural formulas are given.

Chapter 3, "Sugar in the Body," deals with the metabolism of sugar (glucose) within the body. The vital needs of a person for glucose are stressed. The input of glucose to the blood and the output of glucose from the blood, as well as the storage facility for glucose in the body, are described. The various effects from eating starch or sugar are pointed out.

Chapter 4, "Carbohydrate: From Mouth to Liver," takes us on a journey from the plant kingdom through the human digestive tract to the liver, a major organ having to do with the metabolism of sugar in the body. Digestion and absorption of carbohydrate is discussed in some detail. Maldigestion of starch is noted as a not infrequent clinical disorder.

Chapter 5, "Insulin Enters The Picture," tells the reader about this important anabolic (tissue-building) hormone produced by the pancreas. The production, release, and actions of insulin are discussed.

Chapter 6, "The Fate of Glucose," describes the chemical pathways of glucose breakdown in the body for the purpose of obtaining energy. Becoming a bit more biochemical, this chapter deals with these processes: glycolysis, the Kreb's Cycle, glycogenesis, glycogenolysis, and gluconeogenesis. The importance of vitamins and minerals to these functions is pointed out. The role of chemical balance and imbalance at the gate to the Kreb's Cycle, as well as in the body as a whole, is emphasized.

Chapter 7, "Reflection," is a pause. This "seventh inning stretch" briefly reviews from the beginning what we know about carbohydrates in the body. This chapter also places the human body in perspective as a being of soul and worth as well as a complex chemical entity.

Chapter 8, "Interplay of Hormones: A Delicate Balance," provides the reader with awareness of the delicate state of hormonal balance that occurs from moment to moment in the human body. Hormonal "fine tuning" is accomplished by interlocking relationships which are described in this chapter. An important diagram is provided which shows the overall picture of hormonal relationships in the body as they are known. Estrogen therapy is discussed. Adrenocortical insufficiency is described and separated into the severe pathological condition (Addison's disease) and a less severe functional variety. Information about a number of clinical cases is given. The use of adrenocortical extract as a therapeutic agent is discussed.

Chapter 9, "The Pancreas and Insulin," elaborates on the structure and function of the pancreas, the organ which produces insulin. The vital importance of pancreatic secretions in the digestion of food is pointed out. The isles of Langerhans within the pancreas, and the beta cells within the isles are given particular attention as the source of insulin. The nature and actions of insulin are given, and a discussion of hyperinsulinism is included. Of particular interest in this chapter are figures showing lowered blood sugar after the intake of glucose or food.

Chapter 10, "Diabetes Mellitus," discusses this common disorder of metabolism that is increasingly more common in our population. The symptoms are reviewed and the diagnostic features given. Two principal types of diabetes are identified as insulin-secreting (usually maturity onset) and insulin-deficient (usually juvenile). The basic problem underlying diabetes is discussed in detail, including animal research. The complications of diabetes are noted. Sugar in the urine is explained. Three stages of diabetes are enumerated: clinical, chemical, and prediabetes. Each stage is defined and discussed. The characteristics of the infant of the diabetic mother are noted. Drugs that induce a diabetic state are identified. A section on the treatment of diabetes includes the oral hypoglycemic drugs. The effect of pregnancy on blood sugar is discussed. The chapter ends with a promising outlook for the future in view of the progress that can be made with life-style change and alteration of body chemistry. In addition, the advent of the "artificial pancreas" and pancreatic beta cell transplants is noted.

Chapter 11, "The Glucose Tolerance Test (GTT)," is devoted

to a description of this major diagnostic tool for detecting dysfunctions in carbohydrate metabolism. A glucose tolerance test (GTT) of 3 or 4 hours duration is most often used to detect diabetes, but a 6-hour test is necessary to exclude low blood sugar or combined states of diabetes and low blood sugar. The GTT evaluates a person's ability to manage a load of sugar (glucose) which has been given to him by mouth or vein. Possible hazards of the test are not overlooked. Details of administration of the GTT are described and the use of a symptom checklist suggested. Criteria for the diagnosis of diabetes are given, and variations of the GTT are discussed.

Chapter 12, "Glucose Tolerance Curves," acquaints the reader with the graphic representation of the glucose tolerance test. Glucose tolerance curves, according to Dr. J. W. Tintera (M.D.) and Dr. Hugh Powers (M.D.) are presented. A number of the author's case studies are presented with a glucose tolerance curve for each.

Chapter 13, "The Natural Tolerance Test (NTT)," describes this procedure for investigating glucose equilibrium in the body following the ingestion of food. The profound effect of some foods on some individuals is emphasized. In the illustrative case studies, there is considerable discussion about blood-sugar levels in response to food.

Chapter 14, "Low Blood Sugar (Hypoglycemia)," is a high point of this book. The definition of low blood sugar is addressed head-on. Representative views of the academically oriented physician, the functionally oriented clinical physician, and lay individuals are set out. Absolute and functional categories of low blood sugar are given. Emphasis is given to hypoglycemia as a nonspecific manifestation of various underlying conditions. Symptoms and signs of low blood sugar are presented.

The continuum between clearly abnormal and optimal blood-sugar levels is stressed. The importance of full medical evaluation (differential diagnosis) for identification of underlying disorders is advised. Criteria for separating out functional hypoglycemia from more desirable blood-sugar patterns on the GTT are given.

A historical perspective is rendered, including the remarkable story of Dr. Stephen Gyland (M.D.) and his personal struggle with hypoglycemia. Dr. Gyland's unpublished manuscript on functional hypoglycemia has been reviewed by the author and many of Dr. Gyland's findings are presented.

A section, "Where Do We Stand?" grapples directly with the issue of academic medicine versus the lay health movement. The health food store as a focal spot for dissemination of information is discussed. A plea is made for cooperative study, flexible viewpoints, and preventive medicine.

Fasting blood-sugar levels and blood-insulin values are discussed. The origin of symptoms in low blood sugar is reviewed. A large section, "Hyperinsulinism," describes absolute and relative types. The considerable incidence of hypoglycemia in the American population as well as in various clinical conditions is noted. Those conditions include alcoholism, neuropsychiatric disorders, juvenile delinquency, and asthma. An association of functional hypoglycemia with vicious behavior, irritable children, and hyperactive children is suggested.

A correlation of trace-mineral imbalance with low blood sugar is pointed out. Patterns of trace minerals (as measured in hair) that are indicative of hypoglycemia are given.

The enormously high intake of refined sugar and refined carbohydrates is noted as a major factor in the origin of functional hypoglycemia.

Several classifications of low blood sugar are given. Functional and organic varieties are noted in addition to fasting, fed, and stimulative types. Pseudohypoglycemia, the lowering of blood-sugar levels after blood is drawn from the patient, is discussed.

Treatment of hypoglycemia is set out. The need of the acutely ill patient for sugar is noted; and in the chronic condition, the need to avoid sugar is stressed. The important matter of protein in the diet is addressed. Other nutritional matters and hormone therapy is discussed.

The chapter is closed with the presentation of a number of case studies with GTT's (glucose tolerance tests) and NTT's (natural tolerance tests) illustrating hypoglycemia.

Chapter 15, "What Can Be Done?" points out the improvement that can be obtained in individuals with chemical diabetes. Nutritional measures that result in improved states of health for all persons are considered in detail in a section entitled, "Improving Your Diet." The relationship of sugar to learning is discussed, considerable attention is devoted to overweight as a clinical problem, and an outlook for the future is presented.

Appendix A, "Evaluation of the Six-Hour Glucose Tolerance Test," outlines for physicians and other interested persons a logical and sequential method of analyzing this diagnostic tool. Exemplary graphs of 9 basic types of glucose tolerance curves are furnished.

Appendix B, "Glucose Tolerance Test Curves," provides additional clinical material in the form of brief case studies with accompanying glucose tolerance curves. The reader who wishes to become better acquainted with glucose tolerance curves, in their wide variety, will find this material helpful. Notes on each case provide considerable clinical insight.

Appendix C, "Historical Notes Concerning Insulin: A Story of Doctors Banting and Best; A Living Tribute to Dr. Ross Cameron," is a treasure house for the individual who appreciates the whys and wherefores of matters that we take more-or-less for granted today. Some of the hard work and dedicated investigation of past medical pioneers come alive in this section, which includes photographs of great historic interest. The work of Dr. Banting and Charles Best, the co-discoverers of insulin, is described through the eyes of Dr. Ross Cameron (M.D.), who was a medical student at the time. Appendix C ends with a tribute to Dr. Cameron.

Appendix D, "More on the Dangers of Sucrose," is a succinct statement from a medical publication indicating present concern over the dangers of sugar consumption in affluent societies.

For simplicity, throughout this book I will use masculine pronouns to refer to persons of either sex.

CHAPTER 1

SICKNESS
FROM SUGAR

The small boy lay in his mother's arms wheezing and gasping for breath, wracked by fits of coughing which turned him an ashen color and left him weak. The mother told me that Chris had frequent severe asthma attacks just like this one. He was 2 years old.

I quickly administered Adrenalin by hypo and sat with the young boy as his labored breathing eased off. Chris's look of air hunger changed to one of relaxation as the Adrenalin took effect. Exhausted from his asthmatic labors, the boy fell asleep in his mother's arms as air now coursed easily into his relaxed bronchial tubes.

Then I asked the mother about Chris's diet. She stated that he was a "good eater." The boy's overweight appearance suggested she was telling the truth — at least in one sense.

"Does he eat sugar?" I asked.

"No," she said. "He doesn't like sweets."

"How about colas, candy bars, ice cream, or cookies?"

"No, Doc, he hardly ever gets that stuff."

"Kool-Ade, Hi-C, or Jello?"

"No, we don't buy that for him."

I sat and watched the boy responding so nicely to Adrenalin, a medicine that is a God-send for the asthmatic, a medicine which dilates the bronchial tubes and has as one of its properties the ability to elevate the blood sugar.

Sitting there, silent for the moment, I thought that this was one sick kid who didn't consume a high load of sweets in his diet.

Just then his mother spoke up. "Doctor, he does drink a lot of chocolate syrup."

"Oh? Tell me about it."

"Well, he drinks it all day long. We fill a glass half full of syrup and the rest with milk. That's the only way he'll drink his milk. We go through a couple of cans of syrup a day, but it sure gets him to drink his milk!"

I looked at the boy again. Chris was coughing occasionally to clear the remaining mucus in his bronchial tubes. He was very pale and massively overweight. His tissues had a waterlogged appearance.

Once again, sugar had not failed me. The kid that was sick was also the one who habitually ate a large amount of sugar, in one form or another, in his daily diet.

I instructed Chris's mother right then and there to withdraw all sugar and sugar-containing foods from his diet. I told her that Chris would lose weight and be a healthier boy.

The mother followed these instructions to the letter. And just as predicted, Chris lost weight, became more active, had many fewer colds than he did before, and had no further attacks of asthma.

By-and-large, sick kids *are* the ones who consume a heavy load of sugar — in one form or another — in their diets. The kids who get repeated colds; nose, throat, and ear infections; bronchitis; or skin infections are very often those whose diets are heavy in sugary snack and convenience foods. In my experience, their diets are too often *sickeningly sweet!*

Our culture has rather quietly accepted this condition. Our children continue to watch charming TV commercials which encourage them to eat sweetened refined cereals which supply empty calories. A popular entertainer, Sammy Davis Jr., made a bestselling record, "The Candy Man," extolling the virtues of candy — "satisfying and delicious." Underwear made out of candy actually exists. Kojak, a well-known TV detective, presents a lollipop-sucking model for his viewers.

Television's Lieutenant Columbo is admired for his unerring capacity to outwit others. He takes 3 lumps of sugar in his tea! There's a subtle message there that we can be like him if we take sugar, too.

This Florida anecdote indicates how addicting sweets can be. A young boy was retrieving golf balls from a pond near a golf course. As he scooped a golf ball from the edge of the pond with a net, an alligator grasped his arm and lacerated it severely. The alligator had been fed daily by persons who lived near the pond. Marshmallows were the food given to the gator. Golf balls and marshmallows are about the same color and size. Mr. Alligator liked his sweet so much that he snapped at the boy whom he thought was taking it away!

Prison officials in Tennessee followed a trail of chewing gum wrappers that led to the capture of James Earl Ray (jailed as the alleged murderer of Dr. Martin Luther King Jr.) and his fellow prison escapees. Chewing gum contains sugar unless certified sugar-free.

Candy wrappers were found littering the spot from which Lee Harvey Oswald presumably shot President Kennedy. This raises the question of whether Oswald was a candy addict and possibly whether he had a sugar-related illness such as low blood sugar, or violent behavior associated with sugar "allergy," or nutritional imbalance.

There once was a dictator who ended his life by suicide. Before he died, he was responsible for World War II and unparalleled amounts of human suffering. This man, Adolf Hitler, was a confirmed sweet addict. Albert Speer, Hitler's architect and war-time Minister of Armaments,[1] gives this description in a personal communication:

"It is definite that Hitler devoured sweets of all kinds in an almost greedy way. He himself occasionally laughed at this passion. He took alcohol on no condition, but it is quite correct that one could call him sugar-drunk (zucker-trunken). He had an intractable appetite (lust) for sweets; and although he wanted to refrain from them again and again, he was unable to do so."

What role did sugar play in determining the mental and physical state of this powerful man? We shall never know for sure.

[1] Albert Speer is the author of *Inside the Third Reich, Memoirs,* published in 1970 by The Macmillan Company, 866 Third Avenue, New York, N.Y. 10022.

When I see some hyperaggressive children change into placid kids when sugar is removed from their diets, I wonder about the relation of sweets and Hitler's inhuman acts.

Certainly Albert Speer's lucid description of Hitler's sugar-cravings remind us of ourselves, for most of us in modern society are caught up in the habit of wanting something sweet and habitually indulging this want.

Consumption of sugar by persons in our society is a character-rooted problem, a culturally derived habit of which many persons are mostly unaware.

Some brief case studies will now be presented. They can make us more aware of the sweet condition around us and the effects of sugar on some individuals.

John had been treated with antibiotics once a month for several years. Now in third grade, he had missed 30 to 45 days of school each year due to illness. His mother wished to know if I could advise any treatment for his repeated nose, throat, and sinus infections.

John was a cola-holic. He also consumed several candy bars each day and never ate breakfast.

I informed the mother that John was nutritionally deficient and that he suffered from "candy catarrh." He was given no medications, but his diet was successfully "cleaned up" by elimination of sugar (sucrose) in all forms.

John's school attendance steadily improved. His repeated infections disappeared, and he required no antibiotics thereafter.

Virginia was a slow learner in the third grade. Her attention span was very short, and she was very hyperactive. It was "impossible" to keep her in her seat in school. Elimination of dietary sugar was followed by a marked reduction in hyperactivity and a marked increase in attention span. Consequently, Virginia had no difficulty staying in her seat in class when it was appropriate to do so. Whenever she "slipped" and ate sugar, she became hyperactive, irritable, and "couldn't think." Virginia's teachers were able to tell, by her diminished classroom performance and her out-of-seat behavior, whenever she had eaten sugar.

A 2-year-old child died a few hours after drinking a bottle of pancake syrup. Laboratory tests confirmed the cause of death as extreme high blood sugar and an acid condition of the blood brought on by drinking the syrup.

The boy had undergone surgical repair of a lye-damaged esophagus four months earlier, and a segment of his colon had replaced the damaged esophagus. The doctors believed that the colon facilitated absorption of sugar from the syrup and prevented vomiting of the sweet mass of fluid.[2]

Bob, a 3-year-old son of a nurse, was seen because of aggressive, destructive behavior with short attention span and hyperactivity. He rattled doors, kicked walls, struck out at other children, and failed to respond to discipline at home and at nursery school. Bob also had frequent bouts of coughing, wheezing, and middle ear infections.

The mother was instructed to remove all sugar (sucrose) from the boy's diet. Although she believed this to be an unusual treatment and completely unnecessary, she was persuaded to go along with it.

A few weeks later the mother returned, stating that Bob was "a different child." His attention span was markedly improved, he now responded to discipline at home and at school, and he was much kinder to other children around him. In addition, as the child had changed, the mother noted that she herself had improved in her attitude toward the child.

The boy's mother further stated that on several occasions Bob had obtained sugary foods from neighbors. On each of these occasions there was a relapse to aggressive, destructive, hyperactive behavior. Whenever he ate vanilla ice cream, yellow cake, Hershey's Kisses, or peppermints, there followed a one-half to one day period of irritable behavior. Moreover, the boy's irritability lasted longer when the intake of sugary food was greater in amount.

In order to find out if sugar was actually the culprit, I had the mother give Bob a large spoonful of table sugar by itself. A rather hectic day followed, with the typical striking-out, hyperactive, poorly controlled behavior on Bob's part. Once again, sugar had done its thing.

[2]Case reported in *Journal of the American Medical Association*, Vol. 22, No. 7, May 13, 1974, page 817.

Ralph was a 63-year-old man with chronic inflammation of the prostate gland. Despite the efforts of urologists and general physicians, he continued to have pus in his urine. Multiple antibiotics had not altered the prostate swelling and urinary pus. He was considered to have a chronic condition that he would just have to live with. When Ralph gave up the use of honey in his diet, the prostate swelling subsided, and his urine became free of pus and remained so.

A new patient came to town and was admitted to the hospital with bronchopneumonia. As the young infant screamed irritably in his oxygen tent, his mother prepared successive bottles of cola, liquid jello, and sugar water for him to drink. The mother turned to me and asked if I knew any reason why the infant had been ill all winter and couldn't get well despite multiple antibiotics. I explained that there were many reasons for recurrent illness but a likely probability was that the illnesses were sweet-related.

The parents had never entertained the possibility that diet and nutritional status might be an important factor in the infant's health. They were glad to be informed about dietary improvements, including the elimination of sugar. The infant was fed fresh, whole foods that did not contain added sweets, and nutritional supplements were given.

No further illnesses occurred, hence the lad required no further antibiotics. As a spin-off benefit, Mom lost considerable poundage, and Dad's disposition improved as the family's dietary habits underwent considerable change.

A 20-year-old man was the son of a prominent family in town. He had been a school dropout from the age of 13, when he had suffered the first of many bouts of recurrent psychotic mental illness. He had been hospitalized many times and had been under the care of several different psychiatrists. His diagnosis was paranoid schizophrenia. His family noted that he had immediate outbursts of violent behavior after eating sweets such as a candy bar, cake, or cola drinks.

When he was placed on a diet free of sugar, other refined carbohydrates, and food additives, he had no further mental illness. Thereafter, he was able to successfully use his energies in positive channels for rebuilding his life.

Most of these individuals were or could have been treated with medications to alter their symptoms and behavior. Undoubtedly, some successful results would have been obtained. Certainly, too, psychological counseling for individuals and families is helpful. My book, *Allergy, Brains, and Children Coping*,[3] indicates the great importance of allergy treatment in many clinical disorders.

No one approach has all the answers, but each approach is likely to have some answers. Wise are the patient and therapist who are aware of many avenues of help. Wiser still are the patient and therapist who travel nutritional avenues toward health. The first construction block in the nutritional avenue is the avoidance of sugar.

SWEET AND DANGEROUS

Sugar in your diet — or in the diet of your parents and grandparents — may be the single most important nutritional factor having to do with how you feel and how you function. Sugar is truly an extraordinary substance that most of us have taken for granted for too long. At the turn of the century, the average person ate 10 to 20 pounds of sugar per year. The present per capita consumption of sugar in the United States is about 120 pounds per year. This is equivalent to an average of 30 teaspoons of sugar per day for every U.S. resident! Most persons now eat their weight in sugar (or sugar equivalent from other refined carbohydrate) every year. It is no honor that the United States leads the world in the production and consumption of candy.

The publication of Dr. John Yudkin's book, *Sweet and Dangerous*,[4] in 1972 gave the public at large reliable information about the health hazards of sugar.

I believe that Dr. Yudkin's observations are basically correct and that his work will eventually be looked upon as a landmark in investigational research.

[3]Published in 1975 by Johnny Reads, Inc., Box 12834, St. Petersburg, Fla. 33733.

[4]*Sweet and Dangerous* by John Yudkin, M.D. Published in 1972 by Peter H. Wyden, Inc., New York, N.Y.

Although refined sugar (sucrose) is not the cause of all health problems, it comes closer to qualifying for that position than most other single factors. Again and again children change from being sickly to healthy when sugar is regularly omitted from their diets. The fact that excessive sugar in the diet may interfere with the ability of the white blood cells to engulf and kill bacteria probably has much to do with the frequent infections that plague many sugar-eaters.

Sugar in the diet is intimately related with vaginal ill health. Many women have recurrent inflammation of the vagina with annoying discomfort, itching, and vaginal discharge. Normally, the vagina has an acid pH, and this acidity inhibits the growth of pathogenic bacteria. In vaginitis, the pH is often found to be alkaline rather than acid. Doctors frequently recommend a dilute vinegar douche to combat such alkalinity. Sugar in the diet is associated with an alkaline vagina, with overgrowth of harmful bacteria and fungi. The patient with diabetes mellitus who "spills" sugar in the urine is often plagued by recurrent fungal (yeast) overgrowth in the vagina. The elimination of refined sugar from the diet and the avoidance of blood-sugar elevation helps maintain an acid vagina with an absence of inflammation.

Some persons react to sugar with alterations in day by day behavior, whereas others seem not to be affected. It is estimated that perhaps some 30% of the population may show no *evident* untoward effects from the sugar they eat in their own lifetime. It is probable, however, that most persons are eventually affected adversely in regard to health and length of life. Dr. Linus Pauling (Ph.D.) includes sugar restriction in the diet as one of his key recommendations for adding 15 to 20 years to one's lifespan. In addition, a high sugar intake may well have a detrimental effect on subsequent generations. Fifty-four percent of our population ultimately experiences a heart attack. Evidence is accumulating to indicate that sugar, in addition to cholesterol, is a major factor in the origin of heart attacks. Elevation of the triglyceride blood fats is a risk factor in coronary heart disease. Triglyceride levels which are elevated commonly return to normal when a sugar-free diet is instituted.

There is considerable evidence to indicate that primitive tribes are healthier in the long run when they remain primitive tribes. The word *Eskimo* is derived from Algonquin and means "eater of raw meat." The Eskimo who has made contact with civilization no longer eats raw meat. He is exposed to a mass of sugar and other refined carbohydrates: chocolates, gumdrops, potato chips, sodas. These rapidly absorbable carbohydrates, largely empty calories and fiber-free, stimulate the production of insulin, triglyceride fat storage, growth hormone, glucocorticoids, and catecholamines. The civilized Eskimo is rapidly becoming one of us with dental decay, obesity, diabetes, gastrointestinal disorders, and cardiovascular ailments. Other studies indicate that primitive natives in Africa and Australia suffer the same fate as Eskimos when exposed to the "benefits" of civilization.

Dr. Emanuel Cheraskin (M.D.) of the University of Alabama Medical School has pointed out that our country is preoccupied with disease rather than health. Dr. Cheraskin emphasizes mistakes in living that lead to disease. Chief among these mistakes is the consumption of refined carbohydrate, of which sugar is the best example.

Dr. Cheraskin has pointed out that in 1962 - 1963, 38% of 17- to 24-year olds had one or more chronic diseases such as heart disease, cancer, rheumatoid arthritis, diabetes, or blindness. He indicates that the figure is rising by more than 1% each year. If that trend continues, Dr. Cheraskin notes that everyone will be chronically ill by the year 1997.

In our country, we have the best medical care in the world — for the intensive care patient. Crisis medicine is, indeed, excellent. Nevertheless, the life expectancy of the male in our country is going down. The mortality rate of newborns is considerably higher than in some other nations. We are spending more and more money on disease and very little on health. One very practical thing that we can do today is to cut down on our intake of refined carbohydrates, especially sugar. At the time this book was written, this would benefit one's pocketbook as well as his health.

This book is concerned with sugar itself. Nevertheless, it must be remembered that sugar may not be eaten alone but is often taken in association with an assortment of other chemicals and foods. Many sugary foods (candy bars, soft drinks, etc.) contain artificial coloring, flavoring, or other additives. Such a flood of sugary chemicals (a "chemical feast") may be responsible for altered health. Dr. Ben Feingold's book points out that artificial colors and flavors may cause hyperactivity in children. (*Why*

Your Child Is Hyperactive by Ben F. Feingold, M.D. Published in 1975 by Random House, New York, N.Y. The Feingold Association of the United States is a group dedicated to Dr. Feingold's principles. Address: 759 National Press Building, Washington, D.C. 20045.) I agree with his observations, and the thesis is being proven in scientifically acceptable formats.

The elimination of sugar from the diet may thus promote health because of the absence of sugar, but also because many other possibly injurious chemicals are simultaneously avoided. In addition, a sugar-free diet is more likely to include a wider variety of nutritious foods than a sugary diet. These foods do not flood the system with sugar that is rapidly absorbed.

The ordinary person has no need for the presence of refined sugar (sucrose) in his diet. Sugar (glucose) for body metabolism is made quite nicely by the body from other carbohydrate (complex carbohydrate), fat, and protein in the diet. Because of this, it is quite possible to get along fine day by day without *any* dietary source of sugar (sucrose). The very rare individual with an unusual metabolism or illness might require sugar as part of a specific treatment program, but the usual individual in our society would probably do well to immediately reduce or eliminate sucrose from his diet. As far as is known, any form of sucrose (raw sugar, brown sugar, etc.) is suspect. Sugar which is truly raw, however, does contain significantly more minerals than the ultrarefined product, white sugar, which is used in so many foods and which fills our sugar bowls. Perhaps the most nourishing sources of sugar are blackstrap molasses and fruit.

Many readers at this point may be saying to themselves, "If sugar is not necessary in the diet, why do I crave it so? It seems impossible for me to get along without it, because I'm so miserable when I don't have it."

The matter may go far back to the eating habits of your mother — what she ate before your birth and what she offered you and allowed you to eat after your birth. The woman who eats sweets during her pregnancy may program the developing child within her to crave sweets after birth. A fetus will more avidly swallow sweetened amniotic fluid than it will the nonsweetened variety.

Refined sugar is a beast. It tenaciously engages a human being and doesn't want to let go once it gains a foothold. One creates a need or craving for sugar by the very act of eating sugar. When the body is programmed to expect sweet in the diet every few hours, it is uncomfortable when it is not present. The blood sugar rises quickly to a high level after sugar is eaten. A sense of satiety takes place — as though a large meal had been eaten — only to be replaced in a few hours with an empty feeling and hunger — sweet craving — when the blood sugar drops from high levels to lower levels.

As the cycle continues, fat cells become overloaded with fat that is trapped and unable to be metabolized. Water may be retained. Overweight and sluggish physical activity occur, and the person feels increasingly a prisoner within his own body and to his own dietary habits.

Sugar replacement of vital nutritious food results in vitamin and mineral inadequacy. One of the minerals that often becomes deficient is zinc. Zinc deficiency may be associated with altered taste sensation. It may be difficult for the person with zinc deficiency to appreciate the taste of food. He may eat foods containing enormous quantities of sugar and only be aware of a slightly sweet taste. The same can be said for salt.

The craving of our population for sugar-coated cereals, salty potato chips, highly spiced hot dogs, and sickeningly sweet soft drinks may be closely related to a dulling of taste. The culturally induced programming with sweets may itself be responsible, as well as mineral, vitamin, and enzyme deficiencies. Refined white sugar is the essence in empty calories. Sugar provides calories but no protein, no enzymes, no vitamins, no minerals, and no fiber.

Desirable nutritious foods are deficient in the diet of the sugar-eater, because sweet is an addicting taste. Satisfying "the sweet tooth" is a pernicious habit easily learned but difficult to undo. Sugar addiction is as difficult to leave behind as cigarette or alcohol addiction. The penalties for not breaking the habit are as severe as they are for continuing addictions that interfere with a balanced life.

It may be more than coincidence that smoking and sugar addiction are two of our major problems in public health. According to William Dufty in his book, *Sugar Blues,*[5] next to food processing, the tobacco industry is the biggest sugar customer in our country.

An average of 5% sugar is added to cigarettes, up to 20% in cigars, and as much as 40% in pipe tobacco. Dufty points out that flue-cured tobacco contains as much as 20% natural sugar by weight and that sugar is added to air-cured tobacco during the blending process. Tobacco with a high sugar content produces a strongly acid smoke, whereas low-sugar tobacco yields a smoke that is weakly acid or alkaline.

Dufty cites information on British, French, American, and Oriental cigarettes and their relationship to lung cancer. High-sugar smokes are correlated with high rates of lung cancer, and low-sugar smokes are found with low rates of lung cancer. Dufty's chapter describing this is appropriately entitled "Reach for a Lucky Instead of a Sweet?"

Refined sugar is not a dietary necessity — although it *seems* to be needed in cultures that have been regularly accustomed to it. Cravings for sugar are induced. They are learned. Sweet craving is an *acquired* trait, not a physiological need. The sweet taste — from baby food to tooth paste — has been a regular part of our American heritage. In addition: as children, most of us have been rewarded for good behavior with some kind of sweet. No wonder we tend to turn to ice cream or candy when we're feeling low!

There is suspicion that the fructose component of sugar may be the real trouble-maker in the production of ill health. Whether this is so remains to be proven. Fructose has been implicated, however, as a source of triglyceride fat in the blood, and there is growing belief that elevation of triglycerides in the blood is a high-risk factor for heart and blood vessel disease. J. Daniel Palm, however, believes that fructose is "nature's tranquilizer" and the ideal "crave-control" food. He indicates that the exchange of fructose for other dietary carbohydrates maintains blood sugar and prevents the stress of hypoglycemia (low blood sugar).[6]

The total answer is not yet available, but the conclusion can

[5] Published in 1975 by Chilton Book Company, Radnor, Pa.

[6] "Benefits of Dietary Fructose in Alleviating the Human Stress Response" by J. Daniel Palm, Department of Biology, St. Olaf College, Northfield, Minnesota. Published in the American Chemical Society Symposium Series, Number 15, *Physiological Effects of Food Carbohydrates*, 1975, pages 54 - 72.

be made that refined sugar adds nothing to one's diet except calories. Frequently, a great deal of misery accompanies those calories.

Because of the high price of cane and beet sugar in our current economy, less expensive sweets are actively being sought by food producers. There is, of course, no source of income for the individual who suggests that it might be better to do without sweets. Manufacturers continue to avidly search for inexpensive ways to satisfy the sweet tooth of the citizens of our culture. An extra-sweet variety of corn syrup has been promoted to replace sugar. It is a high-fructose product. It seems unlikely that this material would be any more healthful than sugar itself, and it is distinctly possible that it may eventually be shown to be more hazardous to health.

Recent controversy about saccharine centers on whether it should be banned as a cancer-producing substance. The point has simply been missed. Saccharine — or any other sugar substitute — should not be used at all. Saccharine programs the taste for sweet, and as such is generally undesirable. Its use delays progress in breaking sweet addiction.

What about the use of sugar in the treatment of the sick? For many years sugar in a sterile watery solution has been given by intravenous drip to countless numbers of sick persons in hospitals. The use of intravenous sugar solutions, often in combination with salt and other electrolytic chemicals, has been a major contribution to successful anesthesia, surgery, and medical treatment of serious disease.

The sugar used in intravenous solutions is nearly always dextrose (glucose). It supplies a source of calories for the patient who is often unable to eat. In its watery solution it also provides a vehicle for the administration of other chemicals that may need to be given by vein.

Dextrose (glucose) given intravenously has a good "track record." When properly used, it has seldom if ever been associated with adverse effects. Dextrose derived from corn sources might theoretically be responsible for unwanted symptoms in patients who are sensitive to corn products. However, I have never been aware of such a happening.

The apparent lack of harmful effects from intravenous dextrose could indicate that reasonable amounts of dextrose in the diet may not be particularly harmful. Dextrose, a simple sugar, may be quite a different "beast" in its effects upon the body than the more complex sugar known as sucrose. Unfortunately, it is sucrose that abounds in the dietary habits of our culture.

At the time this book was written, modern hospitals, along with all the good they do, are most often serving their patients food that is high in sucrose content. Unfortunately the general diet in a hospital, unless specifically ordered otherwise by the physician, suffers from the same "sweet tooth" that has a strangle-hold on the population at large.

HONEY AND MOLASSES

Honey is manufactured by honeybees from the nectar of flowers. It has been used in folk medicine from ancient times. Honey has been recommended as an antiseptic, cathartic, digestive aid, tonic, sedative, blood builder, cough suppressant, and mucolytic substance. It was used for its sweetening properties long before sugar was known. Honey enjoys a considerable reputation among athletes as a substance to restore energy after exertion or to boost energy prior to activity.

Honey exists because bees (like humans) are enamored of sweets. Bees do not collect nectar unless its sugar content is above 15%. At times when no flowers are in bloom, bees from different colonies will fight over honey and rob one another of the sweet.

Bees perform a valuable service in gathering nectar by spreading pollen from flower to flower. This is presumably nature's reason for the existence of honey.

The nectar of flowers is the raw material for honey. What difference is there between flower nectar and the honey of bees? What part does the bee play in effecting the change of nectar to honey?

As far as science is aware, nectar and honey are remarkably the same. The type of flower from which honey is derived is closely and directly related to the honey formed from its nectar. Dr. Marion Cole (D.O.) indicates that bees fed on sugar instead of nectar produce a "honey" that is made of sugar.

Apparently, the bee is more of a mechanical carrier than a metabolic or chemical factory in the production of honey. The bee hunts, finds, and gathers nectar and then returns it to the hive. The nectar is then concentrated by the evaporation of water therefrom.

Nectar is a watery solution of fructose, glucose, and sucrose, with the water content being 50 to 80%. Bees concentrate it to make honey with a water concentration of 16 to 18%. The vitamins and minerals in honey are present initially in the nectar from the plant. The minerals in nectar are all derived from the soil in which plants grow. Obviously, therefore, honey will vary according to the mineral resources in the soil.

Nectar is ripened into honey by the removal of moisture, but another change also takes place. The major portion of the sucrose sugar in nectar is inverted (altered) to its constituent sugars: fructose (levulose) and glucose (dextrose). This takes place in the bee's body under the influence of enzymes in the saliva or in the worker bee's honey "bag" or stomach. Starches and sucrose in the diet of humans must undergo a similar inversion (digestion) in the human gut to convert complex carbohydrate into simple sugars for absorption (see Chapters 3 and 4). In honey, this digestion has already taken place within the bee. Honey is, thus, a *predigested* liquid source of carbohydrate for man.

A crucial matter is whether or not it is important for persons to consume a predigested carbohydrate food. Clinical experience with readily absorbed predigested liquid protein suggests that any relief we can give to the work of the gut may be important for many overstressed persons in our culture. Undoubtedly, some persons do not require a predigested liquid source of carbohydrate in their diet. Many more persons may improve their function by eating reasonable amounts of honey. When an individual has inflammation in his body or when he has hypoglycemia (low blood sugar), the use of honey may need to be curtailed.

The constituents of honey (strained or extracted) include the following (figures derived from varied sources):

water 17%
carbohydrate[7] 82%
calcium . . . . 5 mg./100 gm.

[7] Glucose (dextrose), 31 - 40%; fructose (levulose), 34 - 39%; sucrose, 1%; and others.

phosphorous . . . .   6 mg./100 gm.
iron . . . .   .5 mg./100 gm.
sodium . . . .   5 mg./100 gm.
potassium . . . .   51 mg./100 gm.
chromium . . . .   29 micrograms/100 gm.
pantothenic acid (vitamin B_5)   55 micrograms/100 gm.
thiamine (vitamin B_1) . . . .   4.4 micrograms/100 gm.
riboflavin (vitamin B_2) . . . .   26 micrograms/100 gm.
nicotinic acid (vitamin B_3) . . . .   110 micrograms/100 gm.
pyridoxine (vitamin B_6) . . . .   10 micrograms/100 gm.
biotin . . . .   .066 micrograms/100 gm.
folic acid . . . .   3 micrograms/100 gm.

Honey also contains copper, manganese, silica, chlorine, aluminum, and magnesium. There may be other trace elements present in small quantities.

Note that the sucrose content of honey is only 1% (although this varies a little with different honeys). Note also the relatively high amounts of glucose and fructose, the immediate breakdown products of sucrose.

Glucose and fructose are absorbed intact into the body without undergoing digestion. Persons who cannot tolerate table sugar (there are such persons) because they lack the enzyme sucrase (needed to split sucrose into glucose and fructose for absorption) may tolerate honey. Those with impaired starch digestion may also do better when they eat honey, although it is more important for these persons to adequately treat their digestive disorder.

Many persons in our society are deficient in the mineral potassium. For these persons, honey might be of benefit because the amount of potassium in honey is substantial.

Honey contains more vitamins and minerals than does refined sugar, which is devoid of everything except calories. When honey is filtered, it loses most of its vitamins. The amount of chromium in honey (29 micrograms per 100 grams) is substantial compared with refined sugar, which contains practically none. Chromium is an important trace mineral whose deficiency is linked with diabetes and arteriosclerosis.

Darker honeys generally contain more minerals than honeys of a lighter color. The reputed health-giving effects of honey may be primarily related to its mineral content.

Honey in the crude, unpurified state also contains pollen. It is possible but unproven that eating this pollen could have a favorable effect on an individual's health. Many persons have pollen allergies. Eating the small amount of pollen present in honey could conceivably desensitize a person to pollen allergy.

Some athletes consume pollen in tablet form, believing that it enhances their physical prowess. This, too, may not be as far-fetched as it might initially sound. Certainly, however, bee pollen therapy has not as yet been proven to be an effective health aid. It should never be allowed to take the place of more established measures for the promotion and maintenance of health. Nevertheless, consider these comments:

Pollen is the dustlike male reproductive material of plants. Ovaries (seeds) are the female counterpart. Seeds contain large amounts of vital nutritive material, and it is conceivable that pollen does the same. For bees, pollen does provide essential proteins necessary for the rearing of their young.

Baby bees grow up on royal bee jelly (high in the B vitamin, pantothenic acid), honey (carbohydrate source), and pollen (protein source). Whether humans should likewise do so is problematical; but certainly pantothenic acid, essential protein, and *complex* carbohydrate are building blocks for human health.

There are a number of enzymes in honey, among which are amylase, invertase, catalase, peroxidase, lipase, and glucose oxidase. Some individuals believe these enzymes account for the (supposed) beneficial effects of honey on health. Some organic acids, proteins, amino acids, and a derivative of chlorophyll called xanthophyll are also present in honey in small amounts.

Although honey has a slightly acid reaction itself, when metabolized in the body it leads to an alkaline state. For those individuals whose body fluids are predominantly acid, the use of honey might be helpful. There are, however, excellent dietary regimens available to alkalinize the body without the use of honey.

Honey is about twice as sweet as refined sugar. Thus, when substituting honey for sugar in cooking, only half as much need be used. Honey absorbs moisture from the air, therefore foods cooked with it tend to remain moist. They also brown faster than foods made with refined sugar.

Some persons believe that honey raises the blood sugar in a more gradual fashion and over a more prolonged time than refined sugar does. This seems unusual since honey contains glucose and fructose, sugars that are directly absorbed into the body without need for digestion. It is said that when honey is taken, glucose passes swiftly into the blood but fructose is more slowly absorbed. This supposedly maintains a steady level of blood sugar, not elevated to a level that stresses the body. I do not know whether this is true or not. Perhaps honey's sugars are absorbed rapidly but do not cause the wide metabolic swings that follow the ingestion of refined sugar. The difference may lie in the potassium and other minerals that are present in honey as well as in the relative lack of sucrose.

Applied Kinesiology indicates that refined sugar induces muscle weakness but that pure honey does not.[8] This may well be true. This Kinesiology evidence, however, is not to be construed as proof until Kinesiology, itself, can be separated from the belief-systems of the person who performs muscle testing.

There are some 4,000 types of honey in the United States, depending on the flower source from which honey has been made.

Tupelo honey is obtained from the tupelo trees (gum trees or swamp trees) of North Florida. It is a honey with a distinctive flavor. It is said to contain sugar in a certain physico-chemical configuration (levo rotating crystal) that does not stimulate insulin as greatly as other sugars. Tupelo honey contains 46 to 49% fructose, 23 to 25% glucose, and 5% sucrose. The higher fructose content has been thought by some to be a health aid, because fructose may enter into the body's metabolism without requiring insulin. I have, however, already suggested that fructose itself may not be healthy in large amounts. Because tupelo honey contains more fructose than glucose, however, it seldom develops sugary crystals (granulation or candying effect). Some individuals and some members of the healing arts firmly believe that tupelo honey is distinct from other honey in being able to sustain blood sugar for longer periods at more optimal levels. Tupelo honey has become the dean of medicinal honeys, at least in the southeastern United States. As far as taste is concerned, sourwood honey is king.

[8]Kinesiology: the study of movement (muscle function).

In the state of Florida, honey prepared for commercial use is required to be filtered. Why should this be so? As far as I know, there is no disease transmitted by crude honey. Presumably, filtering is carried out for the same reason that rice is milled and polished — to make it clear, presumably eye-appealing, and a "pure, undirty" substance, a product of man's "improvement" on nature.

In the preparation of honey for commercial use, honey is heated. Why is it heated? To make it thin and free-flowing so it can more easily be filtered.

Whenever honey is used, it seems advisable to use the crude, unheated, and unfiltered product, hopefully produced in areas in which the soil is rich in mineral content.

The health movement known as Natural Hygiene has taken a stand against the use of honey by human beings.[9] In its view, honey is a food for the animal, bees, and not for man. This, of course, fits in with its vegetarian philosophy. It also considers that honey is a preserved substance, since formic acid is added to honey by bees and serves the function of a preservative.

Natural Hygienists also point out that honey is sweet and, as such, combines poorly with other foods. Stimulating properties of honey are attributed to the sugars and acids therein.

Molasses is usually obtained as a by-product in the manufacture of sugar from sugar cane. It is a thick liquid obtained after boiling and separation from the sugar crystals. Further boiling produces various grades of molasses. The molasses left after the third extraction is known as blackstrap molasses. Molasses contains 55 to 70% carbohydrate and 25% water. An 8-ounce cup of blackstrap molasses contains[10]

calcium	2,052 mg.
phosphorous	252 mg.
iron	48.3 mg.
potassium	8,781 mg.
thiamine	.33 mg.

[9] *Dr. Shelton's Hygienic Review*, Vol. XXXVIII, No. 10, June 1977, pages 221 and 222; section entitled "Your Probing Mind" by V. V. Vetrano, B.S., D.C.

[10] These figures were obtained from the *Natural Health Bulletin*, May 23, 1977. Published by Parker Publishing Company, Inc., Route 59A at Brookhill Drive, West Nyack, N.Y. 10994. The data is used with their permission.

```
riboflavin . . . . . . . . . . . . . . .    .57 mg.
niacin . . . . . . . . . . . . . . . . . . .   . 6 mg.
inositol . . . . . . . . . . . . . . . . .    450 mg.
pyridoxine . . . . . . . . .  810 micrograms
pantothenic acid . . .  780 micrograms
biotin . . . . . . . . . . . . .  48 micrograms
```

Blackstrap molasses has long been touted as a health-maintaining substance. Science has found some reasons why this may be the case. Molasses contains 50 times as much calcium and 10 times as much phosphorous as honey. Like honey, molasses is rich in chromium (22 micrograms per 100 grams) compared with refined sugar, which contains practically none. In addition, blackstrap molasses contains 8 times as much zinc as refined honey or sugar.

According to Mary T. Goodwin, Public Health Nutritionist,[11] strained or extracted honey is only slightly better than white sugar in regard to vitamin and mineral content. Brown sugar has significant amounts of minerals, and blackstrap molasses is comparatively high in its content of minerals and vitamins.

Let us remember, however, that each of these substances — honey, brown sugar, and molasses — is primarily a sugar-containing substance. Dr. John Yudkin (M.D.) has stated in his lectures that honey is just another source of sugar. There is considerable belief that the supposed health benefits of honey may just be wishful thinking.

It appears to be an offsetting kind of situation in which the minerals and enzymes of honey may be beneficial but the sugars undesirable. The important question about the effect of sucrose versus fructose and glucose in the body has not yet been answered. The evidence appears to indicate that refined sucrose is undesirable and that fructose and glucose may be so in large amounts. I would guess that almost any sugar would be harmful if it were present in the body in excess of need. It will certainly be harmful if present without the vitamins and minerals needed to metabolize it, as is the case with refined sugar.

My clinical experience indicates that the use of blackstrap

[11]*Better Living Through Better Eating* by Mary T. Goodwin. Published in 1974 by Montgomery County Health Dept., 611 Rockville Pike, Rockville, Md. 20852.

molasses can be a significant health aid because of the content of minerals. It can be useful as a sweetening agent for homemade bread, bran muffins, and granola cereals. Dark crude honey may also be helpful in regard to mineral input, but more clinical problems are associated with the use of honey than with molasses. Perhaps this is so because honey is overused: "If a little is good, more is better." Infections of the prostate, middle ear, bronchial tubes, and skin as well as arthritis flareups have been seen to disappear when the use of honey has been discontinued. Molasses, because of its strong taste, is less often used to excess.

Honey may be a health-promoting factor for some individuals and not for others. Honey may also be a health-promoting substance in appropriate quantities. Until the time, however, that a definite health-promoting effect can be demonstrated, it is probably wise to avoid the use of large amounts of honey. Those individuals with inflammation in the body such as arthritis, skin disorders, and recurrent infections, and those with elevated blood fats may profit by the elimination of honey from the diet. For the person in need of a dietary source of carbohydrate, reasonable amounts may be well tolerated. Each person may have his own tolerance limits.

Most hypoglycemic persons should distinctly avoid honey — at least in the early stages of their treatment. It is true that a hypoglycemic may, at some stage, profitably use honey as a substitute for refined sugar. This can be a helpful step toward establishing proper dietary intake. Many hypoglycemics improve further, however, when they forego honey for other non-sweet foods.

A person who has hypoglycemia (low blood sugar) can be "wiped out" by the use of sugar or honey. He may become severely fatigued and experience the feeling that the world has dropped out from under him. This often occurs after an initial "lift." The person who does poorly with honey may not recognize that honey is the cause of his problem, because it seems to improve him — initially — after he eats it.

Perhaps the strongest point against eating honey is that it may make a person more susceptible to eating more honey and other less desirable sweets. If one can resist the cultural forces that perpetually urge one to partake of soft drinks, gum, candy, and other refined carbohydrates, and if one can eat only a small quantity of honey, then its presence in the diet may not be objectionable for those who are free of inflammation and hypoglycemia.

Aside from the question of the actual value or risk of pure honey, there is the unfortunate question of whether the honey that one eats has been "cut" with sugar. Unscrupulous persons with magnified economic interests make their honey go further by adding sugared water to it. Unfortunately, the only way to detect such a nefarious practice is to have one's honey analyzed for sucrose content. This is a rather impractical measure for the ordinary citizen. If one is a honey-eater, he had best be sure of his source, have his honey analyzed, or produce his own.

Honey's positive contribution to health most likely is related to the minerals therein. If one is not deficient in minerals, then the use of honey may not be a positive health measure. To sum it up, honey offers the negative effects of excess sugar and the positive effects of minerals and perhaps pollen and enzymes. What it does for or to an individual depends on his personal body chemistry and upon that person's belief about this interesting sweet.

It is wise to avoid the use of all artificial sweeteners (cyclamate and saccharin) until much more information is available about their long-term effects. Newer amino acid sweeteners also need to be carefully evaluated before they are included in the diet.

The sugar-alcohol sweetening agents, sorbitol and mannitol, are extensively used by the food processing industry. They, also, need careful scrutiny for long-term, cumulative, and additive effects in the body when they are used alone and in combination with other chemicals. The same can be said for xylitol, a new agent promoted in gum and toothpaste to combat tooth decay. (Favorable nutrition, an absence of refined sugar in the diet, and optimal local hygiene are the backbone of dental health.)

HIDDEN SUGAR

Persons who "pour on" the sugar — in their coffee or tea, or on their cereal, fruit, or vegetables — are obvious to themselves and to others. Less obvious are those chronic sugar consumers who do not take their sugar from the sugar bowl, but who, nevertheless, end up with a high sugar load from foods which are commonly eaten as part of one's everyday diet. These persons may not even be aware that they are consuming sugar. How often does one realize that most bread, as it is commercially prepared, contains about 8% sugar? Table 1 indicates the content of sugar in some popular food items of our culture.

TABLE 1
SO YOU THINK YOU DON'T EAT MUCH SUGAR?

Approximate amounts of refined sugar (added sugar, in addition to sugar naturally present) hidden in the popular foods.

Food Item	Size Portion	Approximate Sugar Content[1]
BEVERAGES		
Cola drinks	1 (6 oz. bottle)	3½
cordials	1 (¾ oz. glass)	1½
ginger ale	1 (6 oz. glass)	5
highball	1 (6 oz. glass)	2½
orangeade	1 (8 oz. glass)	5
root beer	1 (10 oz. bottle)	4½
Seven-Up*	1 (6 oz. bottle)	3¾
soda pop	1 (8 oz. bottle)	5
sweet cider	1 cup	6
whiskey sour	1 (3 oz. glass)	1½
CAKES AND COOKIES		
angel food	1 (4 oz. piece)	7
apple sauce cake	1 (4 oz. piece)	5½
banana cake	1 (2 oz. piece)	2
cheese cake	1 (4 oz. piece)	2
choc. cake (plain)	1 (4 oz. piece)	6
choc. cake (iced)	1 (4 oz. piece)	10
coffee cake	1 (4 oz. piece)	4½
cup cake (iced)	1	6
fruit cake	1 (4 oz. piece)	5
jelly roll	1 (2 oz. piece)	2½
orange cake	1 (4 oz. piece)	4
pound cake	1 (4 oz. piece)	5

[1]In teaspoons of granulated sugar.
*Trade name.

TABLE 1 — *Continued*

Food Item	Size Portion	Approximate Sugar Content[1]
CAKES AND COOKIES — *Continued*		
sponge cake	1 (1 oz. piece)	2
brownies	1 (¾ oz.)	1¾
choc. cookies	1	1½
Fig Newtons*	1	5
gingersnaps	1	3
macaroons	1	6
nut cookies	1	1½
oatmeal cookies	1	2
sugar cookies	1	1½
chocolate eclair	1	7
cream puff	1	2
donut (plain)	1	3
donut (glazed)	1	6
CANDIES		
average choc. milk bar	1 (1½ oz.)	2½
chewing gum	1 stick	½
choc. cream	1 piece	2
butterscotch chew	1 piece	1
chocolate mints	1 piece	2
fudge	1 oz. square	4½
gumdrop	1	2
hard candy	4 oz.	20
Lifesavers*	1	⅓
peanut brittle	1 oz.	3½
CANNED FRUITS AND JUICES		
canned apricots	4 halves + 1 tablespoon syrup	3½
canned fruit juices (sweet)	½ cup	2
canned peaches	2 halves + 1 tablespoon syrup	3½
fruit salad	½ cup	3½
fruit syrup	2 tablespoons	2½
stewed fruits	½ cup	2
DAIRY PRODUCTS		
ice cream	⅓ pt. (3½ oz.)	3½

[1]In teaspoons of granulated sugar.
*Trade name.

TABLE 1 — *Continued*

Food Item	Size Portion	Approximate Sugar Content[1]
DAIRY PRODUCTS — *Continued*		
ice cream cone	1	3½
ice cream soda	1	5
ice cream sundae	1	7
malted milk shake	1 (10 oz. glass)	5
JAMS AND JELLIES		
apple butter	1 tablespoon	1
jelly	1 tablespoon	4 — 6
orange marmalade	1 tablespoon	4 — 6
peach butter	1 tablespoon	1
strawberry jam	1 tablespoon	4
DESSERTS, MISCELLANEOUS		
apple cobbler	½ cup	3
blueberry cobbler	½ cup	3
custard	½ cup	2
French pastry	1 (4 oz. piece)	5
fruit gelatin	½ cup	4½
apple pie	1 slice (average)	7
apricot pie	1 slice	7
berry pie	1 slice	10
butterscotch pie	1 slice	4
cherry pie	1 slice	10
cream pie	1 slice	4
lemon pie	1 slice	7
mince meat pie	1 slice	4
peach pie	1 slice	7
prune pie	1 slice	6
pumpkin pie	1 slice	5
rhubarb pie	1 slice	4
banana pudding	½ cup	2
bread pudding	½ cup	1½
chocolate pudding	½ cup	4
cornstarch pudding	½ cup	2½
date pudding	½ cup	7
fig pudding	½ cup	7
Grapenut* pudding	½ cup	2

[1]In teaspoons of granulated sugar.
*Trade name.

TABLE 1 — *Continued*

Food Item	Size Portion	Approximate Sugar Content[1]
DESSERTS, MISCELLANEOUS — *Continued*		
plum pudding	½ cup	4
rice pudding	½ cup	5
tapioca pudding	½ cup	3
berry tart	1 cup	10
blancmange	½ cup	5
brown Betty	½ cup	3
plain pastry	1 (4 oz. piece)	3
sherbet	½ cup	9
SYRUPS, SUGARS, AND ICINGS		
brown sugar	1 tablespoon	**3
chocolate icing	1 oz.	5
chocolate sauce	1 tablespoon	3½
corn syrup	1 tablespoon	**3
granulated sugar	1 tablespoon	**3
honey	1 tablespoon	**3
Karo* syrup	1 tablespoon	**3
maple syrup	1 tablespoon	**5
molasses	1 tablespoon	**3½
white icing	1 oz.	**5

[1]In teaspoons of granulated sugar.
*Trade name.
**Actual sugar content.
SOURCE: The Chemins Company, 2430 Mesa Road, Colorado Springs, Colorado 80904, and is used with their permission.

Sugar has been compared to a highly combustible fuel which burns with an intense heat but quickly dies out. The stimulating effect of sugar on the pancreas, without providing nourishment other than calories, is shortly followed by a slump. The individual seeks relief by consumption of more sugar, but eventually the pancreas and the body chemistry may become unbalanced or hypersensitive.

One of the chemicals known to become depleted in the body is chromium. Chromium-deficient refined sugar taken into the body leads to a rise in blood chromium with a net chromium loss through the kidneys. The long-term effect of repeated sugar intake may be diabetes and atherosclerosis, conditions known to be associated with chromium deficiency and high-sugar diets.

In actual fact, it is probable that a cluster of dietary habits is responsible for the poor health that is associated with high sugar intake. Excessive intake of total carbohydrate; excessive intake of refined carbohydrate; protein, vitamin, and mineral deficit; insufficient raw food; and a lack of roughage (fiber) in the diet are such factors. Their detrimental effect is heightened in the individual who is underexercised and who lacks proper physical and mental conditioning.

SUGAR AND POOR HEALTH

The list of health disorders associated with heavy sugar consumption in the diet is long. It includes the following:

1. Dental disease: Tooth decay (caries) and gum disease (pyorrhea).
2. "Candy catarrh": Recurrent or chronic congestion of the mucous membranes.
3. Diabetes mellitus: "Sugar diabetes."
4. Obesity and overweight: Infancy and adulthood.
5. Low blood sugar and related categories of less than optimal blood sugar.
6. High blood pressure and low blood pressure: When the blood sugar rises, the blood pressure rises; when the blood sugar declines, the blood pressure drops.
7. Underweight and malnutrition.
8. Indigestion and other functional gastrointestinal complaints.
9. Atherosclerosis: Heart and blood vessel disease usually associated with fat and calcium deposits in blood vessel walls. Simple sugars are known to elevate blood-fat levels.
10. Heart attacks: The higher the consumption of sugar, the more heart attacks are found.

In addition, there are numerous other conditions which in all likelihood have some relation to the consumption of sugar. These conditions are

1. farsightedness (hyperopia),
2. nearsightedness (myopia),
3. acne,
4. gout,
5. cancer,
6. liver disease,
7. gallbladder disease,
8. peptic ulcer,
9. diverticulitis,
10. constipation,
11. accelerated maturity,
12. reduction of lifespan,
13. functional adrenal insufficiency,
14. recurrent infections,
15. aggressive, irritable behavior, and
16. allergy (hypersensitivity).

It is of interest to note that the medicines used in treating allergic asthma — cortisone, Adrenalin, Isuprel, aminophylline, etc. — all have in common the property of elevating the blood sugar.

DENTAL PROBLEMS

By 60 years of age, thirty-five percent of the American public is without teeth. Many more have nutritional deficiencies associated with inadequate chewing. Sixty-eight percent of Americans between 12 and 17 years of age have some visible sign of periodontal disease. Ninety-five percent of the entire American population has significant dental decay. Teeth are decaying faster than dentists can fix them. Television commercials remind us that children from 5 to 15 are in the cavity-prone years. Poor nutrition in general throughout a lifetime is responsible for much of this. Many primitive tribes have excellent teeth until introduced to the diet of "advanced" civilization. Chronic intake of refined carbohydrate (such as sugar and white flour) is thought to be a dominant factor in causing periodontal disease, cavities, and loss of teeth.

In the 1930's, Dr. Weston Price (D.D.S.) conducted an ambitious study on the occurrence of dental caries among primitive tribes all over the world. He found, for example, only 1% tooth decay among Eskimos still living in a wholly traditional manner, and only 2 to 3% diseased teeth in isolated Swiss villages. Examination of skeletal remains consistently revealed the same low incidence of dental disease. Speaking of modern society, Dr. Price concluded: "No era in the long journey of mankind reveals . . . such a terrible degeneration of teeth and bones as this brief modern period records."[12]

It is also likely that the large number of orthodontic problems in our society may be traced to maldevelopment of facial bones as the result of nutritional inadequacies. The high intake of sugar and other refined carbohydrates is at the core of such nutritional inadequacies.

Dr. Melvin Page (D.D.S.) emphatically believes that sugar adversely affects the calcium and phosphorous levels of the blood and that dental disease is a primary result of these unbalanced factors in the body as a whole. Dr. Page feels that sugar has a devastating effect on body chemistry and that these effects can be detected by measurement of blood sugar, calcium, and phosphorous. His view emphasizes the internal effects of sugar on the teeth, gums, and body as a whole (systemic factors) rather than the local effects of sugar on the teeth. Dr. Page emphasizes that periodontal disease is a systemic disorder with prominent local manifestations in the gums ("arthritis" of the teeth).[13]

Drs. Cheraskin (M.D., D.M.D.) and Ringsdorf (D.M.D.) have pointed out that loss of bone around the teeth correlates with

[12]*Nutrition and Physical Degeneration, a Comparison of Primitive and Modern Diets and Their Effects* by Weston A. Price, M.S., D.D.S., F.A.C.D. Published in 1945 and 1970 by the Price-Pottenger Foundation, Inc., 2901 Wilshire Boulevard, Suite 345, Santa Monica, Calif. 90403.

[13]*Your Body Is Your Best Doctor!* by Melvin E. Page, D.D.S., and Leon Abrams Jr. Published in 1972 by Keats Publishing, Inc., New Canaan, Conn. Also, *Body Chemistry in Health and Disease* by Melvin E. Page, D.D.S., and D. L. Brooks, A. B. Published in 1954 by The Page Foundation, 5235 Gulf Boulevard, St. Petersburg Beach, Fla. Also, *Degeneration/Regeneration* by Melvin E. Page, D.D.S. Published in 1949 by Biochemical Research Corporation, 2810 First Street North, St. Petersburg, Fla. (now the Page Foundation, same address as in preceding reference).

states of low blood sugar as well as with high blood sugar conditions.[14]

In West German schools it has been reported that students with bad grades tend to have bad teeth. It is tacitly assumed that parents who are interested in the dental health of their children are also interested in their schoolwork. A strong case could be made for an equally strong association of bad nutrition with bad grades and bad teeth. The most common feature of bad nutrition is the consumption of sugar and other refined carbohydrates frequently throughout the day.

Cavities in teeth are caused by bacteria which produce acid from carbohydrate (sugars and starches) in the diet. Certain strains of bacteria (streptococcus mutans) are the chief offenders. These organisms form sticky, long-chain sugars (dextrans and levans) from sucrose. These sugars and the bacteria themselves form adherent plaques on the teeth under which the acid tooth destruction occurs. There is also evidence that sugar interferes with the proper movement of fluid into the tooth from the oral cavity as well as from the blood circulation inside the tooth pulp. Steinman has shown that a high sucrose diet reduces fluid transport in the dentin of teeth.[15]

A list of foods high in sugar and other refined carbohydrates, and a list of plaque-forming foods is given in Table 2.

TABLE 2
LIST OF FOODS HIGH IN REFINED
CARBOHYDRATES OTHER THAN SUGAR

Ready to serve breakfast cereals	* Cookies
White bread	* Pies
Pancakes	Rolls and muffins
*Cakes and icings	Sandwich buns and English muffins

*These foods are included in all three categories.

[14] *Predictive Medicine, A Study in Strategy* by E. Cheraskin, M.D., D.M.D., and W. M. Ringsdorf Jr., D.M.D., M.S. Published in 1973 by Pacific Press Publishing Association, Mountain View, Calif.

[15] "The Control of Dental Caries by Manipulating Dentinal Fluid Movement" by Ralph R. Steinman, D.D.S. Published in *The Journal of Applied Nutrition*, Vol. 26, No. 3, Summer 1974.

TABLE 2 — *Continued*
LIST OF FOODS HIGH IN REFINED
CARBOHYDRATES OTHER THAN SUGAR

Biscuits
Saltines
* Graham crackers
Pretzels
Macaroni
Noodles
Spaghetti
* Sweet rolls
* Doughnuts and poptarts
* Coffee cakes
White and instant rice
Flour tortillas
Cream sauces and soups
Candied sweet potatoes
Sweet pickles
Snack foods (cheese puffs,
   onion rings, etc.)
* Ice cream
* Ice milk
* Sherbet
* Fruit, canned or frozen
   in syrup
Sweetened applesauce

* Syrups and sweet sauces
* Jams and jellies
* Candy
* Chocolate
* Jello
* Puddings
* Custards
* Sweetened yogurt
* Instant breakfast
* Breakfast squares
* Hot chocolate
* Milkshake
* Ovaltine
Soft drinks
Kool-Aid
Tang and fruit drinks
Popsicles
Beer
Wine
Hard liquor
Brandy
Cordials

LIST OF FOODS HIGH IN
SUGAR

Sweetened and sugar coated
   cereals
* Cakes and icings
* Cookies
* Pies
Bran muffins
* Graham crackers
* Sweet rolls
* Coffee cakes
* Doughnuts
* Ice cream
* Ice milk
* Sherbet

* Fruit, canned or frozen
   in syrup
Sweetened applesauce
Sweet potatoes, candied
   or in syrup
Chocolate sauce
Sweet and sour sauce
Maple sauce
* Other sweet sauces
   and syrups
* Jams and jellies
* Chocolate
* Instant breakfast

*These foods are included in all three categories.

TABLE 2 — *Continued*
LIST OF FOODS HIGH IN
SUGAR

* Candy, including:
  Candy bars
  Hard candy
  Life savers
  Cough drops
* Breakfast squares
  Sweet pickles
* Sweetened yogurt
* Jello
* Puddings
* Custards
* Milkshake

* Hot chocolate and
  chocolate milk
* Ovaltine
  Kool-Aid
  Tang
  Canned or frozen
  fruit drinks
  Soft drinks
  Popsicles
  Dessert wines and
  cordials

LIST OF PLAQUE-FORMING FOODS

Sweetened and sugar coated
  cereals
Oatmeal
Pancakes
* Cakes and icings
* Cookies
* Pies
* Graham crackers
* Sweet rolls
* Coffee cakes
* Doughnuts
* Ice cream
* Ice milk
* Sherbet
* Fruit, canned or frozen
  in syrup

Fruit, canned (sugar free)
* Instant breakfast
* Breakfast squares
* Hot chocolate
* Milkshake
* Ovaltine
* Jams and jellies
* Syrups and sweet sauces
* Candy
* Chocolate
  Thousand Island dressing
* Sweetened yogurt
* Jello
* Puddings
* Custards

*These foods are included in all three categories.

SOURCE: Dietronics® ©, 19727 Bahama Street, P.O. Box 34, Northridge, Calif. 91324, and is used with their permission.

Experiments using vaccines of the offending strep germs to prevent cavities are being carried out. As yet, it is too early to indicate whether immunity to cavities may be obtained by this technique. It is known that experimental animals can be rendered caries-free by including antibiotics in their diet. Also, caries fail to occur in rats bred in germ-free conditions.

Because of observations such as these, it is concluded that cavities are associated with dietary carbohydrates as well as offending bacteria. A third factor, the strength of the tooth surface, may be improved by fluoride. Fluoride, however, remains a controversial substance. Benefit versus harm has not been established to the satisfaction of a number of investigators and many health-oriented lay persons.

The longer a carbohydrate is in contact with tooth surfaces, the greater the likelihood of cavity formation. Much of this cavity formation is preceded by the accumulation of plaque. The stickier carbohydrates have the greater cavity-producing effect. Caramels, for example, are worse than soft drinks in this regard even though there may be the same amount of sugar in each.

Sucrose, the sugar in our sugar bowls, is the sugar most implicated in dental decay. It is a rather sticky substance when moistened. Interestingly enough, natives who chew on sugar cane from the fields are said to be quite free of dental cavities. This raises the possibility that unrefined sucrose may be less cariogenic than the refined product.

The frequency with which a carbohydrate food is eaten is also a major factor in tooth decay. When sweets are eaten at mealtime they are less likely to produce cavities than if they are eaten between meals.

Most persons know by now that it is advisable to thoroughly cleanse the mouth, teeth, and gums after eating to rinse away particles of food, particularly carbohydrate. Most of us, however, are negligent about this important health measure.

It is particularly important to cleanse the mouth before bedtime, because at night there is little saliva flow to assist in washing away food particles. I have found that children with lazy tongues and poor facial muscle tone create pools of stagnant saliva in their mouths. Vitamin and mineral deficiency with a high dietary intake of sweets seems to encourage the development of poor muscular action.

A cavity-protecting substance known as sialin is contained in human saliva. It is a peptide, a small molecule made up of amino acids. Sialin acts by entering bacteria to divert harmful acid production into alkali. Sialin in the saliva is apparently over-whelmed by excess sugar in the diet.

The National Institute of Dental Research is conducting a study of xylitol, a natural sugar found in berries, some fruits, leaves, and in birch bark. Xylitol keeps the mouth alkaline instead of acid as is the case when sucrose sugar is taken in the mouth. Xylitol is a sweetener that, like fructose, is utilized in the body without the presence of insulin. In Finland, xylitol has been found to not only prevent cavities but to cure early cavities. The xylitol-containing chewing gum used in Finnish tests is widely sold in Europe. (For further information consult the Xylitol Information Bureau, P. O. Box 80, Cresskill, New Jersey 07626.) Further research is needed on the long-term use of this substance and its safety.

Presently available preventive measures to protect against dental disease include fluoridation,[16] tooth brushing, "swishing" after eating, flossing, and cleaning of the teeth by a dentist or dental hygienist. The latter provides a physical means of removing adherent plaque. Judicious, gentle but thorough use of dental floss is an important preventive measure usually recommended by dentists. Removal of abscessed teeth and appropriate dental repair of defective teeth, as well as preventive orthodontics, will aid in preserving local health in the mouth as well as general bodily health. Even more helpful is staying away from sugar (sucrose) and other sticky carbohydrates. If "poor teeth" run in the family, these measures become even more important.

Strontium, a trace mineral, is said to be an important anti-cavity substance, but we need to know more about its clinical use.

It may be important that we ask ourselves, "What are we washing our teeth with day after day throughout our lives?" The manufacturers of toothpastes and powders do not use sugar in their dentrifices. However, the majority of toothpastes and powders are sweet. Often saccharine is used as a sweetener.

Does the long-term use of a sweet chemical in our mouths have any local or general effect on the body? Does the sweetness

---

[16]In this book, I do not wish to engage further in the controversy about the value and toxicity of fluoride, because it does not come within the scope of this book.

"fool" the body into reacting as though sugar were present? The answers to these questions are unknown at this time.

Available in health food stores are dolomite-containing tooth powders or pastes. They also frequently contain a food-digesting enzyme and chlorophyll.

Dental researchers are presently testing the application of transparent plastic adhesive sealants to the teeth. This appears to be a promising advance in cavity prevention for those who cannot or will not resist the "junk food" of their culture.

One's lifestyle has as much to do with the health of his mouth as it does with the health of the rest of his body and his mind. Dr Cheraskin indicates that daily exercise, for example, measurably tightens the teeth.

The mouth is the portal of entry for food, the mode of exit for vomitus, the storehouse of food-choppers, and an organ of lovemaking. Let us care for it like a treasure through local and general methods that lead to health rather than disease. Chapter 15 details what can be done.

ALLERGY TO SUGAR?

Some individuals become so sensitive to sugar that they appear to have an allergy to it. They may not break out in hives, but they do develop overweight, fatigue, irritability, craving for sweets, and elevated fats in the blood. As is the case with other foods, it may be difficult to detect a food allergy when the offending food is consumed on a regular daily basis. Any symptoms arising from the food may not appear for hours or days after it is eaten, and these symptoms may persist for days or weeks. Dr. William Philpott (M.D.), an orthomolecular psychiatrist and preventive medicine specialist, has emphasized in his lectures the close relationship of food allergy and food addiction.

It is necessary to suspect the possibility of allergy to foods when a person has unexplained symptoms such as upset stomach, fatigue, tension, headaches, rash, etc. Removing the food from the diet for several weeks and then reinstating it in the diet as a challenge is a good test for food allergy. If symptoms disappear during the food elimination and reappear upon reintroduction of the food in the diet, it is likely that the offending food has been discovered.

I recently saw a 36-year-old man because of chronic recurrent mental depression and headaches. His symptoms were worse after eating sweets. He was given an oral load of sugar for a glucose tolerance test (see Chapter 11). Within 5 minutes he developed a cold feeling all over his body, with shaking chills. The cold feeling persisted for 8 hours. Twenty minutes after the sugar load, he developed a severe headache that persisted throughout the day. He became mentally depressed and sluggish in his mental activity.

Such dramatic reactions suggest a hypersensistive or allergic response to sugar. Whether this is a true allergic reaction involving antigen, antibody, and chemical mediators is not known. It is possible that this kind of reaction releases harmful chemicals in the body without invoking the antibody mechanism.

Dr. William Crook (M.D.)[17] has tested for sensitivity to sugar by placing a diluted sugar solution under the tongue. He finds a high reaction rate to cane sugar (sucrose) in children with learning disorders and with hyperactivity. When the test is done, flushing of the skin and hyperactive behavior are often seen to occur in response to the sugar. Dr. Crook points out that sugar is a common cause of disturbed child behavior and a cause of hyperactivity and minimal brain dysfunction. He notes that cane sugar, along with milk, leads the list of allergy-causing foods.[18]

One of the most important factors in producing allergy in a person is the frequency and amount of exposure to the substance. For example, Europeans who live where there is no ragweed pollen do not have ragweed allergy. Infants who consume very large amounts of cow's milk in proportion to their body weight have a high incidence of milk allergy.

Since the dietary consumption of sugar in our country is so high (the average American consumes about one teaspoon of sugar every 35 minutes, 24 hours a day, for a lifetime), it is understandable that sugar may be a frequent cause of allergy or allergylike disorders. This certainly fits in with the observations of Dr. Crook and myself.

[17]Pediatrician at Child Health Centers of America, Inc., Jackson, Tenn.

[18]"Pediatrician Sees Link Between Sugar and Behavior Problems" by William G. Crook, M.D. Published in *Prevention*, January 1975, pages 169 and 170.

It is common experience for persons on no-sugar diets to experience headaches, vomiting, fatigue, or other symptoms when they happen to eat sugar.

In my practice, I have found that it is common for individuals who consume large amounts of sugar to have an elevation of the eosinophils ("allergy cells") in their blood counts. Furthermore, these eosinophil counts often return to normal levels when sugar, other refined carbohydrates, and additives are eliminated from the diet. Further research needs to be made to delineate specific substances in the diet that may be associated with increased levels of eosinophils.

SUBTLE EFFECTS

It may be difficult to become aware that sugar is affecting one's health at all. Because sugar (sucrose) is so abundantly present in the foods that we usually consume, one maintains a diet containing sugar even though he is not consciously aware of it. Unless a person becomes knowledgeable about the many foods that have sugar added to them, he will be unable to eat a diet which is sugar-free. Reading labels is the only way to surely avoid sugar, and even then one can be fooled. Natural, raw, brown, and all other descriptive words before the word *sugar*, do not change the fact that sugar (sucrose) is sugar (sucrose).

If we grow accustomed to a certain level of fatigue, sinus congestion, overweight, bad temper, foul breath, or menstrual irregularity, we will come to accept that particular state as normal. Only when we experience the improved health that can follow a sugar-free diet will we then reset our sights on a new and improved level of healthful living.

A diet free of sugar and well regulated in regard to other healthful substances may be responsible for improved exercise tolerance, appreciation of the viewpoints of others, enhanced motivation to perform in job or school, weight loss, more normal menstrual patterns, etc. In addition, years may be added to one's life because the risk of cancer, heart disease, stroke, obesity, and other maladies may be lessened when sugar is eliminated from the diet and vitamin-mineral deficits restored to normal. (It must be emphasized again and again that a sugary diet is one that is not optimal in regard to many dietary features. The presence of sugar in the diet is a signpost that an individual is not as discriminating in health matters as he might be.)

If one is interested in integrating his mind and body with his environment to the benefit of oneself and his associates, a good place to start is by eliminating sugar from his diet.

A young woman in good health voluntarily decided to eliminate sugar from her diet. Thereafter, she felt well but otherwise noted no difference in her health until one day she "just felt like" having sugar on her breakfast cereal. About six hours after eating the sugared cereal, the girl noticed nausea and blurred and diminished vision in one eye. She then developed a severe migraine headache on one side of her head. This migraine headache was only moderately controlled with a medicine known as Cafergot.

A common cause of migraine headaches is chocolate in the diet. Chocolate contains a high proportion of sucrose. This migraine headache followed the eating of table sugar. The girl refused to eat sugar again as a test because of the fear that a migraine headache would reoccur over the Christmas holiday. However, she ate a piece of chocolate and developed a severe migraine headache.

Whether a reaction such as this is a true allergy is an unsettled question. Nevertheless, many individuals experience reactions (immediate or delayed) to sugar-containing foods, which in our present state of ignorance are termed allergy.

NATURAL VS. REFINED SUGAR

Does sucrose from a plant in nature have the same effect as sucrose from the sugar bowl? The answer to this question is not clear at this time. My experience, however, leads me to believe that refined sugar, prepared from sugar cane, corn, or beets, gives more adverse effects in humans than does sucrose consumed in its natural fruit or vegetable sources.

Very probably, refined sugar is a more dangerous substance than sugar in whole foods because of the ready availability of large quantities of refined sugar. It is less likely that one would eat large quantities of sugar cane, beets, corn, or fruit at one sitting than it is that one would use large amounts of sugar from the sugar bowl.

The matter of natural versus refined sugar in the diet must be looked at from this standpoint of quantity of sugar eaten. Much of the undesirable effects of refined sugar may be because it is in such widespread use in our culture and thus the absolute amounts eaten are so great.

Refined sugar could also be harmful to health because of hydrocarbons and other chemicals contained in refined sugar which would not be present in the unrefined plant sucrose.

Bleaching agents, colors, flavors, and preservatives may all be responsible for adverse effects when a food containing them is eaten.

In addition, the very act of refining may alter the substance being refined so that it produces maladaptive reactions in the individual consuming it. It is not uncommon to find an individual, for example, who can eat fresh corn-on-the-cob but who cannot tolerate corn syrup or even grits.

Natural vitamins, minerals, and enzymes are contained in the fresh plant source but not in the refined product. There is considerable evidence to indicate that refined sugar in the diet increases the requirements for B vitamins in the body. As others have pointed out, sugar increases the vitamin need and lowers the supply.

There are, however, factors other than sugar itself to consider in regard to refining and processing of food. The element magnesium is considerably reduced in foods that are refined or processed. The same is true for zinc and chromium. Magnesium is necessary for proper energy utilization in the body. Magnesium is involved in 75% of the enzyme reactions in the body. Chromium deficiency is intimately associated with atherosclerosis and diabetes. Zinc is a constitutent of the hormone that regulates blood sugar. Zinc is also intimately involved with copper, histamine, and vitamin B_6 in the body.

Many authors have pointed out the loss in vitamins that occurs when foods are refined, canned, stored, and otherwise prepared for cosmetic appeal and convenience. "Enrichment" of these foods almost always fails to restore the natural complement of vitamins originally in the food. Also, it could never restore the vital factors in fresh food that may be necessary for health but are not yet discovered. It would be haughty as well as dangerous to presume that our knowledge of nutrition is complete.

Some individuals believe they have done enough when they eliminate refined sugar from their diets. In many cases it is highly desirable to also limit the amount of natural sugar that is eaten. Many fruits and juices contain considerable quantities of sugar (sucrose). Further discussion of this matter will be found in Chapter 15.

SUGAR AND SIGHT

Sugar has a great deal to do with the eyes and vision. In diabetics, whose blood-sugar level fluctuates widely, a common troublesome symptom is fluctuating eyesight. It is not unusual for a new diabetic patient to have difficulty in obtaining a satisfactory optical prescription. The nuisance and expense of changing glasses several times is due to the fluctuating state of the body in metabolizing sugar.

Diabetes with its disturbances in blood sugar prominently affects the eyes in other ways. Diabetes is probably the leading cause of blindness in this country. When a physician examines the retina in the back of the eye of a long-term diabetic, he can often see hemorrhages, discharges, narrowed blood vessels, and ballooned-out blood vessels. The diabetic patient who is well-controlled (that is, the patient whose blood sugar is not allowed to remain abnormally high) may experience fewer of these visual complications.

Refined carbohydrates in the diet have been implicated in the origin of nearsightedness. Before the second World War, nearsightedness was on the rise in Japan, where sugar and other refined foods were extensively consumed. Return to a more natural diet during the war years was accompanied by a drop in nearsightedness; but it is now returning, as the nation's affluence has led to a more Americanized diet.

Dr. John Yudkin (M.D.) has performed experiments in animals and humans which lead him to believe that nearsightedness is associated with elevated blood sugar, and farsightedness is associated with lowered blood sugar.[19]

If sugar is fed to a rabbit, the animal becomes nearsighted. This type of observation leads to the belief that there's a lot more to nearsightedness than heredity.

Lack of calcium and/or vitamin D may also be implicated in nearsightedness. According to Dr. Melvin Page (D.D.S.), sugar intake has a distinct effect on calcium, phosphorous, and perhaps hormonal balance.

Nearsightedness is also related to physical inactivity and extensive close visual work. (How many nearsighted plainsmen are there?) I have observed children in the late elementary school ages who become more nearsighted during school months, but whose nearsightedness regressed or went away during the school-free summer months.

One can speculate that sugar is related to nearsightedness in the following way:[20] When sugar is regularly eaten, fluid is retained with it in the body. The eyes are soft-tissue organs and contain considerable fluid. In the presence of heightened amounts of sugar and water, the eyes become heavier (waterlogged) and more easily deformed in shape according to the force of gravity. The eye is not entirely contained in its orbit and is relatively unrestricted in movement in the direction of the cornea. When a child leans forward in his school desk to look at reading and writing material, his "heavy" eyes elongate due to the pull of gravity. The elongated eye is one which sees out of focus and is the hallmark of the nearsighted person. Undoubtedly, other factors are operative. For example, poor nutrition (mineral and protein deficit, etc.) may allow the soft tissues of the eyes to be more easily stretched or deformed.

As pointed out by Dr. Carl C. Pfeiffer (M.D.), stretch marks have an association with zinc deficiency.[21] Rapidly expanding tissues such as breasts, hips, and abdomens in adolescents and

[19]*Sweet and Dangerous* by John Yudkin, M.D. Published in 1972 by Peter H. Wyden, Inc., New York, N.Y.

[20]Background for these ideas was generated as a result of conversation with Dr. Howard Coleman (O.D.) of Rumford, Rhode Island.

[21]*Mental and Elemental Nutrients* by Carl C. Pfeiffer, Ph.D., M.D., and the Publications Committee of the Brain Bio Center. Published in 1975 by Keats Publishing, Inc., New Canaan, Conn.

pregnant females stretch and rupture skin fibers, leaving scars that we know as stretch marks. Might it not be that the rapidly expanding (growing) eye in the zinc-deficient individual stretches excessively to produce the elongated eyeball and the short sight that we recognize as myopia?

When a nearsighted person successfully wears contact lenses for seeing, it is common observation that the progression of nearsightedness halts or considerably slows down. The profession of orthokeratology uses the application of contact lenses early in a child's life to prevent the development of nearsightedness. It appears that the contact lens acts like a compress or pressure bandage to minimize deformation of the eye.

Nearsightedness, with its tendency to occur in families, is probably influenced by factors other than sugar, other refined carbohydrates, and vitamin-mineral inadequacy. Nevertheless, a good starting place for prevention of this problem is the avoidance of sugar.

MOTHER'S INFLUENCE

Everyone knows that mothers have a great deal to do with the way their children turn out. Mothers, indeed, have a profound influence on child development. They manage the home, buy the food, prepare the meals, and often provide discipline when needed.

But mothers have an equally profound impact on child development which starts before the child ever sees the light of day! The mother's diet throughout her lifetime and during her pregnancy may produce long-term effects on the weight, growth, and metabolism of her offspring! The mother's diet — and her mother's before her — may have a great deal to do with the health of her child as well as the development of nearsightedness in the child.

A woman who is in nutritional deficit at the time that she conceives is ill prepared to bear and deliver a healthy newborn infant and to care for it after birth. It has been estimated that superior versus average nourishment in the mother-to-be in-increases the likelihood that the offspring will be a perfect example of good health by fourfold. A high sugar intake in the adolescent years (and all the nutritional problems that go along with such a diet) does not prepare a young woman well for the weighty tasks of pregnancy, childbirth, and child-rearing.

In respect to pregnancy itself, Dr. Robert L. Davis (D.Sc.) and others have provided us with some excellent animal experiments that illustrate the profound effects of sugar.[22] These experiments indicate that the composition of a mother's diet can have later effects on the development of her offspring, even when these offspring are subsequently fed an adequate diet! Feeding sugar (sucrose) to the mother has been shown to exert a geneticlike effect in controlling the weight and disposition of glucose throughout the life of the offspring. This is the case despite the fact that there is no difference in food intake between experimental and control groups.

In experimental animals, feeding sugar to pregnant mothers produces an imprint made in utero, controlling the growth potential and certain metabolic pathways of the offspring.

In Dr. Davis's experiments, the offspring of sucrose-fed mothers were significantly heavier than those whose mothers were fed their regular food without sugar, and their bodies had a higher fat content. The female offspring were especially heavy. Thus, we see that the maternal diet strongly influences the development of obesity in the offspring — even though the offspring eat no more food than lean control animals! It may be that sucrose in the mother's diet stimulates liver enzymes which activate fat production.

Throughout these experimental studies, the fasting blood-sugar levels of the sucrose offspring averaged about 23 mg.% higher than controls. In addition, their glucose tolerance tests[23] exhibited diabetic traits.

Measurement of the blood fats showed that the prebeta lipoprotein and cholesterol levels of the sucrose animals were much higher than those of control animals.

Dr. Davis has also pointed out human studies on the island

[22]"The Effect of Maternal Diet on the Growth and Metabolic Patterns of Progeny (Mice)" by Robert L. Davis, Sylvia M. Hargen, and Bacon F. Chow. Published in *Nutrition Reports International,* Vol. 6, No. 1, July 1972; and "Long Term Effects of Alterations of Maternal Diet in Mice" by Robert L. Davis, Sylvia M. Hargen, Frances M. Yeomans, and Bacon F. Chow. Published in *Nutrition Reports International,* Vol. 7, No. 5, May 1973.

[23] The glucose tolerance test is a procedure done to establish the presence of diabetes or low blood sugar. It is extensively described in Chapter 11 of this book.

of Formosa, which have reached the same conclusions as the animal experiments that I have just mentioned.

This work should have a sobering effect on every teenage girl who glibly consumes cola drinks, cookies, candy bars, pie, and ice cream. If only young persons knew the importance of proper nutrition long before they started their families! Perhaps you, the reader, can become an educator for this cause. Your effort can significantly contribute to the health of generations yet to be born. Educate so we will not have to medicate.

THE STEADY STATE (HOMEOSTASIS)

There is a growing realization that disease is a breakdown in the body's steady-state arrangement. When the status quo (homeostasis) is interrupted, adverse stress calls upon the body's restorative mechanisms to correct the situation. Excessive usage of these corrective mechanisms may result in their abnormal function, and, at a later date, biological damage to body tissues.

Repetitive ingestion of simple sugar stresses the body's metabolism, setting into action many compensatory chemical and nervous actions. In this way, sugar consumption may upset homeostasis, interfere with endocrine and autonomic nervous system balance, and predispose to disease.

Kinesiologists[24] believe that sugar interferes with the body's energy state. Those who practice Applied Kinesiology believe that the ingestion of refined sugar significantly interferes with the strength of muscles in the body — and that this takes place within a second or two after sugar is eaten. Kinesiologists demonstrate that a strong muscle can become a weak muscle almost instantaneously when sugar is taken in the mouth.

Dr. Emanuel Cheraskin (M.D., D.M.D.) has shown that omission of refined carbohydrates (such as sugar and white flour products) from the diet yields a better homeostatic body mechanism; that is, fewer and less abrupt fluctuations in the blood glucose (see "References," pages 489 and 490).

What about the quick energy effects of sugar? It is indeed true that sugar (sucrose) in the diet raises the blood sugar very quickly. In fact, it does so more quickly than any other food. This

[24]Kinesiology: the study of muscle function and movement.

is really more of a curse than a blessing. *If* the blood sugar has previously been abnormally low, then this rise in blood sugar will, indeed, provide the quick energy lift that the individual needs — at least temporarily. If, however, the blood sugar has been normal or high, then producing a higher blood-sugar level will have no beneficial effect at all and may, indeed, produce undesirable effects. Sleepiness, storage of fat, waterlogged tissues, and loss of minerals are some of these undesirable effects.

The temporary blood-sugar elevation that eating sugar provides is often a double-edged sword. *Raising* the blood sugar by eating is followed 2 to 3 hours later by a *lowering* of blood sugar. What goes up must come down. This lowering in many persons becomes more and more difficult to avoid and perhaps more severe as the dietary intake of sugar increases. This matter will be discussed in more detail in Chapters 9 and 14.

Sugar can drop the blood-sugar level just as quickly as it can raise it. *In some individuals, sugar in the diet actually brings about an initial lowering of the blood-sugar level.* Such a drop may come precipitiously, producing nervous symptoms such as jitters, rapid heart action, sweating, etc. Or, the blood sugar drop may come about more slowly, bringing with it cerebral symptoms such as sleepiness, confusion, hallucinations, coma, or even convulsions. Examples of these situations will be given later on in this book.

Interestingly, the perceptual distortions brought on by low blood sugar may be indistinguishable from the classic symptoms of schizophrenia. Schizophrenia, in each of its several forms, is the most common type of mental illness and is probably the least understood. It is commonly associated, however, with disturbances of glucose homeostasis.

When blood sugar rises above normal, certain endocrine glands in the body are called upon to cope with the elevated sugar. After they have done their job, these glands turn off or reduce their function. Then, an opposing set of glands (counter-regulatory factors) play their part in trying to bring the blood-sugar level to a desirable steady position.

At the same time, calcium and phosphorous levels in the blood are fluctuating, and other chemical reactions and counter-reactions are taking place. The whole matter is something like a pendulum set in motion, swinging first to one side and then the other until stability is regained.

In a particular individual, the release of the regulatory and counterregulatory hormones may not be properly timed, and quantities of the hormones may be inadequate or excessive. Because these hormones exert effects on the entire body as well as on the blood sugar, their repetitive release, overrelease, or underrelease may have profound effects on the whole organism. Wide swings in this homeostatic mechanism are generally undesirable and lead to disease over months and years.

All the cells of our body may suffer the consequences of overactivity, underactivity, or faulty timing of hormone secretion. It all starts, very often, with our civilized habit of eating refined sugar. Sugar is a readily available substance eaten in great quantities because man's technology has made it possible for him to have this substance without eating the plant from which it came. As Dr. Melvin Page has said, "Man can be considered an upstart, because he so readily presumes that technical progress improves his life."[25] Readily available refined sugar produced by the technology of modern man appears to be a prime source of dietary pollution, rapidly limiting man in his health and longevity. The effect of sugar on generations past may be even now wreaking degenerative havoc on ourselves and our children.

At this point, it must be stressed that a diet lacking in refined carbohydrates is not synonymous with a diet lacking in carbohydrate. Some individuals, at some phases of their lives, may need a diet low in carbohydrates. At many other times, however, a diet high in natural whole carbohydrates[26] may be health-producing. There is evidence indicating that a low fat, low protein, high complex carbohydrate diet can reverse or prevent the deposition of atherosclerotic placques.[27]

Further along in this book, I will discuss the metabolism of sugar in the body and tests for sugar disorders. Considerable

[25]"Man — The Upstart" by Melvin E. Page, D.D.S., and Bruce Pacetti, D.D.S. Published in *Journal of Applied Nutrition*, Vol. 26, No. 1 and No. 2, Spring 1974, pages 11 - 16.

[26]Potatoes, barley, whole-grain rice, beans, etc.

[27]"High Carbohydrate Diets: Maligned and Misunderstood" by Nathan Pritikin. Published in *Journal of Applied Nutrition*, Vol. 28, No. 3 and No. 4, Winter 1976.

attention will be devoted to diabetes and low blood sugar. A descriptive evaluation of the standard glucose tolerance test will be set forth, and a description of the "natural" tolerance test will be given. Finally, I will present ways for the reader to prevent or modify existing situations that threaten his health.

CHAPTER 2

SOME
DEFINITIONS

Before I proceed further to explain sugar in the body and the illnesses associated with it, the reader must understand clearly what sugar is.

The constitutents of the food that we eat can be divided into three categories: protein, fat, and carbohydrate. A reemphasis on dieting with a low carbohydrate diet[1] popularized the word carbohydrate. As the word suggests, carbohydrates are food substances that consist of carbon and hydrate (water, hydrogen, and oxygen).

Sugar is a carbohydrate.

Carbohydrate, in a practical sense, consists of *sugar* and *starches*. It is the primary fuel source in the body. Plants, for the most part, are composed of carbohydrates.

Green plants manufacture carbohydrates under the influence of sunshine through photosynthetic action. As they do so, they bind (fix) the sun's energy in carbohydrate molecules. Some of the carbohydrate plant material eaten by animals and man is nondigestible carbohydrate (cellulose, lignin, etc.), but the sugar and starch carbohydrate of the plant is available to man for use in his body.

[1] *Dr. Atkin's Diet Revolution* by Robert C. Atkins, M.D. Published in 1972 by Bantam Books, Inc., 666 Fifth Avenue, New York, N.Y. 10019.

The simplest carbohydrates are known as *simple sugars* (monosaccharides). Simple sugar is the fundamental unit or building block of all carbohydrates. When eaten in the diet, a simple sugar is absorbed into the body without digestion.

The most important simple sugar is glucose (dextrose), since it is the form of sugar that circulates in our blood as a readily available fuel. This is the chemical formula of a simple sugar such as glucose:

$$C_6H_{12}O_6$$

These are some different ways of representing the structure of a glucose molecule:

What now do we know about sugar? It is a carbohydrate. One form of sugar is known as simple sugar. Glucose is a simple sugar, and it is the primary fuel for our bodies. It exists in nature but is also derived from more complex sugars and starches.

Now, our story goes on. A more complex form of carbohydrate is represented by the *double sugars* (disaccharides). They are found extensively in nature. As the term implies, they are made up of two simple sugars. Sucrose (cane sugar) is a double sugar and is the sugar that we know as refined sugar and which is found in our sugar bowls. Nearly always when a food has sugar added to it, the sugar is sucrose. Sucrose is frequently used as a food additive because of its highly sweet taste. Since it is a highly refined food, a devitalized food, it keeps for long periods without spoilage.

This is the chemical formula of a double sugar such as sucrose:

$$C_{12}H_{22}O_{11}$$

When we eat the double sugar, sucrose, a chemical action (digestion) in our intestines breaks sucrose down into its two component simple sugars, glucose and fructose (levulose) (see Figure 1). These simple sugars are then absorbed into the body.

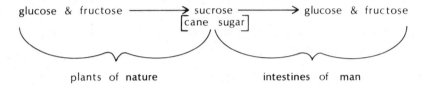

glucose & fructose ⟶ sucrose ⟶ glucose & fructose
[cane sugar]

plants of nature   intestines of man

Figure 1. Sucrose

Nature provides us with supplies of sucrose in plants such as sugar cane and sugar beets. These plants have been grown in abundance and then extensively refined to satisfy the "sweet-tooth" of modern man. In the digestive tract of man, sucrose is broken down into its basic component sugars.

What now do we know about sugar? It is a carbohydrate. Two simple sugars, glucose and fructose, make up the double sugar, sucrose. Sucrose is refined sugar. It is very sweet and is consumed in large quantities in the average person's diet whether he realizes it or not. Sucrose is digested (broken down) in our intestines into its constitutent simple sugars, and these are absorbed into the blood.

Most of the sugar (sucrose) that we use is obtained from sugar cane or from sugar beets. It may also be derived from corn or from maple syrup. Regardless of these sources, the ultimate project, refined sugar (sucrose), is identical. Each person in the United States takes into his body more than 100 pounds of refined sugar every year. On the average, this amounts to eating nothing but sugar every fourth day! Much of this sugar is contained in soft drinks, baked goods, confectionaries, canned fruits and jellies, and ice cream. Prepared foods have sugar added to them so the consumer may not be aware he is eating sugar at all unless he reads the fine print on the label. Prepared meats, salami, sausage, cereal, catsup, canned and frozen vegetables, salad dressings, and canned and frozen fruits are more than likely to contain sugar.

Even more complex forms of carbohydrate are formed when simple sugars are linked together into long chains. These *compound sugars* (polysaccharides) include starch within plants and starch within animals, known as glycogen. Glycogen plays an important role in glucose metabolism in man. It is the storage form of carbohydrate warehoused in man in the liver and muscles. Glycogen is rapidly broken down to glucose when there is a demand for fuel in the body.

The chemical formula of a compound sugar is as follows (the n indicates the number of simple sugars linked together; it is usually a large number):

$$(C_6H_{10}O_5)n$$

Figure 2 shows in schematic fashion the structure of the carbohydrates.

a simple sugar: glucose, e.g.

a double sugar: sucrose, e.g.

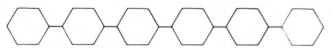

a compound sugar: starch and glycogen

Figure 2. Structure of sugars and starch

This is an illustrative diagram to show the basic structure of sugars and starch.

The simple sugar has one component unit, the double sugar has two components, and the compound sugar has many components.

In summary, what can we now say about sugar? Sugar is the primary fuel substance of our bodies. Sugar is carbohydrate and occurs naturally in plants. It exists as simple sugars, double sugars, and compound sugars.

Glucose, the sugar in our blood, is a simple sugar.

Sucrose, the very sweet substance used so commonly in our diets, is a double sugar which contains glucose as well as fructose.

Starch from plants and "starch" (glycogen) in our bodies are compound sugars. Plant starch requires greater digestion and hence is absorbed more slowly from the intestinal tract than the less complex sugars such as sucrose and glucose.

CHAPTER 3

SUGAR IN THE BODY

Now that we know something about sugar, let us examine more closely the principal sugar of the human body, glucose. This sugar circulates in our blood and is the major fuel substance used by the cells of our body for the production of energy. Glucose, also known as dextrose, is a simple sugar containing six carbon atoms.

The brain and the cellular elements of the blood depend upon glucose for their energy requirements. Other tissues (muscle, for example) prefer glucose as a fuel but can burn fatty substances as an alternate source of energy. During starvation, the brain switches from its reliance on glucose and also develops the ability to use fatty breakdown substances (ketones) as a fuel supply. Figure 3 shows glucose and fat as major fuel substances, contributing to the development of energy for body function.

In the body at rest, the brain consumes about two-thirds of the total circulating glucose supply. Most of the remaining third of the glucose supply goes to the skeletal muscles and the red blood cells. In vigorous exercise, there is redistribution of blood to provide the muscles with additional glucose and oxygen.

The human brain requires about 100 to 145 grams of glucose per day. Reserve supplies stored in the liver amount to less than 100 grams. The remainder is obtained from glucose manufactured in the liver or from the dietary supply. At night the body ma-

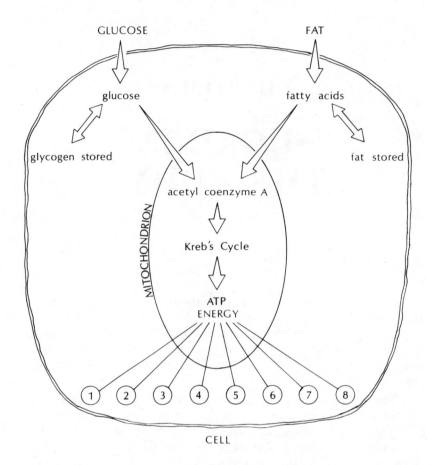

Figure 3. Metabolism within the cell

The two major metabolic fuels, glucose and fat, once inside the cell, give rise to acetyl-coenzyme A which enters the Tricarboxylic (Kreb's) Cycle to produce energy via adenosine-triphosphate, ATP. This ATP-energy is the power source for

    1. amino acid transfer into the cell,

    2. protein manufacture on ribosomes,

    3. DNA synthesis in the nucleus,

    4. change of fatty acids to fat,

    5. fat breakdown to acetyl-coenzyme A,

    6. glycogen manufacture,

    7. phosphorylation of glucose, and

    8. preservation of the integrity of the cell (for example, maintenance of sodium-potassium ratio).

chinery is actively working to provide material for the production of glucose. This occurs predominantly in the liver. The glucose is produced from protein and fat by a process known as gluconeogenesis.

It is necessary for the body to constantly provide an adequate amount of glucose in the blood in order that vital life processes can proceed. This is accomplished by a number of complex interlocking reactions. The proper working of these processes provides *homeostasis*, the steady state in regard to the body's chemical function.

It is thus the aim of the body to sustain a reasonably steady level of glucose in the blood. At all times, the body works toward this purpose: to right the disturbances which push blood sugar too high or too low.

The body's regulation of blood sugar is dependent on two principal factors:

1. the input of sugar to the blood and
2. the output of sugar from the blood.

INPUT

Sugar is supplied to the blood from the intestines when dietary carbohydrate is digested and absorbed (see Figure 4). If more carbohydrate is eaten, the blood sugar will rise higher after a meal. The height and rapidity of blood-sugar rise are significant factors leading to overactive insulin response, fat buildup, and subsequent lowered blood sugar. When a substance is rapidly absorbed into the blood, the blood-sugar levels may reach high peaks before the regulatory efforts of the body can come into play. This would be something like starting your car and accelerating to 60 miles per hour immediately — the shock can be terrific.

Our country is currently experiencing an "epidemic" of coronary-artery-bypass surgery. This procedure is done for persons whose heart arteries are obstructed and narrowed. In this surgery, catastrophic brain damage may occur when sufficient oxygen is not delivered to the brain. Such brain damage is greatly worsened when blood-sugar levels are high. This is so because high levels of blood glucose give rise to high levels of lactic acid, and lactic acid accumulation is toxic to the brain.

Stroke patients, infants with respiratory disorders, and others with poor oxygenation of the brain may well be at risk of brain damage if blood-sugar levels are high. Physicians may need to be particularly thoughtful about administering glucose to patients, especially when there is no precise need for it.

Alcohol and sugars are the most concentrated and rapidly absorbed carbohydrate substances. Protein, fat, and starch (complex carbohydrate) are more slowly processed and absorbed. One hundred percent of dietary starch is ultimately changed to glucose, whereas 57% of protein and 10% of fat ultimately find their way to this primary fuel source.

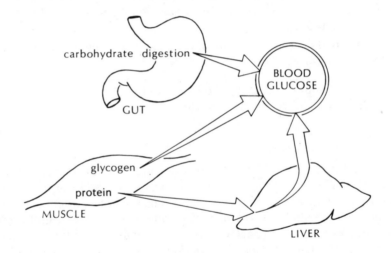

Figure 4. Adding glucose to the blood
The glucose circulating in the blood is derived from the gut, the liver, and indirectly from the muscles.

In order to provide for an emergency supply of power, the body in its wisdom has supplied a storehouse for glucose. Glucose itself cannot be stored in cells, hence the body synthesizes simple sugar back into the complex carbohydrate, glycogen, which is stored in the liver and in muscles. This backup supply can feed glucose to the blood when dietary supplies of fuel are lacking. The storehouse is such an important feature to the body that it is given top priority — it's filled up pronto as soon as glucose becomes available from the gut.

The liver glycogen can give rise to glucose directly when it is needed, but muscle lacks the enzyme necessary to change muscle glycogen into glucose. Use of glycogen in muscle produces lactic acid, which can be built up indirectly into glucose.

OUTPUT

Sugar is removed from the blood when it is metabolized (used up) by the tissues of the body for the production of energy. In addition, blood glucose may be lowered when the glucose storehouse is resupplied (see Figure 5). On occasion, glucose may be lost to the body by spillage into the urine. This cause of sugar outflow is seen in diabetes mellitus (sugar diabetes) and in rare cases of kidney disorder (renal glycosuria). Ordinarily, however, in the normal, healthy individual, glucose is readily conserved for the body by kidney function.

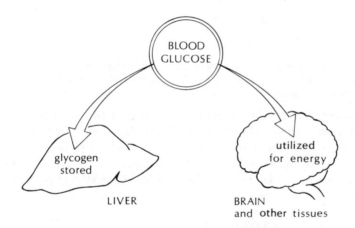

Figure 5. Removing glucose from the blood
Filling up the stores of glucose as glycogen in the liver and utilization of glucose in the tissues for energy needs deplete blood glucose.

Figure 6 illustrates some of the functions that are known to take place within the body at various levels of blood-glucose concentration.

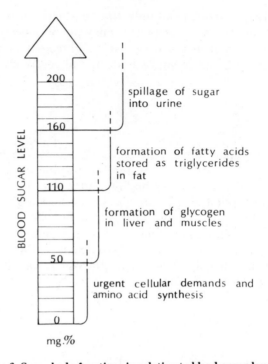

Figure 6. Some body functions in relation to blood-sugar levels
The optimal level of blood sugar could vary for each individual but appears to be between 80 and 90 mg.% for most persons.

Recent information raises the question of the relationship of sugar levels in the blood and in the brain.[1] Glucose concentrations in the brain and blood may not be proportionate to one another. This can be of great importance, especially for individuals with the clinical condition, hypoglycemia (low blood sugar) (see Chapter 14).

THE GLUCOSE STOREHOUSE

The storehouse for glucose in the body is a remarkable organ, the liver. When glucose is absorbed from the intestine, it flows directly to the liver. The cells of this unique organ, the largest organ of the body, soak up glucose like a sponge. This is so be-

[1]"Blood Glucose: How Reliable an Indicator of Brain Glucose?" by Jean Holowach Thurston. Published in *Hospital Practice*, September 1976, pages 123 - 130.

cause glucose can freely enter the liver cells without the presence of insulin. In many other areas of the body, insulin must be present before glucose can gain entry into cells.

Once inside the liver, glucose is changed to its storage form, glycogen (animal starch), if the glycogen stores are low. If the glycogen storage banks are full, or if the glucose load to the liver is extensive, glucose flows out of the liver into the general circulation (see Figure 7).

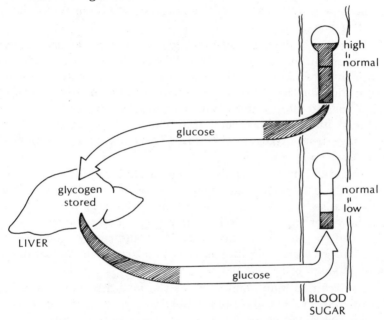

Figure 7. Role of the liver in maintenance of normal blood sugar
When the blood sugar is higher than normal, glucose is stored in the liver as glycogen. When blood sugar is lower than normal, hormonal action causes glucose release from its storage form in the liver to restore the blood-sugar level to normal.

Whenever blood levels of glucose are low, the liver gears up its chemical machinery to mobilize glycogen by converting it to glucose. This output of glucose to the general circulation from the liver is a vital factor in maintaining the steady state (homeostasis) within the body.

Whenever blood sugar rises, blood pressure and body temperature rise. Similarly, when blood sugar falls, blood pressure and body temperature tend to fall. One can see that fluctuations in blood-sugar levels are accompanied by significant changes in other basic physiological processes.

One of the major functions of the liver is the maintenance of a normal level of blood sugar. In fact, the liver is the major glucose-forming organ in the body. The storage of glycogen enables it to do this, but the liver is also the primary site for the manufacture of new glucose from protein (gluconeogenesis). Further information about this vital liver function will be provided later on in this book.

There is a disease known as glycogen-storage disease. As the reader might guess, this is a disorder in which glycogen is unable to be changed back to glucose. Persons with this congenital disorder have low blood sugar because of the failure of their livers to supply the blood with glucose. Glycogen-storage disease is due to the absence of an enzyme necessary to effect the change from glycogen to glucose. There are several different varieties.

As noted, glycogen is also formed and stored in the muscles of the body, but this muscle glycogen is not broken down into glucose, because muscle lacks the enzyme necessary for this. Lactate, however, an end product of muscle metabolism, may be converted to glucose in the liver. Thus, indirectly, muscle glycogen can give rise to glucose. Muscle glycogen is important in the physiology of muscular use.

The trained individual, such as the long-distance runner, is able to conserve his glycogen supply to a much greater degree than his untrained counterpart. Thus, the trained athlete can engage in more strenuous activity for a longer time than the untrained individual.

The skin of the body may also play a role in glucose homeostasis. When sugar is given to a person, the skin-sugar level rises and remains high for a longer time than the blood-sugar level. In diabetic patients, this is magnified. This may have something to do with the fact that diabetic patients are very prone to develop skin infections. Sugar, lingering in skin tissues, may interfere with the effectiveness of white blood cell phagocytes, as they ingest and attempt to destroy invading bacteria. If one is trying to alleviate skin infections, it may be quite helpful to eliminate sugar from the diet.

The starch in a potato which is eaten is changed to glucose by digestion. This digestive process permits a relatively slow absorption of glucose. The glucose derived from the potato arrives at the liver at a rate which the normal liver can handle. Because glucose flooding into the general circulation is minimal, if any, the blood sugar rises little, if any. In the process, the liver has had its glycogen banks restored, so glucose can be released to the blood *when the body needs and demands it.*

This contrasts considerably to the chain of events that occurs when sugar (sucrose) itself is eaten. Sugar, being very rapidly digested and absorbed, floods the liver with an amount of sugar that cannot be processed by the liver. As a result, the blood-sugar level rises considerably and brings forth body reactions which attempt to restore the blood sugar to a normal level. When sugar is eaten, the body gets a load of sugar *which it hasn't asked for.* It is as though the liver is putting out sugar even when the body does not need it, and when the body has not demanded it.

A coordinated endocrine (hormonal) mechanism is necessary to restore blood-sugar levels *when the body has demanded a glucose supply from the liver.* When the glucose load is presented to the body — *without bodily demands for it* — the endocrine mechanisms are apt to become faulty in their coordinated function. Eating sugar when it is not needed is like "crying wolf." The picture of an alarm state is presented when nonesuch exists.

Although starch, a complex carbohydrate, tends to increase blood sugar in a relatively trickling fashion, excessive quantities may exceed the blood-sugar level needed to fill the liver's glycogen banks. When this happens, the excessively elevated blood sugar evokes insulin secretion. This, in turn, stimulates the formation of fat, and a significant lowering of blood sugar may subsequently come about. Refined corn starch appears to be responsible for significantly more clinical problems in this regard than starch from other foods such as potatoes, beans, etc.

The result of chronic repetitive sugar intake may well be faulty insulin release, with the resultant chemical picture of low blood sugar, diabetes, or both. It also appears that in some cases, one's body tissues can overreact to normal amounts of secreted insulin, when sugar intake is large and frequent.

CHAPTER 4

CARBOHYDRATE: FROM MOUTH TO LIVER

We eat to live. Ever since the first one-celled organism ingested a food particle in its environment, we have taken nourishment into our bodies for survival. Our lives intimately revolve around our food supply. Our daily movements to job, to store, etc., have as their ultimate purpose the acquisition of food. By eating organic matter around us, we unleash for the chemical factories within us the energy of the sun, which has been stored in plants and animals (see Figure 8).

The energy production within our bodies must be accomplished smoothly and steadily without jerks and false starts. Moreover, energy must be constantly available for emergency circumstances that might arise. Body ecology, the state of affairs within us, must provide for a relatively steady internal state so that energy production can constantly proceed.

When humans were fashioned, their intestinal tracts were designed to handle food material provided in the form of plants as well as animals. For this reason, man is known as omnivorous.

Protein, fat, and carbohydrate are obtained by man from the food in his diet.

That constituent of food known as carbohydrate provides us with the most readily available energy of all foodstuffs. Carbohydrate (starches and sugars) in the United States diet accounts

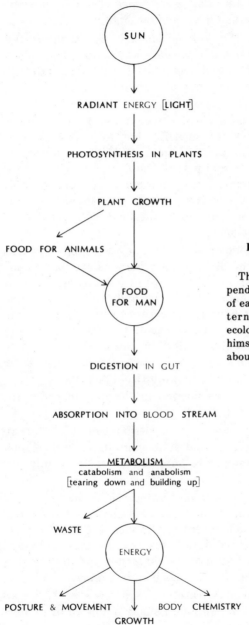

Figure 8. Process of energy production in man

The well-being of a person is dependent upon the efficient function of each of these steps. Problems external to man (environmental ecology) as well as problems within himself (body ecology) may bring about an impaired energy supply.

for some 40 - 65% of the total calories consumed each day. In some areas of the world, this figure may be even higher.

Four factors explain the widespread use of carbohydrates. First, carbohydrates — especially sweets — are addicting. They satisfy the "sweet tooth," and that satisfaction engenders a need to be satisfied again and again and again.

Related to that is the second factor. Carbohydrates are temporarily filling and satisfying. Never mind what happens later! Sweets provide instant gratification for persons who, childlike, wish to feel good *now* — again, and again, and again. Carbohydrates fit the lifestyle of persons immersed in an "instant society" and who live primarily for the moment. After all, if the TV turns on the moment I turn the switch, why shouldn't I feel good *now?* The wide availability of *refined* carbohydrates does nothing to help this situation.

Third, carbohydrate foods are cheap. They are generally less expensive than others. (The diets of poor persons are particularly high in sweets and starches. As a result, low-income families tend to be deficient in protein. This is especially so if the population is inland and does not have ready access to fish. Low-income persons who are nutritionally deficient waste time, money, and energy because of inefficient function or illness. A more nutritious diet might be temporarily a little more expensive, but the long-term dividends would provide a multifold payoff.)

Fourth, carbohydrates are needed in the diet in order for protein to be properly metabolized.

Reference has already been made to the studies of Nathan Pritikin that lead one to believe that a diet high in natural, whole, unrefined, complex carbohydrate may be most favorable for health.[1]

We eat carbohydrate in our diet as starch or sugar, and it ends up in our liver as simple sugars. This change takes place between the time that food enters our mouth and reaches our liver. Let us examine in more detail this remarkable "alchemy" of nature in which complex substances (starches and double sugars) are changed to simpler substances (single sugars).

About 60% of the carbohydrate in the diet is starch. We already know that starch is a compound sugar made up of numer-

[1]See Chapter 1, page 60.

ous glucose molecules bound to one another.[2] The end product of the digestion of starch is glucose. This dietary starch as well as double sugars in the diet must be broken down to their constituent simple sugars before absorption into the body can occur.

Starch digestion is apparently considered a very important survival mechanism by the body. The very first part of the digestive tract, the mouth, is provided with a saliva enzyme, amylase, that starts the digestion of starch. In the small intestine, amylase from the pancreas continues this digestive breakdown of starch to less complex molecules. (Oddly enough, human infants are often deficient in their ability to digest starch.)

Up to this point, starch digestion has taken place within the interior cavity of the mouth or gut. When the double sugar, maltose, and other starch fragments are produced from the starch, the scene of digestion shifts to the intestinal epithelial (lining) cell. In, or on these cells, maltose as well as other double sugars from the diet are digested to produce simple sugars.

The two principal double sugars (disaccharides) in the diet are sucrose[3] and lactose (milk sugar). Sucrose makes up at least 30% of the dietary carbohydrates; and in some individuals, it is much higher. Lactose accounts for about 10% of the dietary carbohydrates for adults.

The reduction of double sugars to simple sugars in the intestinal cells takes place because of the presence of enzymes within these cells. These enzymes are specific for each double sugar (see Figure 9).

The result of carbohydrate digestion and absorption is the entry of three simple sugars (monosaccharides) into the bloodstream. Glucose, galactose, and fructose are these three simple sugars. They are almost completely absorbed from the small intestine in the normal person. A greater degree of sugar absorption occurs in the first part than in the lower part of the small intestine. In deficiencies of the adrenal gland cortex, thyroid deficiency, and in states of B vitamin deficiency, there is decreased

[2]Starch has two components: one is called amylose and makes up 20% of starch, and the other component, amylopectin, makes up 80%.

[3]Sucrose has not always occupied such a prominent position in the diet of man, however. It is only within the last 200 years that a ready supply of sucrose (refined sugar) has permeated developed and developing societies.

absorption of sugars. Under normal conditions, fructose is absorbed more slowly than glucose or galactose, and it does not appear that there is competition between fructose and glucose for absorption.

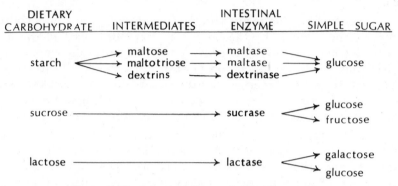

Figure 9. Breakdown of dietary carbohydrate in the intestine to simple sugars

Glucose, a simple sugar, is derived from starch, maltose, sucrose, and lactose. Fructose and galactose, other simple sugars, are derived from sucrose and lactose respectively.

(Figure 9 is modified from a figure by G. M. Gray in the journal, *Gastroenterology*, Vol. 58, 1970, pages 96 - 107, and it is used with the permission of the author and the Williams and Wilkins Company, Baltimore, Md.)

When the intestine has a high content of glucose in it, some glucose may diffuse into the blood. Following an ordinary meal, however, the glucose concentration in the intestine is probably less than that in blood, yet glucose is rapidly absorbed into the blood. In fact, glucose is absorbed into the blood even though the blood glucose may be very high, as in the case of diabetics. In order for this to occur, some kind of glucose "pump" exists in the intestinal cell wall. The nature of this pump is not known at this time.

Fructose enters the intestinal epithelial cells and is there converted to glucose for the most part. Further change to glucose occurs in the liver. Thus, blood in the general arterial circulation is virtually free of fructose.

The blood in the general circulation of the body is virtually free of sucrose, also, because of its rather complete alteration to glucose and fructose before absorption.

There is growing clinical suspicion that fructose may account for the harmful effects of excess sugar (sucrose) intake in the body. It may be that large amounts of sucrose in the diet may bring about changes in the ability of the gut to alter fructose to glucose. Although this is entirely speculative, it appears that somehow sucrose itself, or fructose, can be associated with harmful effects in the body not found with glucose. Perhaps in some persons, these sugars gain entry to the blood without change and then are toxic.

Although the three simple sugars (glucose, fructose, and galactose) are different chemical compounds, each readily enters into the metabolic pathways that use sugar for the production of energy.

The double and simple sugars involved in the digestion of carbohydrate are summarized in Figure 10.

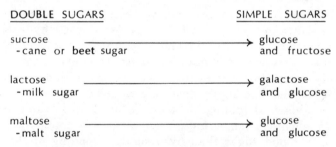

Figure 10. Double and simple sugars of importance in the digestion and absorption of dietary carbohydrate

When there is a deficiency in intestinal digestive enzymes, diarrhea, gas, and acid (burning) stools may result. Intestinal enzyme lack may be a congenital disorder or an acquired disability. A deficiency of the milk sugar enzyme (lactose) is not unusual in adults, especially in the black race. Such a deficiency in lactose produces an intolerance for milk, since milk is the source of lactose (milk sugar) in the diet.

Infants who sustain a bout of infectious diarrhea may also become milk-intolerant because of damage to the cells of the intestinal wall. Such infants continue to have diarrhea and skin-scalding stools long after their intestinal infection is gone, unless milk (lactose) is removed from their diet. A reliable lactose-free infant formula, CHO-Free Formula Base,[4] is available to feed infants with lactose deficiency. This preparation may also be used for individuals with sucrose-enzyme (sucrase) deficiency or maltose-enzyme (maltase) deficiency, since it contains no sucrose or maltose. Usually glucose (dextrose) or honey is added to the CHO-Free Formula Base to bring the calorie content up to an adequate amount. It is worth noting that commercial soy-milk preparations, used to feed individuals with milk allergy, usually contain sucrose.

When a person fasts (abstains from eating) there is a decrease in the activity of the intestinal enzymes. The addition of glucose to the diet will restore all enzymes to normal activity. If one consumes sucrose or fructose, the intestinal cells are stimulated to produce the sucrase enzyme. Since sucrose contains fructose, it appears that fructose is the agent which induces the development of the sucrose enzyme.

The interrelations of carbohydrate, fat, and protein metabolism are shown in overview in Figure 11. When carbohydrates are eaten together with protein and fats, they must compete for absorbing surfaces in the gut and, hence, they are absorbed into the body more slowly than if they were eaten alone. Except in rare situations, it is highly desirable to absorb dietary carbohydrate in a slow "dribbling" fashion rather than in rapid bursts. There is much less homeostatic change brought about by a gradual rise in blood sugar than by a sudden flooding of the blood with sugar. For this reason, it may be advisable to consume carbohydrate foods in association with protein and fats.

On the other hand, one may wish to consider the views of Natural Hygienists.[5] They believe that proper food combining enhances digestion and is favorable to health. Because the salivary amylase enzyme is active in a weak alkaline media, they believe that acid foods and protein should not be eaten with starch.

In some individuals, it may be wise to severely limit carbohydrate in the diet rather than merely to avoid sugar. Such nu-

[4]CHO-Free Formula Base, Syntex Laboratories, Inc., Stanford Industrial Park, Palo Alto, Calif. 94304.

[5]*Food Combining Made Easy* by Herbert M. Shelton. Published in 1951 by Dr. Shelton's Health School, San Antonio, Tex.

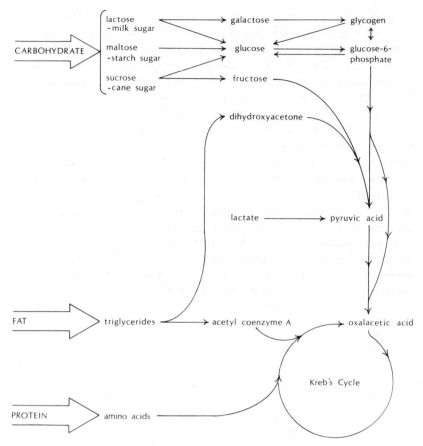

**Figure 11. Simplified diagram of carbohydrate metabolism
showing link-ups with fat and protein**
All three substances (carbohydrate, fat, and protein) must undergo digestion
in the gut and absorption through the gut wall before they enter into the metabolic scheme shown here.

tritional change should be carried out only on the advice of a
physician.

It should be noted that, so far, the pancreatic hormone, insulin, has not been involved in any way with carbohydrate entering the body. That's to come.

The simple sugars absorbed from the gut are transported
to the liver by the portal vein leaving the gut. In the liver, the

nonglucose sugars may be changed to glucose. Glucose may be stored as glycogen (if the storehouse is low), released to the blood for circulation to the tissues of the body (if it is needed), or broken down for energy. It should be remembered that sugar is the most concentrated carbohydrate that we eat. Because of this, it is easy to overload the body's metabolic machinery when one consumes table sugar (sucrose). A similar overload could occur in the individual who gorges himself on juices or fruits which contain considerable sugar. This is much less likely, however, than spooning out considerable amounts of refined sugar from a bowl.

The digestion of starch is accomplished, as has been pointed out, by digestive enzymes. Since our bodies do not always work perfectly, especially considering what we do to them, it is not unusual to find that a considerable number of individuals have faulty digestion of starch. A test for starch in the feces of an individual can indicate faulty starch digestion. In such cases in which excessive dietary starch is passed out of the body in the feces, an oral supplement can be provided which will provide starch-digesting enzymes.[6] Proper food combining may also be helpful.

[6]Enzyme supplements to aid protein digestion and fat digestion are also available and are usually given along with the starch-digesting enzymes. These substances can be of help in persons with allergies that presumably are due to incomplete protein digestion.

CHAPTER 5

INSULIN ENTERS THE PICTURE

A major accomplishment of the body is the changing of complex dietary carbohydrate to the final common product, glucose. This, however, is only the prelude of bigger and better things to come.

Glucose is of very little use to the body as a fuel when it remains in the blood. It is necessary for glucose to gain entrance to the interior of cells before it can be processed for energy.

Insulin is a protein hormone produced by the pancreas. Insulin is necessary for the metabolism of glucose in the body. The body in its internal wisdom "knows" that carbohydrate requires insulin so it can be processed for energy. Accordingly, carbohydrate entering the gut causes the release of some agent (the gastrointestinal signal)[1] which influences the pancreas to secrete insulin into the blood. The rise in blood sugar which follows absorption of carbohydrate is thus anticipated by the body.

Imbalances in the gastrointestinal signal, the timing of this signal with insulin release, and the absence or overworking of the signal may explain many of the findings in diabetes or states of low blood sugar.

[1]It is known that the gastrointestinal hormones — gastrin, secretin, and pancreozymin — stimulate insulin secretion. For this reason, glucose given by mouth stimulates more insulin release than glucose given by vein.

Other stimuli to insulin secretion may be elevated blood sugar, some amino acids (leucine, arginine), chemical substances known as ketones, growth hormone (somatotropin), adrenal-cortical hormones, obesity, and psychic stimuli.

The levels of minerals in the body also have an important influence on insulin production. Particularly important are calcium, zinc, potassium, and chromium.

When glucose is absorbed by the gut, it travels in the bloodstream until it reaches the liver. There it freely enters the liver cells. Glucose likewise travels freely from the blood into the red cells of the blood and into the cells of the brain, kidney, and intestinal wall. In all these areas, glucose freely courses into the interior of the cell without special help. In all other cells of the body (fat and muscles, for example), glucose is denied entry unless insulin is present to unlock the gate into the cell. Whenever insulin *is* present, glucose will enter the fat cells and bring about a build-up of fat within the cell.

This, then, is a principal action of insulin: to provide a means of entry for glucose into cells. One can thus see why insulin lowers the content of blood sugar in the blood.

Insulin secreted by the pancreas courses along with absorbed food materials in the portal vein until it reaches the liver. The great importance of the liver in the body economy is shown by the fact that about half of the insulin is utilized within the liver (see Figure 12). Insulin has two effects in the liver: to encourage the formation of glycogen and to inhibit the production of glucose from protein. The overall effect is to decrease the amount of glucose leaving the liver.

There is evidence that the action of insulin works directly on the cell membrane. Insulin may serve to open pores in the cell through which glucose molecules can travel. One can readily understand that an increased number of cells in the body (for example, in obesity) would require an increased amount of insulin in order to allow normal glucose intracellular metabolism.

Special locations on the cell surface known as insulin receptor sites may determine the effectiveness of insulin in facilitating glucose entry into cells. The biochemical makeup of insulin is another factor that may determine whether insulin is operational or not.

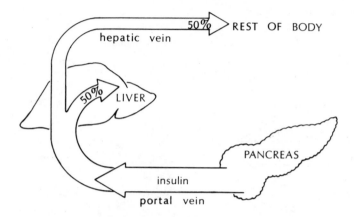

Figure 12. Dispersal of insulin from the pancreas
The liver and pancreas are outgrowths from the gut. Their function is intimately involved with that of the gut. Insulin is secreted in response to events which take place in the gut and then travels to the liver and all other body tissues to exert its actions.

Once inside cells, glucose enters into a chemical reaction in which phosphate is added to the glucose molecule. The substance, glucose-6-phosphate, is formed.

$$\text{glucose} \longleftrightarrow \text{glucose-6-phosphate}$$

This phosphorylation of glucose "locks" glucose in the cell, because the cell membrane is impervious to phosphoric acid compounds such as this. This is an indispensable step for the subsequent metabolism of sugar. Once glucose is phosphorylated, it has been prepared so it can proceed in various pathways of metabolism within the cell.

In this phosphorylation reaction, the phosphate is obtained from a compound known as ATP (adenosine triphosphate). The reaction is catalyzed (facilitated) by the enzyme hexokinase. Insulin also enhances or promotes the hexokinase reaction which is responsible for the coupling of phosphate to glucose.

Thus, we can see that insulin is an essential element in getting glucose involved in intracellular metabolism (see Figure 13).

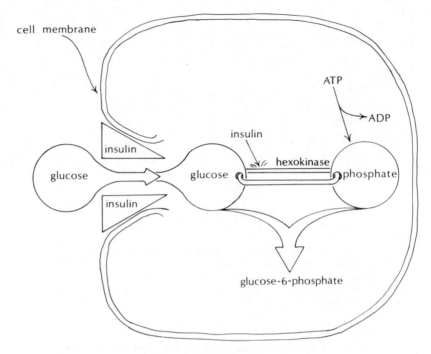

Figure 13. Actions of insulin

Insulin permits glucose to gain entry to cells; and once therein, insulin keeps glucose there by encouraging its binding with phosphate.

Actually, the relationship of glucose to the phosphorylated glucose is more involved than has been presented. The addition of phosphate, however, to glucose "captures" the glucose molecule within the cell. This reaction is encouraged by three enzymes: hexokinase, glucokinase, and the synthetic action of glucose-6-phosphatase. Only when phosphorylated glucose is split (hydrolyzed), producing free glucose, may it then leave the cell.

The formation of glucose from protein is accomplished largely in the liver. It is known as *gluconeogenesis*. The process starts with certain amino acids known as glucogenic (glucose forming) amino acids. Insulin inhibits gluconeogenesis.

In summary, then, insulin brings about a fall in the blood sugar as a result of

1. an increase in the use of glucose by tissue cells and

2. a decrease in glucose output to the blood from the liver.

Further on in this book I will tell you more about this most amazing hormone, insulin.

At this point I will summarize some factors at work in the body which regulate the level of glucose in the blood. They are

1. the amount of available insulin,
2. the amount of glucose formed from protein and fat (gluconeogenesis), and
3. the amount of glucose production from glycogen (glycogenolysis).

In the next chapter, I will tell you more about what happens to glucose in the body.

CHAPTER 6

THE FATE
OF GLUCOSE

As we have seen, glucose is the major fuel for the cells of the body, and ordinarily it is the necessary fuel for brain and blood cells. Because of the importance of this simple sugar in the body economy, I will further describe the fate of glucose in the body.

The factors that sustain an effective level of glucose in the blood as well as those which tend to lower the glucose level are as follows:

Factors That Lower Blood Sugar	Factors That Sustain Blood Sugar
Liver function (glycogen formation)	Liver function (glycogen breakdown and gluconeogenesis)
Fasting	Eating
Exercise (prolonged)	Glucagon
Insulin	Anterior pituitary hormones: growth hormone, ACTH
	Adrenal hormones: glucocorticoids, adrenalin
	Obesity

Notice that the liver can contribute to elevation of the blood sugar or, when glucose stores are being rebuilt, the liver can contribute to lowering blood sugar.

When glucose is "burned" by the body to provide energy, it is broken down in a series of sequential chemical steps. This is known as *glycolysis*. This process of glucose degradation takes place in the cytoplasm[1] of cells. It has already been explained that insulin is necessary for the very first steps in this chemical sequence. Insulin also activates phosphofructokinase, a key enzyme in glycolysis.

At least ten chemical steps are involved in the glycolytic breakdown of sugar for energy. In this process, the 6-carbon molecule, glucose, is eventually changed into two smaller compounds (fragments), each containing three carbon atoms.

Glycolysis is the basic skeleton of metabolism. From an evolutionary standpoint, it is an ancient biochemical process. Since life arose under oxygen-free conditions, it is not unexpected to find that glycolysis takes place in the absence of oxygen (anerobic).[2]

Glycolysis is also known as the Embden-Meyerhof pathway of metabolism. Some 10 - 20% of the energy produced in the breakdown of glucose is obtained in anerobic glycolysis. The "lion's share" of energy is produced by the further breakdown of the small chemical fragments in the Kreb's Cycle.

The Kreb's Cycle is an aerobic[3] process and is also known as the tricarboxylic or citric acid cycle.

In evolutionary development, the Kreb's Cycle is a latecomer. Its oxygen-using biochemical steps are later embellishments added on to an already functional anerobic metabolism.

Mitochondria are sausage-shaped organelles located in the cytoplasm of the cell (see Figure 14). The chemical reactions which make up the Kreb's Cycle take place within these intracellular bodies. Because of the large amount of energy released in the Kreb's Cycle, the mitochondria are known as the "power plants" of the cell. In the Kreb's Cycle within the mitochondria, most of the energy belonging originally to phosphorylated glucose is released and captured in the form of high-energy phosphate bonds.

[1]Cytoplasm: the protoplasm contained within the cell membrane and which surrounds the nucleus.

[2]Anerobic: without oxygen.

[3]Aerobic: with oxygen or oxygen-consuming.

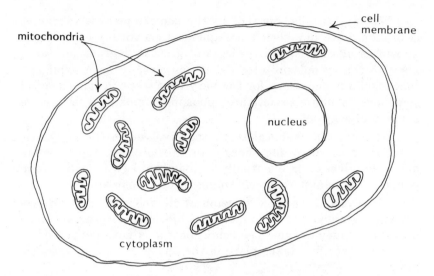

**Figure 14. Diagram of a cell with mitochondria
("power plants") in the cytoplasm**
    Within the mitochondria, the all-important energy-producing Kreb's Cycle
takes place.

    In actuality, the successive oxidative reactions in the Kreb's
Cycle produce a pool of a chemical substance known as reduced
nicotinamide adenine dinucleotide (NADH). This substance is a
principal reactant (substrate) in the process known as oxidative
phosphorylation. In this process, high-energy phosphate bonds
known as ATP (adenosine triphosphate) are formed from ADP
(adenosine diphosphate) and from the energy liberated in the
oxidation of NADH by the mitochondrial electron transport chain.
Flavoprotein, coenzyme Q, and the cytochrome enzymes are also
involved in oxidative phosphorylation.
    For those readers who enjoy chemical equations, here is the
overall formula representing the changes that take place as glucose is metabolized through glycolysis and the Kreb's Cycle:

$$C_6H_{12}O_6 + 6O_2 + 38ADP + 38P \longrightarrow 6CO_2 + 6H_2O + 38ATP$$

One can recognize that 1 molecule of glucose sets in motion a
series of steps that results in the formation of 38 molecules of
high-energy phosphate substance (ATP).

The metabolism of carbohydrate in the body is of central importance in understanding the relationship of body metabolism in general. Some of the important areas of interaction with fat, protein, and nucleic acids are shown in Figure 15.

Throughout the metabolic pathways for carbohydrate metabolism, vitamins and minerals play important roles. In fact, we can say that vitamins and minerals are fundamental elements in metabolism. The mineral magnesium, for example, participates in more than three-fourths of the enzyme reactions in the body! Interestingly enough, magnesium is also the key mineral which is involved in photosynthesis in plants. Vitamin B_6 is involved in a host of metabolic reactions. The essential nutrients of the body are closely linked one to another. Lack of any one vitamin or mineral may affect the body requirements of others. Multiple insufficiencies are apt to develop when the body is under stress: for example, in growth, illness, pregnancy, or lactation. Those persons who do not eat properly or who do not absorb their food are particularly prone to develop vitamin and mineral deficiencies with interruption of normal metabolism. In the United States, the land of plenty, it is not unusual to find individuals with vitamin or mineral deficiencies due to faulty diet or exaggerated needs.

The function of the Kreb's Cycle is highly dependent upon adequate vitamins and minerals. Copper, for example, is involved early in the cycle, magnesium further along, and iron still later.

Whenever glycogen is formed, two molecules of potassium are deposited in the glycogen for every molecule of glucose. Because of this, potassium deficit could come about when glycogen is formed in large amounts. This is especially evident if the diet is deficient in potassium. Potassium deficiency in the diet is becoming an increasing health problem, as more and more refined carbohydrates are consumed.

The next two paragraphs will deal with the function of vitamins in metabolism. The terminology in these two paragraphs is somewhat technical. Skip the next two paragraphs if you are not interested in these biochemical aspects.

Vitamins are the precursors of coenzymes in the body. Coenzymes combine with protein substances (apoenzymes) to form enzymes (holoenzymes). An enzyme combines with a substance (substrate) and catalyzes the formation of a product.

A. Glycogenesis is the formation of glycogen from glucose.

B. Glycogenolysis is the breakdown of glycogen to glucose.

C. Glycolysis (Embden-Meyerhof pathway) is the breakdown of glucose to pyruvic acid, oxalacetic acid, and acetyl-coenzyme A. In the absence of oxygen, lactic acid is formed from pyruvic acid.

D. Gluconeogenesis is the new formation or buildup of glucose from fats or proteins.

E. The pentose shunt is an alternate route of glucose breakdown. There are other alternate paths such as the polyol pathway and the glucuronic acid cycle which are not depicted here.

F. Fat synthesis. This process is promoted by insulin and high blood-sugar levels.

G. Protein synthesis. This process is promoted by insulin.

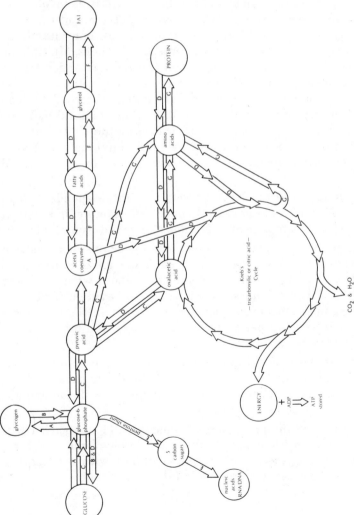

Figure 15. Simplified schematic diagram of carbohydrate metabolism and its relationship with fat, protein, and nucleic acids

Thiamine (B_1) functions in aldehyde-transfer reactions. Pyridoxine (B_6) is useful in transferring amino groups and in decarboxylation of amino acids. Riboflavin (B_2) aids hydrogen-atom (electron) transfer. Cobalamin (B_{12}) is involved in transfer of methyl groups and hydrogen atoms. Folic acid assists in transfer of carbon fragments. Niacin or nicotinic acid (B_3) is active in hydrogenation (electron transfer). Nicotinamide dinucleotide has already been mentioned as a product of the Kreb's Cycle activity and as a vital factor in oxidative phosphorylation. Ascorbic acid (C) is a reducing and chelating agent. Pantothenic acid (B_5) aids transfer of acyl groups. The vitamin biotin activates carbon dioxide and assists in its transfer from one substance to another. The reader who wishes further information about the role of vitamins in metabolism should consult *Review of Physiological Chemistry* by Dr. Harold A. Harper (Ph.D.). Published in 1975 by Lange Medical Publications, Los Altos, California.

Now I will use alcoholism to illustrate the interrelationship of a vitamin with glucose.

The chronic alcoholic may develop a neurological disease involving a defect in memory and ability to learn. This disorder is known as the Wernicke-Korsakoff syndrome. It is due to a specific deficiency in thiamine (vitamin B_1). The administration of thiamin to the alcoholic with this disorder is curative and may change an individual from a noncontributing derelict in society to a person who is able to remain a functioning member of the culture in which he lives.

When a nutritionally-deficient alcoholic is treated, it is often done in a hospital emergency room where the patient may have been taken because of "whiskey fits," simple drunkenness, falls, intolerable social behavior, or delirium tremens (the D.T.'s). Often, in such cases, the patient is treated with an intravenous solution containing glucose. If glucose alone is given, the chronic alcoholic may develop the Wernicke-Korsakoff syndrome, precipitated by the administration of glucose without thiamine.

Because of the vast importance of the Kreb's Cycle in producing the energy-for-life processes, a more detailed outline of this cycle is presented in Figure 16.

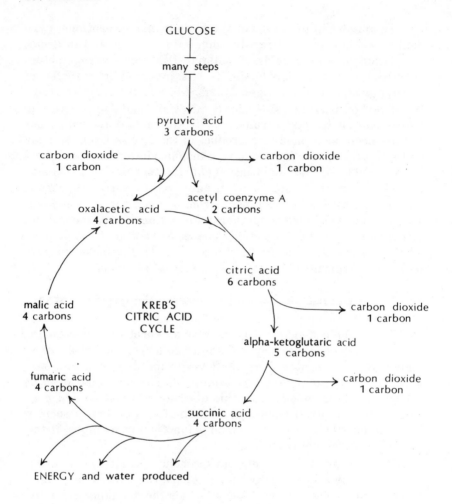

Figure 16. The Kreb's Cycle

The 2-carbon compound acetyl-coenzyme A enters the cycle by condensing with the 4-carbon compound oxalacetic acid to form the 6-carbon compound citric acid. With each turn of the cycle, carbon dioxide, water, and energy are given off, and oxalacetic acid is regenerated. This cycle is an aerobic (oxygen consuming) process which occurs inside the mitochondria of cells.

This diagram has been simplified for instructive purposes. In actuality, reduced nicotinamide adenine dinucleotide (NADH) is produced in the course of citric acid oxidation. This pool of NADH serves as substrate for a process known as oxidative phosphorylation. The end result of the oxidation of NADH is the formation of water and energy which is stored as ATP (adenosine triphosphate).

In order to further appreciate the crucial interactions that occur when the smaller chemical fragments enter the Kreb's Cycle, Figure 17 should be consulted. It has become increasingly clear that a balance or equilibrium should exist at the gate to the Kreb's Cycle. This concept has been elaborated by Dr. George Watson (Ph.D.),[4] who engages in psychochemical research and treatment.

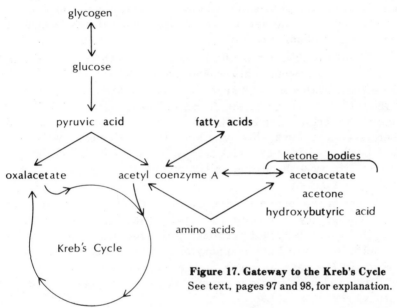

Figure 17. Gateway to the Kreb's Cycle
See text, pages 97 and 98, for explanation.

As is the case in the body as a whole, balanced function is desirable in the metabolism of foodstuffs. Excessive reliance upon any one type of food may be dangerous. Disturbance of chemical balance constitutes a disruption of homeostasis. When this occurs, the body's function becomes less and less efficient.

Metabolism proceeds best when carbohydrates, fats, and proteins assume a balanced role in the production of energy and growth. An example of this may be the relationship of carbohydrate and fat at the gateway of the Kreb's cycle.

When carbohydrate is not being utilized normally, fatty acid degradation increases. This brings about the increased produc-

[4]*Nutrition and Your Mind* by George Watson, Ph.D. Published in 1972 by Harper and Row, Publishers, Inc., 10 East 53rd Street, New York, N.Y. 10022.

tion of acetyl-coenzyme A ("acetate") with no corresponding increase in the compound known as oxalacetate. The excess "acetate" may be unable to proceed into the Kreb's Cycle, because
there is insufficient oxalacetate with which it can combine. The
excess "acetate" may lead to the formation of ketone bodies (acetone, etc.), and the Kreb's Cycle may be in a state of "starvation." Let us recall that the brain and the blood cells are dependent upon proper glucose metabolism for continued function. In
diabetic acidosis, the severe lack of insulin produces failure of
glucose metabolism with ketosis.

A similar starvation of the Kreb's Cycle with lack of energy
production may occur when carbohydrate is being used in excess of fats. In this situation, excessive oxalacetate may be formed
in relation to acetyl-coenzyme A. An imbalance again may occur.
Again, the Kreb's Cycle may not proceed at a normal rate to
liberate energy for bodily function. Many Americans who consume large quantities of "quick-energy" foods complain of chronic
fatigue. It may well be that in the midst of plenty their Kreb's
Cycles are starved.

Sugar in the form of glucose is a vital core substance for the
body. The ability to maintain an optimal level of blood glucose
for cellular metabolism determines to a large degree one's state
of health. In Figure 18, a summary of glucose homeostasis is presented. In the case of sugar in our bodies, not too much, not
too little, and just the right amount at the right time are crucial
watchwords.

Most often, conditions of illness are found to be associated
with a greater than needed supply of glucose to the body. In regard to obesity, Dr. George E. Schauf (M.D.) has pointed out that
the proportion of fat, carbohydrate, and protein in the diet is a
crucial determinant of whether food is stored as fat or used to
make essential enzymes, hormones, and cells.[5] Whenever enzymes,
hormones, and cells of the lean body mass are in short supply,
metabolism is at a low ebb. As a result, it becomes progressively more and more difficult to rid the body of accumulated fat.

Dr. Schauf has summarized some of the biochemical events
that take place. Glucose is necessary for fat formation and stor-

[5]"The Caloric Theory Does Not Apply to Obesity" by George E. Schauf,
M.D. Published in *Journal of the International Academy of Preventive Medicine,*
Vol. III, No. 2, December 1976, pages 33 - 41.

Figure 18.
Glucose homeostasis

Notice that ordinarily glucose is the necessary fuel for brain and blood cells. Other body tissues preferentially use glucose but may use ketones for fuel when glucose is in short supply.

The liver manufactures ketones from fats when glucose metabolism is blocked.

Gluconeogenesis, the new formation of glucose from protein (amino acids) and fat (glycerol), is carried out predominantly by the liver and is a major source of blood glucose. Glycogenolysis, the breakdown of glycogen to sugar, is the other source of blood glucose from the liver.

Notice that the kidney also performs gluconeogenesis, although to a much smaller extent than the liver.

(This figure is slightly modified from one kindly supplied by Dr. John I. Malone (M.D.), College of Medicine, Dept. of Pediatrics, University of South Florida, Tampa, Florida.)

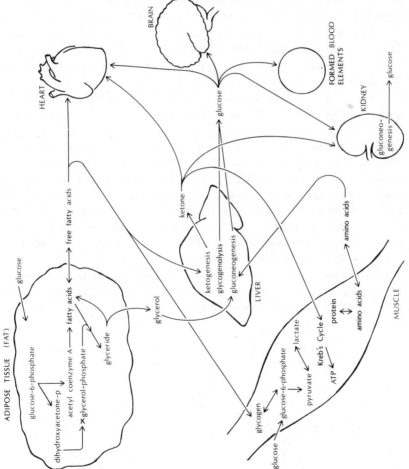

age. It supplies acetyl-coenzyme A, a 2-carbon fragment that is built into fatty acids. Glucose also supplies hydrogen ions for NADPH (dihydronicotinamide adenine dinucleotide phosphate), a substance that is necessary for fat formation. Insulin facilitates the conversion of glucose to fatty acids.

Fatty acids from food that is eaten or those made from glucose within the body must undergo a chemical change known as esterification (the making of an ester — see "Glossary") before they are stored as triglyceride fat within the cell. A substance known as alpha-glycerophosphate accomplishes esterification, and it is a breakdown product of glucose. Therefore, as Dr. Schauf points out, glucose must be available before either endogenous (internal) or exogenous (external) fats can be stored!

Fuel stores for emergency purposes in the body are available in the form of glycogen. Once the glycogen stores are filled, excess carbohydrate in the diet produces glucose which stimulates insulin release, and storage fat (triglyceride) is formed.

Many individuals in our society today visit their physicians and find that they are overweight and that their triglyceride blood fats are elevated. In nearly all cases, the blood triglycerides can be lowered by reduction or elimination of the following dietary substances: refined sugar, other refined carbohydrates, honey, juices, sweetened liquids, fruits, alcohol, and sometimes starchy vegetables and grains. In addition, avoidance of artificial indoor illumination with increased outdoor activity is a helpful measure in triglyceride reduction. In general, the lower the level of blood triglycerides, the better an individual feels and the fewer symptoms of illness he has.

For most of us, eating a wide variety of whole, nutritious foods and avoiding dietary floods of sugar allow our bodies to carry on the business of maintaining satisfactory chemical equilibrium.

CHAPTER 7

REFLECTION

The biochemistry of carbohydrate metabolism can become very complex. The material presented in this book has been selected, abbreviated, and summarized in an attempt to present the main themes as well as some of the important details as they relate to clinical disorders of carbohydrate metabolism. Readers wishing further details should consult standard texts (see "References," pages 489 and 490).

Let us pause at this point and review again what we know about carbohydrates in the body.

Carbohydrates are plant products which harness the energy of the sun. They are known as starches and sugars. They can be described as simple, double, or compound. Digestion of starch in the cavity of the gut is a slower process than is the digestion of the double sugars in the wall of the small intestine.

Glucose, a 6-carbon sugar, the eventual end product of carbohydrate digestion, circulates in the blood to supply the cells of the body with fuel (see Figure 19). It is stored in the liver as glycogen and utilized in the cells with the help of the pancreatic hormone insulin. The cortex of the adrenal gland supplies corticosteroid hormones to counteract the effect of insulin on the blood sugar. Adrenalin, glucagon, and growth hormone are also available in the normal individual as counterregulatory influences on blood sugar.

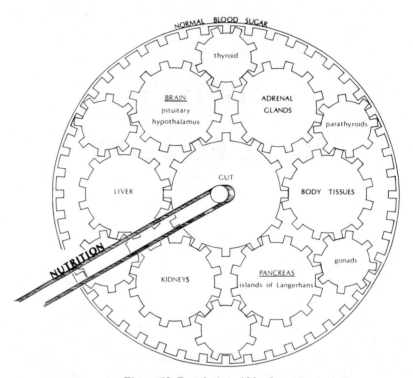

Figure 19. Regulation of blood sugar

A delicate balance in these factors determines the normality of the blood sugar. Nutritive input powers the whole system. The central importance of the gut is shown. Important organs are shown as interlocking cogs. Some cogs are not labeled, indicating that important determinants may yet be discovered.

Glucose may undergo various paths in its breakdown for energy, but the predominant route is by anerobic glycolysis to pyruvic acid,[1] a 3-carbon fragment.

This 3-carbon pyruvate may be changed to a 2-carbon compound, acetyl-coenzyme A, or to a 4-carbon compound, oxalacetate, by deletion or addition of 1 carbon atom, respectively.

These 2 compounds, acetyl-coenzyme A and oxalacetate, join together to form the 6-carbon compound, citric acid. Thus commences, within the mitochondrial power plants of the cell, the

[1]An acid in the body is often described in terms of its combined state, or "salt." Thus, *pyruvic acid* may be listed as *pyruvate*. For purposes of this book, the terms *pyruvic acid* and *pyruvate* may be considered interchangeable.

aerobic Kreb's (citric acid) Cycle, which is responsible for the "lion's share" of energy production for body chemical functions.

The final stage of energy capture actually occurs in the process known as oxidative phosphorylation. This series of biochemical events involves NADH (the reduced form of the coenzyme nicotinamide adenine dinucleotide), oxygen, and other substances to yield ATP (adenosine triphosphate), high-energy phosphate bonds.

Oxygen is consumed during this process of tissue respiration. In order for the entire program to work correctly, a heart must pump one's blood; oxygen must gain entrance to the lungs; blood must carry oxygen through open blood vessels to the tissues; muscles must be used; food must be obtained, digested, and absorbed; vitamins and minerals must be available in full supply; toxic chemicals must be absent; and waste products must be effectively removed from the body cells and the body as a whole. The lean body mass — the bones, joints, muscles, organs, and connective tissue — must be well nourished and habitually used at a level that develops and maintains efficient function.

The biochemical reactions involved in the metabolic pathways require adequate quantities of protein enzymes, mineral catalysts, and vitamin cofactors.

While all this biochemistry goes on at cellular and subcellular levels, the whole entity that we know as a person must find a satisfying niche in his environment. Moreoever, the person must, in order to thrive, become a contributing member and a significant force in his particular culture. He must experience stimulation, identity, and security; he must be important to some other member of his tribe or group; and he must have goals or objectives for which he strives. When an individual ceases to have goals that require some effort to attain, he usually withers away and dies.

The human body is a marvelous chemical and mechanical machine. It has also been provided with abstract reasoning ability and a soul. It is unfortunate that both the machine and the soul are so often neglected by their owners. It is hoped that this book, explaining carbohydrates, carbohydrate metabolism, and their relationship to health, will lead many readers to better health through more favorable diets, improved lifestyles, and an appreciation of human worth.

The time has come for enlightened citizens to care for themselves as well as they do their domestic animals and automobiles. By so doing, man's inhumanity to man may well be relegated to the past, as man and his body chemistry come into balance.

The following illustration is a part of the next chapter, Chapter 8. It contains information that will help you to better understand that chapter.

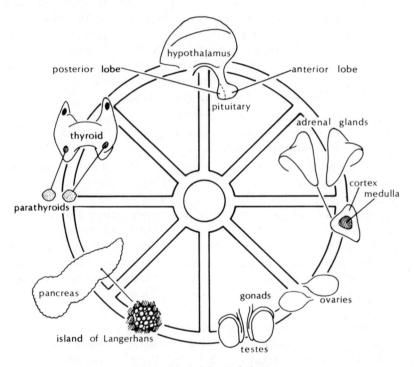

Figure 20. A delicate balance

Every human function is the result of a delicate balance between agonist, antagonist; secretion, nonsecretion; action, inaction; and other factors. In this diagram, the endocrine glands of the body are shown as vital contributors to the wheel of life. The text of this chapter discusses many of the hormonal factors that determine human function.

CHAPTER 8

INTERPLAY OF HORMONES: A DELICATE BALANCE

The hormone system of the human body is very much like a symphony orchestra. For proper performance, both the body and the orchestra depend on the accurate function of each individual member. Whether it's harps or hormones, one mistake may bring about altered performance from other members of the group, with an overall discordant result.

The endocrine glands which produce the major hormones of the body are depicted in Figure 21. It is entirely possible that additional endocrine glands will be discovered in the future. Organs that are currently thought to have no endocrine function may be found to produce hormonal secretions delivered directly into the bloodstream.

There is an especially delicate balance in the body between insulin and the hormones of the adrenal gland (cortex) and the anterior pituitary gland (see Figure 22). The overall effect of insulin is antagonistic to the effect of the glucocorticoid hormones of the adrenal and to the growth hormone of the anterior pituitary. Insulin lowers the blood sugar; but hydrocortisone from the adrenal and adrenocorticotropin hormone (ACTH) and growth hormone from the anterior pituitary tend to elevate the blood sugar. A decrease in insulin has the same action on blood sugar as does an increase in the adrenocortical hormone or the

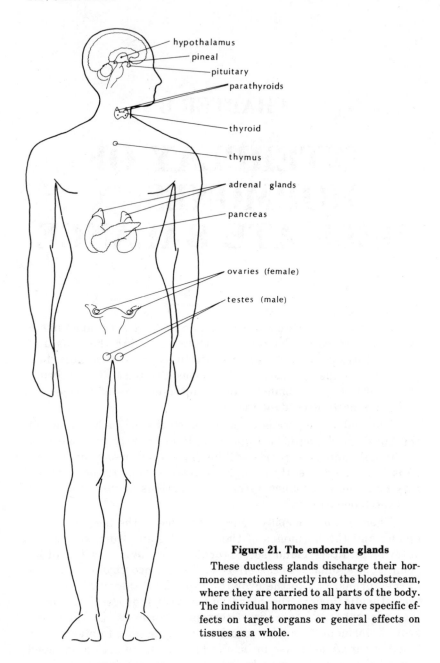

Figure 21. The endocrine glands

These ductless glands discharge their hormone secretions directly into the bloodstream, where they are carried to all parts of the body. The individual hormones may have specific effects on target organs or general effects on tissues as a whole.

pituitary hormone. It has long been known that diabetes produced by surgical removal of the pancreas can be relieved by
removal of the pituitary or adrenal glands.

Body metabolism involves processes which break down tissue (catabolism) and those that build up tissue (anabolism). The
hormones of the adrenal gland and the pituitary gland are predominantly those which encourage tissue breakdown. Insulin is
the prime substance in the body responsible for tissue buildup.
Because of this, the role of insulin is much greater than just a
regulator of sugar metabolism in the body. The proper rates of
formation of protein and fat, reflected in the normal growth and
repair of tissues, depend upon an appropriate supply of insulin.

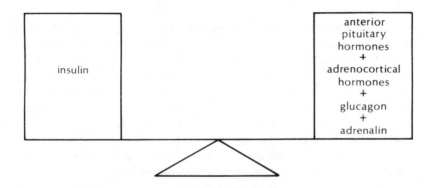

Figure 22. Hormonal balance
In the normal individual, a state of delicate hormone balance occurs, and
carbohydrate metabolism is normal. In the diabetic, insulin deficiency, insulin ineffectiveness, or relative insulin deficit occurs with impairment of carbohydrate
metabolism. Experimentally, removal of the pancreas (producing a diabetic
animal) can be cured by surgical removal of the anterior pituitary or the adrenal
gland.

From a clinical standpoint, obesity could be added to the scales on the right
side of the diagram, because it is often an important factor in "tipping the scales"
toward diabetes. The liver could be represented on both sides of the diagram, as
it acts as a balancing or compensatory factor for imbalances in blood sugar.

As we have already learned, insulin favors the formation of
glycogen (glycogenesis).

The breakdown of glycogen to glucose (glycogenolysis) is
encouraged by the following hormones:

epinephrine (adrenalin from the adrenal medulla),
growth hormone (somatotropin from the anterior pituitary),
thyroid hormones (thyroxine and triiodothyronine from the
  thyroid gland),
adrenocortical hormones (glucocorticoids from the adrenal
  cortex),[1] and
glucagon (the noninsulin hormone from the pancreas).

The breakdown of protein to form glucose (gluconeogenesis) is favored by adrenocorticoids but opposed by insulin.

Directly or indirectly, insulin affects the structure and function of every organ and biochemical function in the body. A direct effect of insulin on receptors in the brain has recently been discovered.

The power of insulin is evident when we realize that it is the only hormone whose major action produces a lowering of blood sugar. It does this by altering the permeability of muscle and fat membranes. Insulin also inhibits the output of liver glucose; inhibits the breakdown of fat; and, as mentioned, stimulates fat, protein, and glycogen formation. In addition, insulin has a reputation in some circles as an anti-infectious substance. Tiny doses of long-acting insulin have been used by some clinicians at the first sign of clinical infection. Others have used insulin topically (on body surfaces) for a presumed healing effect.

Insulin has been thought by some to have a cancer-inhibiting effect. Cancer cells which are grown in tissue culture require sugar. Patients with cancer frequently are found to have elevations of blood sugar. Small doses of insulin given on a daily basis have been used for its supposed cancer-suppressing effect. Certainly, much research needs to be done on this matter.

Epinephrine (adrenalin) and norepinephrine (noradrenalin) are the emergency adrenal medullary hormones released for fight or flight (see footnote number 1). Their release occurs in response to a drop in blood sugar as well as in states of fear, anger, pain, or exertion. When an adequate supply of glycogen exists, the release

---

[1]The adrenal glands are made up of an outer portion, the cortex, and an inner portion, the medulla. The cortical hormones are concerned with stress on a long-term basis and evidence a pronounced circadian rhythm. The medullary hormones are more concerned with stress on a short-term basis and are more related to events of an immediate "fight or flight" nature.

of these hormones brings about a breakdown of glycogen and a prompt rise in blood sugar.

Epinephrine and norepinephrine are the neurochemical transmitters of the sympathetic branch of the autonomic nervous system. They are the only internally released chemicals known to inhibit insulin secretion. It is possible that states of stress in which these chemicals are released in large amounts may be responsible for a diabetic state. Individuals are occasionally seen who "spill" sugar in their urine after frightening or anxious experiences. This is probably due to the release of epinephrine and/or norepinephrine that inhibit insulin and at the same time stimulate release of glucose from glycogen in the liver.

Excess insulin may lead to a deficiency of epinephrine. It is probable that some individuals make insufficient epinephrine and profit by epinephrinelike substances present in tea, coffee, cola drinks, or chocolate.

Thyroid hormone increases the rate of absorption of glucose from the bowel and it also facilitates the entry of pyruvate (3-carbon fragment) into the mitochondria.

The sex hormones, estrogen and progesterone in the female and androgens in the male, are important factors in determining our physical and physiological make-up. Each person, male or female, manufactures some male and female hormones. In some cases, these sex hormones or their effects are imbalanced, and the patient may be helped by administration of the appropriate substance. Premenopausal women who have had their ovaries surgically removed often need replacement amounts of estrogen. The advisability of using estrogen in females of postmenopausal age is less certain, although at times their use is helpful in managing psychic, circulatory, and orthopedic disturbances. At times, the use of progesterone in the female is more helpful than estrogen.

An association of breast cancer with the use of estrogens in the postmenopausal female has been noted. At that time of life, estrogen may bring about relief of undesirable symptoms at the same time that it exerts an undesirable biological effect. Relief of these same symptoms by dietary change and by the use of vitamins E, C, B_6, and certain minerals has been noted in many women. Significant hormonal imbalances in the body are to be avoided whenever possible.

Because men are more prone to heart attacks than women, the administration of female sex hormones to men has been considered. Recent technological advances in the measurements of hormones have, however, indicated that this may not be desirable. Men who have risk factors that are correlated with heart attacks (including elevated blood sugar) have been found to have high levels of estradiol, the major female sex hormone. Deviations from optimal hormone balance are not conducive to health.

Patients with cancers known as osteogenic sarcoma and chondrosarcoma have been noted to have a high frequency of abnormal glucose tolerance tests. Treatment with estrogen and/or progesterone has recently been shown to be associated with the return of these tests to normal at the same time that the cancer has stopped in progression.

Growth hormone (somatotropin) of the anterior pituitary gland has two known effects in the body: to stimulate growth and to elevate blood sugar. Growth hormone can be thought of as an insulin antagonist in regard to its effect on blood sugar. It directly inhibits the action of insulin at the cell membrane. It can cause diabetes, and it is found to be elevated in diabetics. Growth hormone also paradoxically stimulates insulin output by sensitizing the beta cells to release more insulin than they would otherwise do in response to glucose. Interestingly enough, growth hormone is ineffective in the absence of insulin. This emphasizes again the great importance of insulin for the build-up (anabolic) processes in the body.

Patients who are deficient in growth hormone are dwarfed in height. If given growth hormone, they will grow in height; but intact pancreatic and thyroid function is necessary for the growth-promoting effect to be realized.

Growth hormone has been synthesized in the laboratory. It is a 191-amino-acid peptide.

Insulin-induced hypoglycemia, stimulation with the amino acid arginine, and vigorous exercise can all be used as provocative tests to assess the normalcy of growth-hormone production.

Growth hormone has no single target organ upon which it exerts its influence. It produces many of its effects in the body through substances known as somatomedins. These chemically active agents are produced in the liver, kidney, and skeletal muscle as a result of growth hormone stimulation of these tissues. Somatomedin was originally known as the sulfation factor, because it stimulated chondroitin sulfate syntheses in cartilage, a necessary step in bone growth. It is now known to have other functions among which is the true growth-promoting effect of stimulating cell division. A particular kind of dwarf known as the Laron dwarf is unable to manufacture these somatomedin growth substances.

Somatomedin is structurally similar to insulin and may compete with insulin for its binding site in tissues. This may lead to glucose intolerance and diabeticlike GTT's (glucose tolerance tests).

Glucagon, the hormone produced by the alpha cells of the pancreas, rises in the blood whenever the needs of the body for fuel increase. This has been documented in emergency conditions such as severe infection, severe burns, starvation, exercise, diabetes, and severe trauma. Glucagon release may be the body's way of maintaining an optimal glucose supply for brain metabolism in stressful conditions which threaten the supply of blood to the brain. Glucagon also suppresses insulin synthesis in the beta cells of the pancreas. If glucagon is deficient, excessive insulin secretion and release may occur.

Figures 23, 24, and 25 show in diagramatic fashion the effects of some of the hormones on blood sugar.

The endocrine glands secreting their hormones directly into the bloodstream are highly interactive one with another. When a person has one endocrine deficiency, he is very apt to have an imbalance in the function of the other endocrine glands. Sometimes this can only be gauged in clinical terms, because our tests of endocrine function may not be sufficiently sensitive to measure the endocrine dysfunction. The pituitary gland, the "master gland," was formerly thought to be the primary source of stimulation for many of the other endocrine glands. It is now known that the underpart of the brain, the hypothalamus, releases hormones which are carried in blood vessels to influence the pituitary gland. Thus, we can see that the brain itself, at least the hypothalamic portion, is an endocrine organ.

One such hypothalamic secretion is somatostatin. It is a hypothalamic peptide also known as growth hormone release-inhibiting factor. This hormone turns off the pituitary production of growth hormone (somatotropin). Thus, somatostatin can be said to be a growth hormone antagonist. There is also evidence that

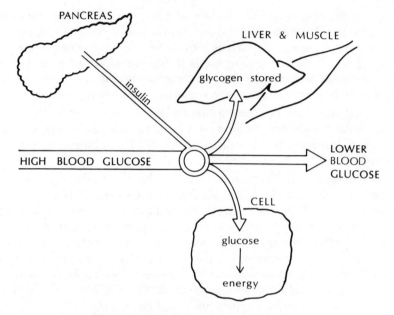

Figure 23. Effect of insulin on blood sugar

Insulin promotes the formation of glycogen and the cellular metabolism of sugar for energy. Insulin also promotes fat and protein formation, although these functions have not been shown in this diagram.

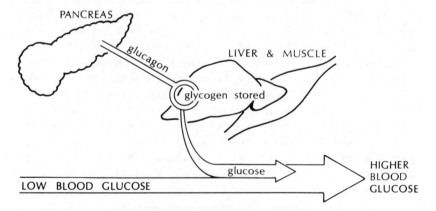

Figure 24. Effect of glucagon on blood sugar

Glucagon is a highly potent pancreatic hormone, which promotes the breakdown of stored glycogen to glucose. Its effect is to elevate the level of blood glucose. Glucagon may be used by diabetics whose treatment with insulin may bring about unwanted episodes of low blood sugar.

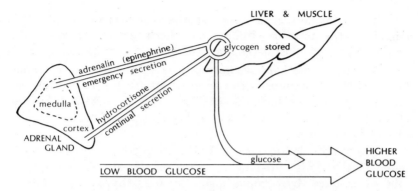

Figure 25. Effect of adrenal hormones on blood sugar

Both the inner and the outer parts of the adrenal gland secrete hormones which encourage the breakdown of glycogen to glucose. The inner secretions, adrenalin and noradrenalin, are released in acutely stressful conditions, whereas the outer secretions, hydrocortisone and the like, are more responsive to situations of chronic stress. The adrenal secretions have a glucose-elevating effect. Adrenocortical hormones counter the effect of insulin.

somatostatin inhibits insulin release by a direct action on the beta cells of the pancreatic islets. It has also been shown to inhibit glucagon, the anti-insulin hormone of the pancreas, as well as prolactin, thyrotropin, and adrenocorticotropin (ACTH), all pituitary hormones. The use of somatostatin, when long-acting forms are developed, may hold promise for the treatment of hard-to-manage "brittle" diabetics. Preliminary research indicates that the use of insulin plus somatostatin provides better control of blood-sugar levels than the use of insulin alone.

Preliminary news reports indicate that Dr. Andrew Schally (Ph.D.), who first detected somatotropin, has discovered an analogue of somatostatin that inhibits glucagon but not insulin.

Somatostatin is the second hormone that has been produced as the result of genetic engineering. In 1977, scientists at the University of California at San Francisco manipulated bacteria in such a way to produce somatostatin.[2] (Foreign genes were introduced into bacteria earlier that year by another team at the same university when they inserted a rat insulin-producing gene into bacteria.)

[2]"Bacteria Synthesize Brain Hormone." Published in *Science News of the Week*, Vol. 112, No. 20, November 12, 1977. Published by Science News, 1719 N Street, N.W., Washington, D.C. 20036.

It is probable that another hypothalamic hormone exists that stimulates the anterior pituitary to release growth hormone. As yet, however, this postulated hormone has not been isolated.

Another hypothalamic hormone is corticotropin releasing factor (CRF). It stimulates the anterior pituitary gland to release adrenocorticotropin hormone (ACTH). This important hormone stimulates the adrenal cortex to release corticoid hormones (hydrocortisone, etc.).

The hypothalamus also produces thyrotropin releasing hormone (TRH). This substance stimulates the anterior pituitary lobe to produce thyrotropin, also known as thyroid stimulating hormone (TSH). It is now realized that the level of TSH in the blood is a most accurate indicator of the adequacy of thyroid function. When thyroid function is deficient, the level of TSH rises to an abnormally elevated level, and it does so before other blood tests of thyroid function become abnormal.

The hypothalamus produces yet another hormone known as LRF. This luteinizing releasing factor stimulates the anterior pituitary to release gonadotropins, hormones that in turn stimulate the gonads to function.

These hypothalamic or brain hormones are clinically important substances which may soon play an important role in the management of endocrine disorders. They appear to be effective when administered to patients by mouth.

The various hormones from the pituitary gland travel in the bloodstream to their various target tissues in the body. These target glands release hormones in response to the pituitary hormones. When the level of the target-gland hormone is high, there may be a feedback effect which shuts off the production of the stimulating pituitary hormone.

Releasing as well as inhibiting hormones exist for growth hormone, prolactin, and melanocyte-stimulating hormone. In the case of the thyroid, adrenal, and gonads, inhibition of the hypothalamic-pituitary stimulation is by means of the secretions of these target glands.

The hormones circulating in the blood are known as first messengers. All hormones in the body appear to exert their particular effect through common substances in cells known as cyclic AMP and cyclic GMP (AMP: adenosine mono-phosphate;

GMP: guanosine mono-phosphate). These chemicals are known as the second messengers. They appear to translate the message of the hormones (the first messengers) into action inside the cells.

The overall situation in regard to hormones in the body is depicted diagramatically in Figure 26. (It should be realized that research developments are proceeding at a rapid pace so that any summary of hormonal balance in the body will be somewhat out-of-date by the time a publication appears.)

The hypothalamus itself is affected by inputs reaching it from other points of the brain. There is reason to believe that the visual input to the hypothalamus is particularly significant. Man is a predominantly visual creature. Drs. Skeffington, Getman, Streff, Greenstein, and other developmental optometrists have for many years emphasized the impact of vision on human function. They have stressed the vital role of vision in the overall development of mind and body, and they have pointed out the interrelations of vision with posture and movement, body warps, and feeling tone. I have witnessed patients with imbalances in body homeostasis who were improved when the stress of visual dysfunction was relieved. It may be that a more favorable hormonal balance is achieved when visual factors are improved. Since vision is a dominant sense in human operations, the detailed function of this system in relation to hormonal balance should be vigorously investigated. One would like to see a joint investigative project for this in which ophthalmologists, developmental optometrists, endocrinologists, and neurologists work in concert.

Intriguing possibilities for therapy arise when one considers the possible role of chorionic gonadotropin as a regulating agent of the hypothalamus. Human chorionic gonadotropin (HCG) is a hormone produced by the placenta in pregnancy. It is present in high amounts when brain development and endocrine glands are forming in the fetus. Dr. A. T. W. Simeons (M.D.),[3] along with Dr. Jonas Miller (M.D.) and a few other physicians have indicated that HCG is helpful in the treatment of some endocrine dysfunctions. It also is effective in many cases of migraine headache.

It is now known that psychological factors can profoundly influence the function of the brain and the endocrine system.

[3]*Man's Presumptuous Brain, an Evolutionary Interpretation of Psychosomatic Disease* by A. T. W. Simeons, M.D. Published in 1960 by E.P. Dutton and Co., Inc., New York, N.Y.

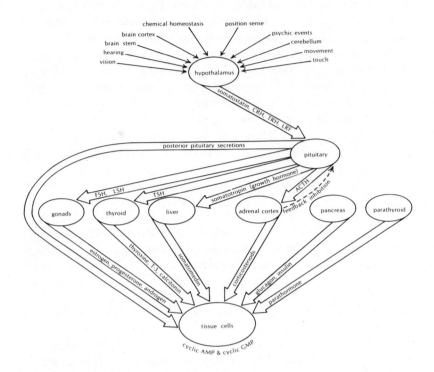

Figure 26. Interplay of hormones

See text, pages 115 and 117, for explanation.

For simplicity, feedback inhibition has only been shown for the adrenal glands. It also occurs for the thyroid and gonadal secretions.

Abbreviations:

CRH: corticotropin releasing hormone
TRH: thyrotropin releasing hormone
LRF: luteinizing releasing factor
TSH: thyroid stimulating hormone
FSH: follicle stimulating hormone
LSH: luteinizing stimulating hormone
ACTH: adrenocorticotropin hormone
T-3:   triiodothyronine

Growth hormone, TSH, FSH, LSH, and ACTH, are anterior pituitary lobe secretions. The posterior pituitary lobe secretions are vasopressin (antidiuretic hormone, pitressin) and oxytocin (pitocin). Pitressin acts upon the kidney to reabsorb water. Pitocin acts upon the uterus to stimulate smooth muscle.

For simplicity's sake, the pineal gland has not been included in this diagram. It is involved in the production of melatonin, a hormone having to do with body pigment and the sex glands. The pineal also secretes serotonin, an important chemical in the nervous system.

Other hypothalamic hormones exist but have not been included in this diagram.

There are some 40 or 50 hormones produced by the adrenal cortex. As yet the exact function of these hormones has not been classified. These hormones include gluco-corticoids, mineral-corticoids, androgens, and estrogens.

Growth-hormone deficiency occurs in psychosocial dwarfism. In this condition, the child is undergrown due to an abused environment. It is a disorder of child rearing in which the parents continuously reject the child. When such a child is removed from the home environment, normal pituitary function rapidly returns and catch-up growth occurs.

A particularly important relationship exists between the pituitary hormone, ACTH, and its target gland, the adrenal cortex. Whenever there is a situation of chronic adverse stress in the body, the hypothalamic-pituitary-adrenal axis brings about an increased response on the part of the adrenal cortex. A high level of corticoid hormones is released into the body and the inhibiting feedback of the pituitary is overridden. Thus there continues to be a high-level secretion of adrenal-corticoid hormones as a response to the stressful condition.

ACTH stimulation and the adrenocortical response constitute a defense mechanism against adverse stress. This system, however, was not apparently designed for long-continued operation. In the clinical situation, continuing states of adverse stress "wear out" the system, so its effectiveness is compromised or its continuing operation becomes a detriment to body function.

Conditions of chronic adverse stress — real or perceived — indicate to the body a need for chronic alertness and readiness for fight or flight. Adrenal secretions pour out and continue to do so. It is as though an alarm signal is stuck in the "on" position. Adrenocortical hormones, responding to the signal of chronic stress, promote tissue breakdown. The body does not tolerate forever the adrenal drive that fosters tissue breakdown at the expense of tissue buildup. As time goes on, body regulation suffers and degenerative disorders appear. Heart attack, diabetes mellitus, high blood pressure, cancer, arthritis, and various gut diseases of civilization may be related in part to disturbed endocrine balance.

In many cases, the adrenal cortex becomes depleted from

constant use and overuse, and it may cease to function as an effective regulatory agent.

The importance of adequate adrenocortical hormones can be appreciated when the body is placed under the stressful load of an oral sugar dose. When a glucose tolerance test (GTT) is carried out (see Chapter 11), the performance of the adrenal corticoids is decisive in regard to the outcome. In the normal response to ingested glucose, the level of hydrocortisone[4] in the blood rises 40% in the first 60 minutes, is declining by 90 minutes, and has reached a baseline level by 2 hours. If hydrocortisone does not rise in this manner, there is increased sensitivity to insulin with resultant lowered blood-sugar levels (hypoglycemia). Those individuals who have deficient production of hydrocortisone in this early phase are seriously out-of-balance in regard to body-hormone function.

The GTT can be looked at "from the other end." In the second phase of the test — that is, beyond 2 hours — declining blood-sugar levels may serve as a stimulus for the release of hydrocortisone and possibly other adrenocortical hormones. When blood-sugar levels dip too low, a counterregulatory adrenal secretion occurs. The result is elevation of blood sugar to normal levels. This occurs because of the immediate effect of adrenal-medullary hormones (epinephrine and norepinephrine) and the more prolonged effect of the adrenal-cortical secretion. The latter tends to maintain blood sugar at an adequate level by 3 mechanisms:

1. increasing the formation of glucose from protein (enhanced gluconeogenesis),
2. increasing the breakdown of glycogen to glucose (enhanced glycogenolysis), and
3. decreasing the peripheral utilization of glucose in the tissues.

ADRENOCORTICAL INSUFFICIENCY

Intact function of the adrenal glands is necessary to sustain life. When hormonal secretions from the adrenal cortex are in-

---

[4]Hydrocortisone is thought to be the principal carbohydrate-regulating hormone secreted by the adrenal cortex. Because of its prominent effects on glucose metabolism, it is known as a glucocorticoid.

sufficient, serious disturbances in body homeostasis occur.

Adrenocortical insufficiency may be primary — that is, due to dysfunction of the adrenal cortex itself — or secondary — due to a failure of the hypothalamic-pituitary system to stimulate the adrenal cortex to proper function. I have already mentioned the functional depletion that may occur under states of chronic adverse stress.

Adrenocortical insufficiency is characterized by symptoms of weakness, fatigue, nausea, vomiting, dizziness, poor appetite, pallor, low blood sugar, and perhaps states of allergy. Pigmentation of the skin, disturbance of tooth eruption, and growth disorder may be seen. Alopecia (hair loss) may eventually be shown to be associated with adrenocortical disorders. Mineral imbalance may occur. Sodium depletion as measured by a low sodium level in the blood is a hallmark of adrenocortical insufficiency. Elevation of the potassium level in the blood is likewise characteristic. Calcium disturbances also occur.

In hair-mineral analysis, elevated calcium and/or magnesium levels with depressed sodium and potassium or depressed sodium with elevated potassium are consistent with hypoglycemia and/or hypoadrenocorticism.

Most cases of pathological adrenocortical insufficiency have no apparent cause. They are known as idiopathic in origin. It is conceivable that these cases are due to faulty adrenocortical function in the absence of evident structural disturbance. Nutritional and stressful causes may well have a great deal to do with these cases. Allergic (hypersensitive) adrenalitis is known to be responsible for some cases of adrenocortical underfunction. Antibodies to adrenal tissue may be demonstrable. Idiopathic cases may also be due to allergic or maladaptive causes that do not demonstrate classical antibodies.

The adrenal gland is sometimes injured in the course of infectious disease. In bloodstream infection (sepsis) due to the meningococcal bacterium (Neisseria meningitidis), an acute destruction of the adrenal glands may occur. In this condition, severe hemorrhage takes place in the adrenal glands as the result of a toxic-allergic process (Schwartzman reaction) associated with the infection. This hemorrhagic destruction of the adrenal glands in meningococcal blood infection is known as the Waterhouse-Friderichsen syndrome.

Adrenal hemorrhage also occurs as a result of trauma during childbirth and at times of surgery. Irradiation may also damage the adrenal glands. Leukemia is another cause of hemorrhagic adrenal destruction.

Tuberculosis has been recognized as a principal cause of pathological adrenocortical deficit. Fungal and viral infections may also cause adrenal insufficiency in rare instances.

The adrenogenital syndrome is a group of defects caused by enzyme deficiencies that interfere with the synthesis of hydrocortisone, aldosterone, or both corticosteroids. The adrenogenital syndrome, according to traditional medical interpretation, is the commonest cause of adrenal insufficiency in childhood.

Secondary adrenocortical insufficiency is due to pituitary insufficiency, hypothalamic lesions, or to the exogenous administration of corticosteroid drugs that suppress the normal pituitary output of ACTH.

The disorder known as Addison's disease is a severe pathologic form of primary adrenocortical insufficiency. It may be due to any of the causes already enumerated. In this severe form of adrenocortical deficit, the entire spectrum of problems is manifest. The Addisonian patient is heavily pigmented in scars, creases, and the nipple areas. He may die during an "adrenal crisis" in which shock, sodium depletion, and dehydration occur. Weakness can be profound, and resistance to stress (infections, etc.) is poor. Hypoglycemia may occur. Salt, sugar, fluids, and replacement adrenocortical hormones are effective treatment when appropriately applied.

The Adrenal Metabolic Research Society of the Hypoglycemia Foundation, Inc.,[5] believes that hypoadrenocorticism is a much more prevalent condition than has been indicated. This group believes that hypoglycemia is a very frequent disorder and that the hypoglycemia is only the most prominent manifestation of hypoadrenocorticism, the underlying disorder. Congenital errors of metabolism as well as faulty diet and excess stress are seen by this group as causes of this condition.

Traditional medical views and the beliefs of the Adrenal Metabolic Research Society are at loggerheads in respect to concepts of adrenocortical insufficiency and hypoglycemia. Fur-

[5] 153 Pawling Avenue, Troy, N.Y. 12180.

ther discussion of this problem will be found in Chapter 14 dealing with hypoglycemia.

I believe that under situations of long-term adverse stress, the target organ — the adrenal cortex — may become unable to keep up an adequate production of corticoid hormones. In this case, the individual may fail to provide the adrenal secretions needed to cope with continuing needs. A state of functional adrenal depletion or exhaustion may occur and the person may develop symptoms of weakness, fatigue, and nausea, as well as the entire symptom complex associated with low-level hypoglycemia. A balanced arrangement should exist between insulin and the adrenal-corticoid hormones. In adrenocortical insufficiency, low blood sugar may occur because of relative insulin excess.

It is likely that there is a whole spectrum of disorders involving the adrenal cortex, ranging from Addison's disease on the one hand to an isolated deficiency of certain adrenocortical hormones on the other. The milder varieties may be termed functional adrenocortical insufficiency.

Whether these milder functional varieties are primary or secondary types of adrenocortical insufficiency remains to be seen. I suspect that a large number of them are secondary, associated with minimal brain dysfunction, hypothalamic dysfunction, low-level pituitary dysfunction, allergy, and nutritional deficiencies.

Those physicians who have had experience with hair-mineral analysis know the high frequency of low sodium and potassium content found in the hair of sick patients. It appears that sodium and potassium deficit in hair is a very significant finding, indicating adverse stress not well managed by the individual. These individuals appear to have functional adrenocortical insufficiency. They often crave large quantities of salt and sweets. Most of them respond to treatment with adrenocortical extract, some require allergic desensitizing injections, and nearly all require nutritional improvement and vitamin and mineral supplements. Nearly all profit from psychological counseling.

There appear to be many avenues of intervention that are successful in helping these individuals. In most patients, however, one particular form of therapy appears to be particularly suitable for that individual. Thus, if a patient needs allergy treatment and he only receives counseling, he is apt to continue to have symptoms of distress.

I have observed for many years that there is an association between overdue babies and the later development in these children of allergic conditions. Caucasian babies born significantly past their due date usually have very white skin. Many allergic children have this same type of very white skin; their pale appearance has prompted the term pseudoanemia to describe them. Individuals with pituitary insufficiency often have a very white alabaster appearance of their skin. Could it be that overdue babies and some allergic children have a form of pituitary inadequacy?

Evidence indicates that the postmature infant may be the victim of fetal adrenocortical insufficiency.[6] Prolonged pregnancy with progressive fetal ripening may be due to an absence of a hydrocortisone surge from the fetus. This production of hydrocortisone by the fetus is thought to be the signal for the onset of labor. Could it be that fetal adrenocortical insufficiency is secondary to inadequate hypothalamic-pituitary stimulation? I have called attention to the frequent coexistence of neurological dysfunction and allergy.[7] Is the overdue, postmature infant the earliest example of the neuroallergic syndrome? Answers to these questions await further research.

Dr. David S. Alexander, at the annual meeting of the Canadian Pediatric Society, reported the case of an 8-year-old girl with severe recurrent hypoglycemia.[8] The child was found to be deficient in the production of hydrocortisone because of a deficiency of pituitary ACTH. She was kept well by the administration of adrenal corticoid hormone.

This case shows that an isolated deficiency of ACTH or the hypothalamic-releasing factor of ACTH may occur and be responsible for hypoglycemia.

I suspect that science will eventually discover that brain,

[6]U. Nwosu and others in *American Journal of Obstetrics and Gynecology,* Vol. 121, 1975, page 366.

[7]*Allergy, Brains, and Children Coping* by Ray C. Wunderlich Jr., M.D. Published in 1973 by Johnny Reads, Inc., Box 12834, St. Petersburg, Fla. 33733.

[8]"ACTH Deficiency in Severe Hypoglycemia" by Dr. David S. Alexander, Canadian Pediatric Society. Published in *Pediatric News,* Vol. 8, No. 11, November 1974.

hypothalamus, and pituitary gland dysfunctions underlie or accompany many clinical health problems. Nutritive disorders and toxic states may interfere with the delicately poised neural transmitters as well as hypothalamic, pituitary, and adrenal function. Individuals whose brain function is dysorganized may fail to properly stimulate the pituitary gland to release its hormones in adequate amounts or in proper sequence. Pituitary malfunction, in turn, may result in faulty function of the target endocrine glands as well as diminished efficiency of body tissues in general. Fatigue, obesity, allergy, altered gastrointestinal function, mental depression, and manic behavior may be some of these clinical problems.

Dr. Melvin S. Page (D.D.S.) believes that some cases of manic-depressive illness are associated with a deficiency of pituitary hormone.[9] He uses very tiny doses of pituitary hormone to improve this condition. To the author's knowledge, his claims have not been substantiated by any hormone measurements presently in traditional medical use. Nevertheless, my clinical experience bears out his contention.

When the use of an adrenocortical substance is needed in the treatment of low blood sugar, allergy, inflammatory states, or adrenal insufficiency, I have found that hydrocortisone, dexamethasone, prednisone, or betamethasone have been helpful. In *Allergy, Brains, and Children Coping,*[10] I pointed out the clinical usefulness of these preparations in children with combined states of allergy and neurological dysorganization (the neuroallergic syndrome). Such children probably have cerebral allergy. Sometimes very small doses of adrenocortical hormones are required in treatment. At all times the use of adrenocortical-steroid medications requires close medical supervision. The use of these glucocorticoid hormones should nearly always be a temporary measure until definitive care of allergic, visual, nutritional, and emotional stress can be carried out. Glucocorticoid treatment should not be used at all if simpler treatment measures are effective. My more recent experience suggests that the use of adrenocortical extract (ACE) may, in most cases, be even more helpful than the use of individual proprietary corticosteroids.

[9]Personal communication from Dr. Page to the author.

[10]Published in 1973 by Johnny Reads, Inc., Box 12834, St. Petersburg, Fla. 33733.

Whenever a state of adrenocortical insufficiency occurs, current thinking indicates that there should be an increased level of ACTH in the body. This occurs when pituitary function is intact because of the lack of inhibitory feedback to the pituitary. I have already pointed out the symptoms that may occur due to lack of adrenal-corticoid hormones. Another set of symptoms may be noted as accompaniments of the high-level ACTH in the body. Such persons are likely to be "set in their ways." They manifest perseveration. They have difficulty in changing behavioral sets. When tested with galvanic skin resistance, a measure of reactions in the autonomic nervous system, they demonstrate slowness in extinction of responses. They are prone to escape from frustrating situations rather than trying to deal with them.

The result of this "high-ACTH behavior" is that the person does not readily change course, assimilate new information, or seek out challenging situations. He becomes bogged down in his old way of behavior, which may not be adequate for coping with tasks in the first place. He may also be reluctant to meet situations head on and to struggle with tasks that appear difficult.

It is hoped that a reliable and economically feasible test for blood-ACTH levels will soon be readily available for clinical use.

The adrenal cortex not only secretes glucocorticoids which have effects on carbohydrates but also mineral-corticoids which have effects on the minerals sodium and potassium in the body. Aldosterone and desoxycorticosterone are the chief mineral-corticoids; hydrocortisone and corticosterone are the principal glucocorticoids. Dr. Hans Selye (M.D.) has described the glucocorticoids as anti-inflammatory and the mineral-corticoids as proinflammatory (see *The Stress of Life* by Hans Selye, M.D. Published in 1956 by McGraw-Hill Book Co., New York, N.Y.). An imbalance between these two types of adrenocortical secretions can be responsible tor disease. In my experience, the most common situation is for the glucocorticoids to be deficient.

For some time now, persons outside of the mainstream of academic medicine have claimed that persons with weakness, low blood sugar, flat glucose-tolerance curves, and other symptoms have had low function of their adrenal glands. The concepts of the Adrenal Metabolic Research Society have already been pointed out. Adelle Davis, particularly, wrote and lectured about the importance of diet and vitamin supplements for adequate adrenal function. Japanese prisoners of war who ate nothing but rice had a measurable deficit in adrenocortical function.

When an academic physician examines persons with weakness, fatigue, headaches, etc., he does not usually find the pathologic state of adrenocortical insufficiency, Addison's disease. The physician therefore may conclude that there is no deficiency of the adrenal gland, because he has not commonly been taught that milder degrees of adrenal insufficiency may exist. Oriented to established pathological disease, the physician may fail to recognize disorders of function of this gland which may exist and which may precede structural damage in the body.

The use of the 6-hour GTT can indicate the presence of low blood sugar, a tipoff to the presence of adrenocortical insufficiency. To date, traditional medicine has a "love affair" with one adrenocortical secretion, hydrocortisone. Measurements of hydrocortisone in the blood may be perfectly normal in the patient with significant low blood sugar who responds to the administration of adrenocortical extract. When we are clinically able to measure many more of the adrenocortical hormones, we will be better able to assess the status of adrenocortical function.

It is my belief that there is such an entity as "mild" or subclinical adrenal cortical insufficiency, and that it might best be termed *functional adrenocortical insufficiency*. I believe that it is actually a rather common disturbance.

The hormones of the body, especially those of the adrenal cortex, are made of cholesterol esters, hydroxyl groups, and methyl groups.[11] Since hydroxyl groups and cholesterol are usually present in abundance, the presence of methylating groups becomes a critical factor in the production of these hormones. The source of important methyl groups is dietary. Methionine and choline in the diet are needed in proper amounts on a daily basis to supply methyl groups for hormone synthesis. According to computerized diet surveys obtained from office patients, methionine is one of the substances most commonly lacking in the diet. Vitamin C and the B vitamins are also involved in hormone production. One can readily understand how an inadequate diet can lead to adrenocortical insufficiency, especially when coupled with the

[11]"Hypoglycemia and the Methyl Approach" by Carlous F. Mason, D.Sc., Mato Laboratory, P. O. Box 7006, Riverside, Calif. 92503. (Publication date not given.)

increased demands of stressful living. Enzyme deficiencies in the body are also responsible for disordered hormone synthesis.

The problem is apt to be coincident with other subtle hormone dysfunctions (estrogen, progesterone, thyroid, pituitary, insulin) as well as with nutritional deficiencies, pancreatic dysfunction, gut disorders, and allergic disorders. A functional evaluation of laboratory tests may be necessary to suspect the problem from a laboratory standpoint. Laboratory tests which give values that are low normal, for example, may actually be abnormal for the individual in question.

The treatment of functional adrenocortical insufficiency may be the treatment of an accompanying nutritional or toxic problem, the treatment of a coexistent allergic disorder, the treatment of mental stress, or all together. Appropriate psychological counseling or behavior modification may be crucial in management. Finally, treatment with hormones may be needed. Frequently, all of the preceding treatments are needed to eliminate unwanted symptoms and to promote optimal health.

Further research in endocrinology will undoubtedly delineate more clearly the indications for hormonal treatment so that individuals with functional adrenocortical insufficiency may be restored to a symptom-free state.

I have collected a number of cases of children with learning disorders and health problems who appear to have functional adrenocortical insufficiency, but who do not have clinical Addison's disease. These patients have abnormally low levels of corticoid hormones as measured by the 24-hour urine-excretion test for 17 hydroxycorticosteroids. Their fasting blood sugars are usually low-normal when interpreted in the traditional medical sense (60 to 75 mgm.%). These values are distinctly lower than desirable for optimal function when looked at from a functional viewpoint.

In addition, many of them have an elevated level of eosinophils in the differential white blood cell count. Since one of the effects of corticoids in the body is to suppress the eosinophils, an elevated eosinophil level in the blood is an index of adrenocortical function. None of these patients have had parasites detected in their stool exams (parasites may give rise to eosinophil increases in the blood).

Many of these patients are allergic patients. This leads one to wonder whether their low adrenal function is a result or a cause of their allergic condition. Some observations have shown that the adrenal cortex is functioning below par in patients with allergic asthma. My observations suggest that the allergic problem is often a primary disorder, and functional adrenocortical insufficiency is a result of the longstanding allergic condition. Nutritional disorder must always be considered as an accompaniment, sequel, or cause of allergy and/or functional hypoadrenocorticism.

Many of these patients in my practice have clinical symptoms which resemble those of low blood sugar. These symptoms often disappear with low-dose glucocorticoid therapy, although the symptoms may not have disappeared with tranquilizers, antibiotics, or even antihistamines. Treatment measures other than glucocorticoids are available. Treatment with adrenocortical extract is most helpful as a rule and provision of oral adrenal material may be effective. Although these patients may be helped in this way, it is important to definitely correct the digestive, malabsorptive, infectious, allergic, nutritional, and emotional problems that often exist.

Malnutrition and states of chronic adverse stress are two factors which interfere with the proper function of the adrenal cortex. The most common form of malnutrition encountered in our society is nutritional deficiency from eating the wrong foods.[12] States of mineral toxicity such as lead, mercury, cadmium, and copper accumulation in the body may also be involved in adrenal dysfunction. It is likely that these conditions are associated with the clinical problems of these individuals whom I am describing.

A survey of these children with learning disorders has provided the information shown in Figure 27. Unfortunately, it was not possible to obtain similar determinations of adrenocortical secretion on a control group of children who were not having learning difficulties in school. This certainly should be done.

The average percentage of lymphocytes in the peripheral blood tended to be higher in those individuals with low corticosteroid excretion that in those with normal corticosteroid excretion. The degree of overlap in the groups, however, was too great to permit use of the lymphocyte count to distinguish between the groups.

When the analysis of hair minerals was reviewed and corre-

[12]In Chapter 15, recommendations are given for proper eating.

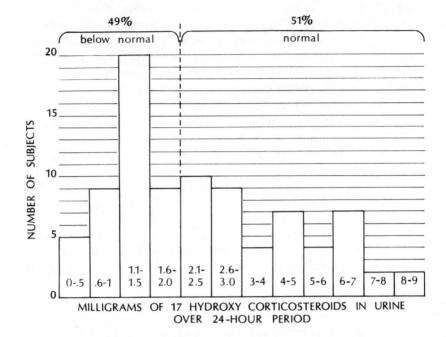

Figure 27. Learning disorders and adrenocortical function

Of 88 individuals ranging in age from 13 months to 46 years,* about half were found to have abnormally low adrenocortical function as measured by determination of 17 hydroxycorticosteroids in a 24-hour urine collection. Sixty-two out of 88 (70%) were below 3 mg. — that is, were below normal or were low normal.

The individuals who had values above 4 mg. tended to be those individuals over 8 years of age.

This large percentage of cases accumulating in the subnormal and low-normal levels would not be expected in a group of persons with normal adrenocortical function.

Sixty-one out of the total 88 individuals (69%) were males. Of the 43 individuals who had values below normal, 34 were males (79%).

*There were 9 adults and 18 children below 6 years of age. These individuals had symptoms of maladaptation to the stresses of living. The remaining 61 were children experiencing difficulty in adaptation to school.

lated with the urinary corticosteroid excretion, some interesting figures were obtained. The level of iron in the hair was found to have an inverse relationship to the level of urinary corticosteroids. Patients with normal levels of corticosteroids had an average level of iron in the hair of 17 parts per million. Patients with lower levels of corticosteroids had an average level of iron in the hair of 25 parts per million.

The ratio of sodium to potassium in the hair was also notable. An average of 1.8 was found in the low-corticosteroid patients and a ratio of 1.25 in those individuals with higher corticosteroids.

Lead and mercury levels in the hair also bore some relationship to urinary corticosteroid levels. Those patients with low corticosteroids tended on the average to have higher levels of lead and mercury.

The most consistent relationship was found when the average ratio of iron to manganese in the hair was compared with the urinary corticosteroid values. The figures are here presented:

urinary 17-hydroxy-corticosteroids (in milligrams/24 hours)	average iron/manganese ratio in hair
less than 2.1	21.50
2.1 - 2.9	18.75
3.0 - 5.9	17.50
more than 5.9	11.75

One can see that the iron to manganese ratio steadily declined as the level of urinary corticosteroids rose. If this finding can be substantiated, the relationship of iron to manganese in the hair could prove to be a helpful index of adrenocortical function.

More intensive evaluation of these patients with learning disorders who appear to have low adrenocortical function should be carried out. ACTH stimulation would be desirable to see if their adrenal-cortex glands are capable of responding to pituitary goading. It would be important, however, when testing with ACTH, to carry on the testing over a prolonged period of time to mimic the *chronic* stress situations that these children encounter in real life. In addition, the status of their hypothalamic and pituitary hormones needs to be thoroughly investigated. Certainly a study needs to be done in which plasma hydrocortisone levels are used rather than urinary corticosteroid levels. If the latter are not reliable indicators of adrenocortical function, the significance of the variations reported here needs to be clarified. Such research is expensive. It is hoped that funds can be provided for further endocrine research on patients such as these.

Several case studies of individuals with evidence of low adrenocortical function will now be presented.

Jimmy was a 7-year-old boy with asthma. He had entered this world vomiting. He had allergic eczematous skin rash as a baby and many allergic reactions to foods in infancy. He had become allergic to several antibiotics.

Whenever Jimmy had severe asthma, the use of cortisone medications seemed to be more helpful than any other treatment measure.

At a time when Jimmy had taken no cortisone preparations for 1½ years, two 24-hour urine tests were obtained for 17-hydroxycorticosteroids. These tests were made to assess the function of his adrenal cortex. A value of 1.5 and 1.3 mg. was obtained. Since normal values for his age are 2 - 8 mg., Jimmy's tests were lower than normal.[13]

Although Jimmy was not having asthma at that time, he was given a small daily dose (5 - 10 mg.) of hydrocortisone (an adrenocortical hormone). Over the next three months, he experienced better health than ever before in his life. He slept peacefully without tossing, turning, or dreaming. He was stronger and did not fatigue as the day went on. There was a "drastic" improvement in his sneezing. The appetite was improved, and weight gain occurred at a rate never before exhibited by this boy. A chronic, allergic skin rash on the elbows and knees disappeared. Most importantly of all to the boy's parents, he did not become ill at times when he previously would have certainly succumbed to illness: after excitement, weather changes, or exposure to cold. Jimmy also became able to tolerate foods which formerly would have made him wheeze, sneeze, break out in a rash, or vomit.

The plan is to leave Jimmy on this small daily dose of hydrocortisone indefinitely or to try him on adrenocortical extract in the future.

The reader should note that this is a special case. Asthmatics, as a general rule, do not require this kind of management. Corticosteroids should be reserved for those cases which are unresponsive to more conservative treatment measures; and when

[13]A creatinine excretion test was normal. The creatinine test serves as a check on the 24-hour urine collection. If it is normal, it indicates that the urine was accurately collected.

they are used, they should be used for the briefest possible period of time. The child under corticosteroid treatment should have careful medical supervision at all times.

Mercelene is a 4-year-old girl whose parents sought help because of her aggressive, violent personality. Uncontrollable temper, hyperactivity, flagrant irritability, and intolerance of any frustration were her hallmarks. She was described as "miserable" since birth, and her problems in self-control were becoming greater as she grew older.

A 24-hour urine specimen was done to measure the output of the hormones of the adrenal cortex. The result of this test was very low, and a repeat test in a different laboratory confirmed the low result.[14]

In view of this evidence for low adrenal-cortical function, Mercelene was placed on an oral dose of a replacement glucocorticoid preparation (hydrocortisone, taken by mouth in a low dose).[15]

The parents noted an immediate change in Mercelene's behavior. She became pleasant, communicable, and easy to manage. Her negative attitude, for the first time in her life, changed to one which was reasonable and which could be worked with. Hyperactivity grew much less, and the child was able to settle down for sleep at night without her usual 1 - 2 hour ordeal of crying, rocking, wandering, etc.

After one month of this "blissful" behavior, I instructed the parents to lower the dose of the hydrocortisone, by cutting it in half. As soon as this was done and for one week thereafter, Mercelene became hyperactive, irritable, negative, and sleepless. As soon as the previous dose was restored, the child again became placid, pleasant, and normally active.

An analysis of the minerals in her hair revealed an elevated level of mercury (28.0 parts per million) and lead (59 parts per million), and levels of sodium and potassium distinctly lower than normal. The child is now being treated to lower the mercury and

[14]The result of the 24-hour urine test for 17-hydroxcorticosteroids was .4 mg./24 hours with a normal creatinine value. The repeat specimen was .5 mg./24 hours with a normal creatinine value. The norms for this procedure in that laboratory for 3- to 6-year-olds are 2 - 6 mg./24 hours.

[15]5 mg. once daily.

lead and to increase the sodium and potassium in her body to more desirable levels. After this is accomplished, she will then be gradually withdrawn from hydrocortisone and followed carefully. It is anticipated that she will do well in the future without need for hydrocortisone medication. If she has further difficulty that appears to be an indicator for corticosteroids, however, adrenocortical extract may be given.

It seems likely that Mercelene was indeed a child whose adrenal cortex was functioning at a lower level than desirable. It may be that elevated mercury and lead in the body were responsible for this adrenocortical insufficiency.

In other patients, the presence of a chronic allergic state has seemed to bring on a state of adrenal depletion, and the treatment of the allergic state has greatly improved the patient's ability to cope with stress.[16] In still other patients, the repair of nutritional inadequacies has been followed by disappearance of hypoadrenocortical states. Similarly, emotional stress has been associated with lowered adrenocortical function, and this altered chemical state has been remedied when appropriate psychological counseling has been provided.

Evaluation of carbohydrate metabolism should be made in patients such as these. At this point, it is not known what the specific relationship is between this degree of adrenocortical insufficiency and sugar metabolism. One would speculate that hormonal imbalance would, indeed, be associated with disordered carbohydrate metabolism. Present experience indicates that glucose-tolerance curves show an inadequate rise or a decline instead of a rise after glucose loading.

The use of corticosteroid medication in the treatment of such cases should ordinarily be viewed as a temporary measure until other more definitive treatement can be carried out.

A 14-year-old boy with chronic hay fever and allergic sinusitis came to me because he was struggling to keep up in school. He was also plagued by recurrent headache, stomachaches, fatigue, and intermittent vomiting. His symptoms had been present for many years.

[16]See *Allergy, Brains, and Children Coping* by Ray C. Wunderlich Jr., M.D. Published in 1973 by Johnny Reads, Inc., Box 12834, St. Petersburg, Fla. 33733.

A 24-hour urine test for adrenocortical hormones indicated low adrenocortical function.[17] The peripheral blood count showed that 17% of this boy's circulating white blood cells were eosinophils ("allergy cells"). Blood tests for thyroid function were low normal. Microscopic examinations of the mucus from his nose revealed a large number of eosinophils ("allergy cells").

Blood-sugar examinations following the boy's usual breakfast ("natural" tolerance test) are shown on the graph in Figure 91, page 262. It is evident that this breakfast resulted in a *lowering* of blood-sugar values. The fasting blood sugar of 74 mg.% and the 6-hour value of 72 mg.% are lower than ideal, although technically they fall near or within the broad range of normal for most laboratory procedures for true blood glucose.[18]

An analysis of the minerals in this boy's hair showed a low sodium level, a very high level of copper and zinc, and lead of 46 parts per million (10 parts per million or less is desirable). Cadmium in his hair was also elevated.

This boy was given vitamins and minerals designed to restore the pattern of hair-mineral balance to normal. He was treated with desensitizing allergy injections for the substances to which he reacted on skin test. Foods to which he was allergic were omitted from his diet. Nutritional improvement measures were instituted.

Within a six-month period, the boy had lost his symptoms of headache, stomachaches, fatigue, and vomiting. He was achieving in school and was participating in the family discussions and projects. His attitude had changed from negative and belligerent to positive and cooperative.

[17]This boy's urinary excretion of a 24-hour urine collection for 17-hydroxycorticosteroids was 1.8 mg./24 hours. (Normal for this age is 2 - 10 mg./24 hours.) The creatinine excretion was 20 mg./kilogram/24 hours. (Normal for this age is 15 - 25 mg./kilogram/24 hours.) The latter test indicates that the 24-hour urine was collected properly. A repeat 24-hour urine test for 17-hydroxycorticosteroids was essentially the same.

[18]In the laboratory which I use, normal values for true blood glucose are 75 - 105 mg.%. In another laboratory, close by, the normal range of true blood sugar is 70 - 110 mg.%. It is imperative that the reader understand that this broad range is merely representative of the population upon which the test was standardized. It does not necessarily indicate the *desired* or optimal blood-sugar concentration. At this time, it appears that a fasting blood-sugar value of approximately 85 mg.% is optimal — that is, most likely to be associated with lack of symptoms and freedom from disease.

The hair-mineral imbalances returned to normal, and the urine test of adrenocortical function rose to a clearly normal level (6.5 mg./24 hours). The eosinophils in the peripheral blood were now 3% (normal).

This boy with low-level insufficiency of the adrenal-cortex gland did not have severe enough involvement to be diagnosed as Addison's disease. Nevertheless, his adrenal cortex was apparently not doing its job. This boy's treatment was followed by return of adrenal-cortical function to normal.

What caused the adrenal suppression or malfunction? Was it the chronic allergic condition? Was it dietary in origin? Vitamin deficiency? Was it the accumulation of lead? Cadmium? Copper? Zinc? Was it hypothalamic or pituitary dysfunction? Was it psychic? At this time, one can only speculate about the factors involved. It may be that each of these causes played a role. There is a strong suspicion that a diet high in refined carbohydrates is particularly stressful to the adrenal cortex. This is especially the case when B vitamins and enzymes in the diet are inadequate.

It is notable that this case was treated successfully without use of corticosteroid medication. As a result of experience with other cases such as this, I would speculate that this boy would have responded quite well to the temporary administration of very low-dosage corticoid medication or adrenocortical extract, properly supervised. Treatment of a child such as this is apt to be markedly improved when we understand the function of the many trace hormones which make up whole adrenal cortex extract.

Now that more precise laboratory tests are becoming available to measure hypothalamic hormones, adrenocorticotropic hormone (ACTH), growth hormone, thyroid stimulating hormone, and others, additional information about cases such as these should be forthcoming.

ADRENOCORTICAL EXTRACT (ACE)

Those physicians who find adrenocortical extract (ACE) to be of clinical use, point out that whole ACE contains 40 to 50 steroid hormones, of which 19 have been shown to be biologically active, according to current criteria. It may be, however, that all of the hormones will eventually be shown to be of importance to the function of the body.

Under normal conditions in which adverse stress is absent, hydrocortisone and corticosterone make up the bulk of the adrenocortical output. Under prolonged stimulation of the adrenal cortex with ACTH (a situation of adverse stress), the trace hormones have been reported to make up 80% or more of the gland's output.

It has also been noted that the action of mixed adrenocortical hormones on the nervous system is greater than that of isolated steroids. In rats who have their adrenal glands removed, there is depressed brain-wave activity, which returns to normal with hydrocortisone or adrenocortical extract. The latter, however, is said to be significantly more effective than the individual hormone.

The use of adrenocortical extract (ACE) in the treatment of hypoglycemia is a controversial subject. ACE is usually given by repeated intravenous injection over a period of weeks or months. ACE was originally used to treat Addison's disease before the synthesis of the cortisone group of drugs. Some physicians hold that ACE is a highly valuable preparation. The orthodox medical view, however, is that adrenocortical extract is obsolete and of little or no value. The reason for this belief is that ACE contains only .1 mgm. of hydrocortisone per milliliter, "an amount hardly capable of raising the blood-sugar level of an average mosquito, much less that of an adult man or woman."[19] Any beneficial effects from ACE injections are often attributed to a placebo effect, but in the author's opinion this is usually not the case.

If a large quantity of hydrocortisone is necessary to counteract the effects of insulin in the body, then these arguments against ACE are quite reasonable. ACE, however, contains many hormones other than hydrocortisone. The beneficial effects of ACE could be on account of these other hormones or conceivably it could be due to the tiny amount of hydrocortisone in correct balance with the other hormones.

I have patients who respond favorably to very small amounts of ACE given by intramuscular injection. These same patients have failed to respond to distilled water given as placebo injections.

---

[19]"Hypoglycemia" by Rachmiel Levine, M.D. Published in the *Journal of the American Medical Association*, Vol. 230, No. 3, October 21, 1974.

If the large number of trace hormones in ACE are considered to be biologically inactive, then the administration of ACE in treatment would not be indicated. On the other hand, if the large number of trace hormones are considered to have biological actions — even though such actions may not yet be identified — then treatment with ACE could be considered a helpful measure.

My experience and that of others[20] leads me to believe that ACE is of considerable value for some patients. The individual who is likely to respond is the person with relative hyperinsulinism unopposed by endogenous adrenocorticoids. The GTT (glucose tolerance test) is usually flat or declines after glucose loading. In some cases, the GTT is clearly normal from the standpoint of absolute values, but the patient experiences hypoglycemic symptoms as the blood sugar changes from peak to lowest values. In other cases, the GTT has been diabetic at some stages, hypoglycemic at others, or combined diabetic and hypoglycemic. Other indications that a patient may respond to ACE are muscle and joint pains and sleep disturbance. The trace hormones contained in whole adrenal cortex extract may have a regulatory action on insulin production and effectiveness. Thus, they may play a vital role in glucose homeostasis. According to Dr. Carlos Mason (D.Sc.),[21] an adrenalcortical hormone controls the amount of insulin made in the pancreas. Dr. Mason also states that an adrenocortical hormone destroys insulin when the blood sugar drops to a normal fasting level, and yet another controls the effect of insulin in allowing glucose to enter the cells.

Surgical removal of the adrenal glands as well as medical inhibition of their function are established modes of therapy for patients with breast cancer. The adrenal is removed or suppressed in order to turn off its production of sterioid hormones, which include adrenal estrogens or estrogen precursors. Suppression of adrenal (cortical) activity is accomplished by giving large doses of a glucocorticoid hormone (dexamethasone). Suppression is probably not as effective as the surgical removal of the adrenal glands.

[20]"Hypoglycemia: The End of Your Sweet Life" by William Currier, M.D., John Baron, D.O., and Dwight K. Kalita, Ph.D. Published in *A Physicians' Handbook on Orthomolecular Medicine* (pages 156 - 160), edited by Roger J. Williams, Ph.D., and Dwight K. Kalita, Ph.D. Published in 1977 by Pergamon Press, New York, N.Y.

[21]Mato Laboratory, P. O. Box 7006, Riverside, Calif. 92503.

From the standpoint of this particular disease process (breast cancer), individual components of the whole adrenal cortex are quite important. Their presence or absence determines to some degree the rate of growth of the tissue. It is likely that similar important effects of adrenocortical hormones are exerted every day on healthy tissue.

Recently in my office, a severely asthmatic child who usually responds well to the preparation Medrol, a glucocorticoid made by Upjohn Company, was found to be unresponsive to it. After the administration of a few doses of whole ACE, the child's sensitivity to Medrol returned.[22]

When individual corticosteroids (hydrocortisone, prednisone, etc.) are used in treatment, it is unusual for the drug to affect the course of the illness for which it is used beyond the time when it is used. Moreover, rebound illness may occur when it is stopped. Also, treatment with hydrocortisone or its derivatives tends to create a need for more and more of the substance in order to obtain relief of symptoms. A potentially life-treatening suppression of adrenocortical function occurs whenever hydrocortisone or its derivatives are used in therapeutic programs that involve pharmacologic doses.

Therapy with ACE, on the other hand, appears to significantly influence the body toward normal function for a considerable time after it is given. In fact, treatment with ACE may anticipate the need for less and less ACE as time goes on. Users of ACE believe that adrenocortical function is improved rather than suppressed.

When the adrenal cortex is low in function, the patient may experience water retention. In past years, this has served as the basis for the water-loading test of adrenocortical function. ACE can be a very helpful treatment for water retention of this origin. A young lady in my practice with severe asthma of pregnancy was treated successfully with repeated intravenous injections of ACE. Not only was her asthma relieved, but her severe fluid retention was remedied. If she missed an ACE injection, severe fluid retention occurred (waterlogging), and she became sluggish

[22]The reader must realize that adrenocortical steroids or ACE are not to be used indiscriminately. Their use should be reserved for cases which do not ordinarily respond to more conservative measures, and their use must be carefully supervised by a physician at all times.

and depressed. After obtaining ACE, prompt diuresis was noted; and her well-being was markedly improved.

Those patients who have sleep disturbances commonly respond well to treatment with ACE. Insomnia, restless sleep, and frightening dreams have all disappeared when an appropriate dose of ACE has been administered. Moreover, some individuals can titrate their need for ACE by the nature of their sleep.

Judge Tom R. Blaine has written an interesting book in which he describes the use of ACE in the treatment of his own case.[23]

ACE needs to be further evaluated as a therapeutic agent and not passed by as an obsolete substance. Furthermore, the individual components of the substance need to be delineated and studied. Optimal doses of ACE also need to be established. My use of the substance indicates that 1 to 5 cc's per intravenous injection (double strength) may be best in children. Adults may require 2 to 10 cc's of the double-strength material. The elderly appear to require less. Dosage for nearly all persons declines over time as improved health results. Intramuscular usage is not as dramatically effective but may be very worthwhile even in small doses. The use of doses given sublingually needs to be evaluated. Oral adrenal cortex material (raw gland) is available but may not have the same effect as that of the injectable material.

At the present time, the Food and Drug Administration has removed ACE from the market and has declared that it can only be returned if "new drug" trials "prove" its value. This, in effect, insists that if adrenal-cortical hormones are indicated for any condition, synthetic adrenal hormones (cortisone and its derivatives) must be used.

The *Homeostasis Quarterly* is a newsletter published 4 times a year by the Adrenal Metabolic Research Society of the Hypoglycemia Foundation, 153 Pawling Avenue, Troy, New York 12180. This newsletter has pointed out (Volume 7, Number 1, 1978) that ACE is beneficial and largely without side effects, whereas cortisone drugs depress the immune system, increase chances of infection, lower resistance, hasten the development of diabetes, and break down the body's integrity. According to the quarterly, ACE *increases* resistance to infection and does not de-

[23]*Goodbye Allergies* by Judge Tom R. Blaine. Published in 1965 by The Citadel Press, Secaucus, N.J. 07094.

press the immune system, hasten diabetes, or break down body tissues. My experience agrees with that statement.

When ACE is needed and properly used, it exerts a nourishing effect in the body, whereas cortisone more often contributes to a breakdown in body tissues.[24] Moreover, the nourishing effect of ACE usually grows more apparent with prolonged use. The unwanted destructive effects of cortisone grow more apparent with prolonged use.

One hopes that legal efforts to reinstate ACE will be successful. Intensive investigation should be carried out to establish the true therapeutic roles of ACE and the synthetic cortisone drugs in the management of human disability.

NUTRITION AND HORMONES

The impact of certain foods on the physiology of certain individuals is at this time not generally appreciated. It seems highly probable that the hormonal status of the body is directly related to our nutritional status. Optimal function of the endocrine glands is closely related to the quality of a person's feeding. The importance of dietary substances providing methyl groups for hormone synthesis has already been pointed out.

Dr. Melvin Page (D.D.S.) of the Page Foundation, St. Petersburg, Florida, has for many years advocated a "basic diet" for individuals who are ill.[25] This "basic diet," among other things, is free of refined sugar and largely free of sugar from fruit, milk, and juice sources. Dr. Page feels that this diet has a beneficial effect on the blood sugar, blood calcium, blood phosphorous, and the hormonal balance of the body. He believes that growth hormone in cow's milk often brings about excessive growth in children and may be associated with the production of cancer in adults. His

[24]In rare cases, cortisone or its derivatives serve nicely as replacements for corticosteroid hormones that are not manufactured in the body because of metabolic error.

[25]*Your Body Is Your Best Doctor* by Melvin E. Page, D.D.S., and H. Leon Abrams Jr. Published in 1972 by Keats Publishing, Inc., New Canaan, Conn. Also, *Body Chemistry in Health and Disease* by Melvin E. Page, D.D.S., and D. L. Brooks, A.B. Published in 1954 by The Page Foundation, 5235 Gulf Boulevard, St. Petersburg Beach, Fla. Also, *Degeneration/Regeneration* by Melvin E. Page, D.D.S. Published in 1949 by Biochemical Research Foundation, 2810 First Street North, St. Petersburg, Fla. (now The Page Foundation, same address as in preceding reference).

clinical success and his respected position among some groups (Southern Academy of Clinical Nutrition, International College of Applied Nutrition) is predicated on this dietary pattern, as well as on the use of minute doses of endocrine hormones (see footnote number 25). His work, which has been scoffed at by many establishment-oriented physicians, should be considered for what it is — a nonproven method of trying to help patients in need and a theory of microendocrinology that may be entirely or partially valid. Two California groups[26] are investigating the Page Method.

For many years, Dr. Page has been normalizing body chemistry and making patients healthier by the administration of microdoses of hormones such as whole pituitary, posterior pituitary, gonadal hormones, insulin, and adrenal cortex. The era is now approaching when biochemistry may isolate specific substances in these preparations that are responsible for favorable change.

Dr. Billy A. Stewart (D.D.S.), Director of Continuing Education at the University of Detroit Dental School, is conducting research on the Page Method and is using its precepts in connection with his program of preventive health. Many other practicing dentists are using the Page concepts to improve the general health of their patients and thus to improve dental conditions.

The University of Pennsylvania School of Dental Medicine is now establishing the Melvin E. Page Oral Medicine Laboratory of the Center for Clinical Research (4001 Spruce Street, A1, Philadelphia, Pennsylvania 19104).

Other persons and groups have been involved through the years in helping sick persons by dietary manipulation. Often, restriction of sugar is a prominent part of the recommended diets. When such diet therapy is advocated as a replacement for sound medical care and treatment, it cannot be condoned. When dietary improvement is used to supplement sound medical care, however, it can benefit the patient and ultimately reduce the need for further medical care.

Although the dietary recommendations of some groups may not be sound, it is likely that sugar restriction is a positive health measure. The careful research of Dr. John Yudkin (M.D.) bears

[26]Naturalism, Inc., P.O. Box 3621, Hollywood, Calif. 90038 and International Academy of Microendocrinology, c/o Dr. Helyn Leuechauer (D.D.S.), 3169 Barbara Court, Hollywood, Calif. 90068.

this out. Dr. Yudkin has found, for example, that the likelihood of a coronary attack is six times as great in a person who ingests 120 pounds of sugar per year as it is for a person who ingests only 60 pounds of sugar per year.

The traditional medical viewpoint toward unusual dietary treatments and the persons who give them has been skeptical. It is certainly desirable that any form of treatment be properly evaluated by unbiased sources and techniques. In such investigation, however, *the response of the individual patient must not be overlooked.* A double-blind investigative study comparing groups of subjects is desirable and proper, but such a method of investigation may not disclose the unique response of *one* individual or a *few* individuals within the groups.

In my experience with the blood-sugar responses of individuals to foods in the diet, I have been amazed to note the profound clinical effects and blood-sugar changes that may take place in connection with eating. The ingestion of food is a powerful event that sets in motion a complex set of chemical reactions and counterreactions within the body.

For an example of the effect of food on body chemical balance, I want the reader to turn to page 265 of this book and examine Figure 94. The blood-sugar (glucose) values of an 8-year-old girl are presented on a graph. The first blood-sugar value, the fasting value, was 56 mg.%. It was obtained just before lunch. The next value, 15 minutes after lunch, was 150 mg.%. The lunch consisted of fried haddock (breaded), green beans, jello, and water. One-half hour after lunch the blood sugar was 172 mg.%. Thereafter, the blood sugars returned toward the fasting level.

This is an *extreme* response of an individual to the ingestion of food. In this case, the blood sugars changed from a generally low (hypoglycemic) level of 56 to a level (150) considerably higher than the fasting level 15 minutes after the meal. The further increase to 172 at one-half hour represents an increase of 116 mg.% within a one-half hour period! Ordinarily, the blood-sugar response after a meal appears to be much less. The rise in this girl represents a distinct diabetic tendency in this individual. The administration of a standard glucose tolerance test (see Chapter 11) would probably indicate diabetes. Research correlating the effect of food and the effect of glucose load on patients is needed. Also, blood insulin, hydrocortisone, and glucagon levels need to be determined on patients such as this and correlated with the blood-sugar values.

We know that the normal patient releases a small amount of insulin within the first 5 - 20 minutes or so after drinking a sugar-test solution. We know that the ingestion of food sets off certain hormonal signals within the gut which influence the digestion and absorption of that food. We need to know much more about these events, about the gastrointestinal hormones and their relationship to types of food eaten and quantities of food eaten. We also need to know more about the long-term effect of dietary intake on hormone production. There is so much yet to do!

CHAPTER 9

THE PANCREAS AND INSULIN

The pancreas, at first thought, is not a very exciting organ. It is hidden away in the rear wall of the abdomen, tucked behind the stomach and cradled in a curve of the upper small intestine (see Figure 28). Perhaps to most of us we know the pancreas better as sweetbreads in the market.

The human pancreas, however, is indeed a most important organ. It controls our destinies more than we realize. The pancreatic secretions, under nervous control as well as being responsive to blood factors and gut factors, can profoundly influence our state of mental and physical fitness.

The pancreas has two principal functions: to aid digestion of foodstuffs (protein, fat, and carbohydrate) in the gut and to preserve a normal level of sugar (glucose) in the blood from moment to moment.

I have already pointed out the important role that the liver plays in regulating blood glucose. The pancreas is equally important. It is understandable that the liver and pancreas are vitally involved in carbohydrate metabolism when we realize that the liver and pancreas arise from the gut itself, as a single bud or shoot, in the embryological development of the organism.

In simplified terms, we can visualize glucose regulation as a function of 4 organs: the gut, the liver, the pancreas, and the adrenal glands.

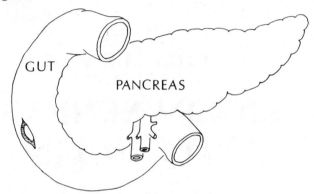

Figure 28. The pancreas

This important gland situated in the rear of the abdomen is nestled in the curve of the small intestine (gut). The pancreatic functions, both exocrine and endocrine, are necessary for sustaining life.

The digestive function in the gut is performed by four secretions of the pancreas. These substances are

1. sodium bicarbonate (to provide an alkaline pH in the upper part of the small intestine),
2. a protein-digesting enzyme,
3. a fat-digesting enzyme, and
4. a carbohydrate-digesting enzyme.

These substances are vital for the complete digestion of foods. They are secreted by the pancreas and carried to the small bowel via a duct from the pancreas. Most of the bulk of the pancreas is made up of cells that produce these digestive aids. This aspect of pancreatic function is termed an *exocrine*[1] function.

The secretion of sodium bicarbonate by the pancreas is perhaps the most important single overlooked process in digestive physiology. The presence of sufficient sodium bicarbonate is necessary to provide the proper *alkaline* pH for the completion of protein digestion begun in the *acid* medium of the stomach.

[1]Exocrine: having to do with the secretions of glands that have ducts, as opposed to endocrine, the secretions of glands that do not have ducts.

Consider the sales of antacids and the consumption of bicarbonate, and one will realize the importance of acid-alkaline balance in digestion! Incomplete protein digestion means protein starvation for the body even though protein intake may be ample.

The production of bicarbonate by the pancreas may be turned down or off by injury, alcohol, inflammation, infection, allergy, etc. The production of the digestive enzymes may also become deficient. Such deficiency may underlie conditions of "protein poisoning," many degenerative disorders, and perhaps cancer.

I suspect that the digestive role of the pancreas in preserving health and overcoming illness is far more important than we have yet realized. We have much to learn about the total role of pancreatic function in human physiology.

THE ISLES OF LANGERHANS

In 1869 Dr. Paul Langerhans (M.D.) identified special cells in the pancreas. These cells were noted to be grouped together in clusters, and have been named the islands (isles) of Langerhans. There is no anatomical duct associated with these cells. They discharge their secretions directly into the bloodstream. The ductless isles of Langerhans constitute an endocrine gland or the endocrine portion of the pancreas gland.

It is now known that the cells of the isles secrete two hormones directly into the blood. Those hormones are insulin (insula = island) and glucagon. They have opposite functions in regard to blood sugar, insulin lowering the sugar level and glucagon elevating it. It is also known that insulin is secreted by the beta cells and glucagon by the alpha cells in the isles.[2] In embryological development, the glucagon-producing cells appear earlier in the pancreas than the insulin-producing cells.

The isles of Langerhans make up only about 1% of the pancreas (see Figures 29 and 30), but their secretions are vital to the life of the individual. It is estimated that there are about one-quarter to one and three-quarter million islets in the pancreas. They are scattered throughout the pancreatic tissue but are

[2]A third type of cell has to do with the hormone gastrin. In Chapter 8, it was pointed out that the hormone somatostatin has also been found in the isles of Langerhans.

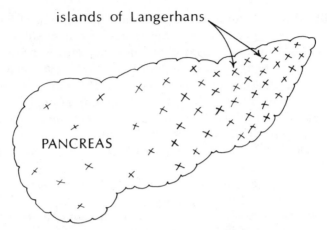

Figure 29. Diagram of pancreas showing isles of Langerhans
The isles make up only a small portion of the bulk of the pancreas. They are more numerous in that portion of the pancreas known as the tail.

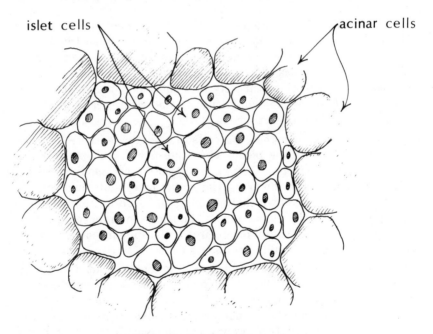

Figure 30. An island of Langerhans
The isle is made up of beta cells (which secrete insulin) and alpha cells (which secrete glucagon). The acinar cells surrounding the isle produce the enzymes used for digestion in the gut.

found in greater numbers in that portion of the pancreas known as the tail. The number of islets is constant throughout the life of an individual unless they are damaged.

In 1889, Dr. Oskar Minkowski (M.D.), a Russian, removed the pancreas from dogs and then saw flies clustering around puddles of the dog's sweet urine. He found that the dogs died after becoming diabetic. It was the absence of the endocrine portion of the pancreas (the isles) that accounted for the sugar in the urine and the excessive urination shown by the dogs before death.

(We, like dogs, cannot live long without our pancreas. Perhaps the great architect of the human body knew this when He placed this vital organ deep in the abdomen, where it is relatively protected from harm.)

Experiments were then made in which a dog's pancreas was transplanted from the abdomen to the skin. The digestive function of the pancreas was eliminated by this procedure, but the dog's blood sugar remained normal. This showed that the pancreas secreted some blood-sugar-controlling substance directly into the blood. Tying the digestive duct of the pancreas was another way to eliminate the digestive function and preserve the endocrine function.

It proved difficult for experimenters to find the "X" substance which kept blood-sugar levels in check. When a pancreas was removed from an animal for study, the digestive enzymes of the pancreas would digest the insulin just as they digested protein in the intestine. Dr. Frederick Banting (M.D.) and Charles Best, however, in 1921 tied off the pancreatic ducts, waited several weeks for the digestive part of the pancreas to degenerate, and only then removed the pancreas. In this way, they were able to successfully obtain the "X" substance (insulin) from the remaining isles of Langerhans. They showed for the first time that elevated blood sugar in diabetic dogs and humans could successfully be treated with this hormone. Since then, the study of the protein insulin has led to a major triumph in biochemistry: the determination of the amino acid sequence in a protein.

Biochemists have even succeeded in synthesizing human insulin in a pure state. The process entails 200 individual reaction stages. (For more insight into the historical events surrounding the discovery of insulin, see Appendix C, pages 449 - 460.)

The functions of insulin in relation to carbohydrate metabolism have already been pointed out. Insulin also has an important effect on growth. Insulin stimulates anabolic (building up) reactions in the body, involving carbohydrate, fat, protein, and nucleic acids. Insulin is necessary for the growth of cells in fetal life and early childhood. The presence of insulin "triggers" the multiplication of cells which account for growth. A second messenger substance known as cyclic AMP, an important chemical mediator in cells, decreases within the cell when insulin is present. This, along with other factors, appears to be a signal to cells to begin the process of cell division.

Diabetics are known to have diminished resistance to infection. One reason for this may be a lack (or relative lack) of insulin, with diminished multiplication of antibody-producing cells.

Some physicians have utilized insulin in very low doses to abort impending infections,[3] although this treatment has not been proven to be effective from a scientific standpoint. To my knowledge, however, neither has it been proven ineffective.

Dr. Noel Maclaren (M.D.) and Dr. Marvin Cornblath (M.D.) have pointed out that the principal role of insulin in the body is to conserve body fuels.[4] They state that insulin is the hormone of feasting, meaning that it is released and exerts its actions in the fed state. They list five actions by which insulin conserves body fuels:

1. discouraging the breakdown of fat,
2. discouraging the breakdown of protein,
3. increasing glycogen storage,
4. augmenting the transport of nutrients (amino acids, glucose, and ions) across body-cell barriers, and
5. growth promotion by increasing cell size.

Insulin is, indeed, the hormone of feasting. As a matter of fact, insulin may be termed *the fattening hormone.*[5] Blood-insu-

[3]"Non-Diabetic Uses for Insulin" by Hal A. Huggins, D.D.S. Published in *Journal of the International Academy of Preventive Medicine,* Vol. 4, No. 1, July 1977, pages 100 - 107.

[4]"Physiology of Diabetes Mellitus" by Noel K. Maclaren, M.D., and Marvin Cornblath, M.D. Published in *Pediatric Annals,* Vol. 4, No. 6, June 1975.

[5]See *Dr. Atkins' Superenergy Diet* by Robert C. Atkins, M.D., and Shirley Linde. Published in 1977 by Crown Publishers, Inc., New York, N.Y.

lin levels are distinctly higher in obese individuals than they are in thin persons. Insulin not only discourages the breakdown of fat, it stimulates the buildup (synthesis) of fat. In our country with its high consumption of sugar, insulin is called forth to metabolize the sugar, and excess fat formation frequently results.

THE BETA CELLS

The insulin-producing beta cells are the most numerous constituents of the isles of Langerhans. They vary in number from 200,000 to 2½ million. The beta cell is richly endowed with something known as endoplasmic reticulum (ergastoplasm). It is a fatty network, upon which is studded a series of particles known as ribosomes. The ribosomes contain ribonucleic acid (RNA) and protein, and are the "engines" of protein-hormone synthesis in the beta cell. It is here on the ribosomes that insulin synthesis occurs (see Figure 31).

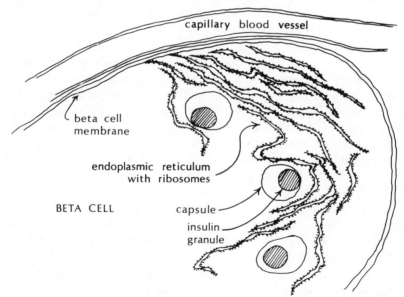

capillary blood vessel

beta cell
membrane

endoplasmic reticulum
with ribosomes

BETA CELL

capsule

insulin
granule

Figure 31. Diagram of the interior of a beta cell
The synthesis of insulin takes place in the ribosomes of the endoplasmic reticulum. Thereafter, insulin is transferred to the Golgi complex (not shown), where it is packaged and stored as granules of insulin. When a signal for insulin need is received, beta cell degranulation occurs, as insulin makes its way out of the beta cell and into nearby blood capillaries.

Insulin is manufactured first by the beta cells as proinsulin.[6] It is actually proinsulin synthesis that is accomplished on the ribosomes of the endoplasmic reticulum. Thereafter, the proinsulin is transferred to an intracellular organelle known as the Golgi Complex. On this Golgi apparatus, the proinsulin is packaged as granules and released into the cellular cytoplasm when needed.

Normal beta cells fill with granules of insulin which are characteristic in their shape from species to species. This characteristic beta-cell granulation can be shown when the insulin-containing pancreas is stained with special dyes. The presence of beta cell granulation shows that reserve insulin is stored and ready for use. Beta cell degranulation (insulin production and release) occurs when a glucose load is given or when an oral hypoglycemic drug such as sulfonylurea is administered.

The exact mechanism of formation of insulin from proinsulin in the beta cell, its movement from the interior to the exterior of the beta cell, and its incorporation into nearby capillary blood vessels await further scientific investigation.

We do know from electron microscopy that the insulin granules are contained in tiny sacs and that these sacs migrate from their site of formation to the surface of the beta cells. Somehow the insulin gets across the sac membrane and the cell membrane to the outside of the cell. Insulin, presumably in solution at that point, gains access to the blood through the walls of capillary blood vessels lying within the pancreas (see Figure 32). The process of insulin release from the beta cells requires calcium in the fluids surrounding the cells. Sodium ions are also involved, as well as cyclic AMP.

The insulin-carrying blood from the pancreas drains to the portal vein, into the liver, and thence to the heart and the general circulation. Within the liver, 50 to 80 percent of the insulin is removed from the blood.

It may be that some patients with sugar-related health problems have difficulties in production, movement, exteriorization, or blood vessel pick-up of insulin molecules. In the future, we may be able to speak of defects in the production, move-

[6]Dr. Donald F. Steiner, M.D., Professor of Medicine at the University of Chicago, the discoverer of proinsulin, has recently identified preproinsulin, a precursor to proinsulin. Thus, insulin is actually manufactured in 3 stages.

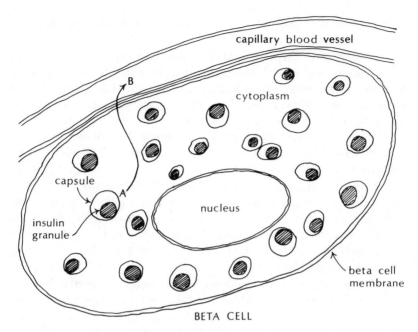

Figure 32. Possible barriers to release of insulin

Defects in the production of insulin may occur because of an absence of beta cells in the pancreas, or because of faulty insulin manufacture and release. Packaged insulin granules must make their way out of their capsules, across cellular cytoplasm, across the cell membrane of the beta cell, across the cell membrane of the capillary blood vessel, and into the blood. As insulin mobilizes from storage form to active hormone in the blood, it must be split from the proinsulin form to insulin itself.

ment, or transfer stages of insulin in the beta cells of the pancreas. The role of the enzyme insulinase also needs to be clarified.

There is evidence to indicate that the insulin output from a beta cell is a pulse lasting only a few seconds.[7] Insulin secretion appears to be a series of intermittent pulses from the beta cells, increasing in frequency and amplitude in response to an elevation in blood glucose.

Nerve endings from both the sympathetic and parasympathetic branches of the autonomic nervous system have been identified in close approximation to the islet cells. This, presum-

[7]"Periodicity of Insulin Secretion" by Paul M. Beigelman, M.D. Published in *Guidelines to Metabolic Therapy*, Vol. 3, No. 3, Fall 1974. Published by the Upjohn Company.

ably, is the mechanism whereby neuromental stimuli influence this endocrine organ.

Activation of the sympathetic nervous system is associated with the release of the neurochemical transmitters epinephrine and norepinephrine. These inhibit insulin secretion. Acetylcholine is the neurochemical transmitter of the parasympathetic nervous system, and it stimulates insulin secretion. The effect of acetylcholine can be blocked by atropine. Atropine and other chemicals similar to it have been used with some success in treating individuals with hyperinsulinism.

Different individuals have different "sets" of their beta cells for insulin production. Physical activity, diet, nervous stimuli, and heredity probably determine the "set" of these cells. The basal (fasting, nonglucose stimulated) insulin secretion is a reflection of the long-term "set" of the islets. It is most readily altered by changes in nutritional input.

Some patients with profoundly low blood sugar are found to have tumors or overgrowth of the beta cells. A related condition is known (nesidioblastosis) in which immature beta cells appear in excessive numbers throughout the pancreas. It may be that these "beta cells" are derived from the digestive (acinar) cells of the pancreas in response to repeated stimulation.

Glucagon, the anti-insulin secretion of the alpha cells of the pancreas, is known to inhibit insulin secretion, to decrease the uptake of glucose by the tissues, and to stimulate glucose formation from protein (gluconeogenesis).

INSULIN

As indicated, insulin is a protein hormone. It is the principal hormone in the body responsible for building up tissues (anabolism).

Insulin consists of 51 amino acids arranged in two polypeptide chains (see Figure 33). Insulin exists as a crystal unit composed of 6 insulin molecules grouped together as 3 pairs of 2 molecule subunits. The mineral zinc is required for the formation of the stable 6-molecule crystal. Unlike the representation shown in Figure 33, the molecule is actually twisted upon itself like a ball of string.

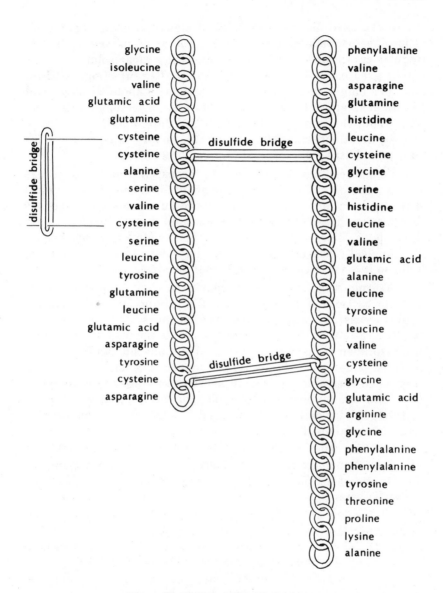

glycine
isoleucine
valine
glutamic acid
glutamine
cysteine
cysteine
alanine
serine
valine
cysteine
serine
leucine
tyrosine
glutamine
leucine
glutamic acid
asparagine
tyrosine
cysteine
asparagine

disulfide bridge

disulfide bridge

disulfide bridge

phenylalanine
valine
asparagine
glutamine
histidine
leucine
cysteine
glycine
serine
histidine
leucine
valine
glutamic acid
alanine
leucine
tyrosine
leucine
valine
cysteine
glycine
glutamic acid
arginine
glycine
phenylalanine
phenylalanine
tyrosine
threonine
proline
lysine
alanine

Figure 33. Diagram of insulin structure

Insulin is a protein hormone consisting of 51 amino acids arranged in 2 polypeptide chains with connecting disulfide bridges. Although not shown, zinc is an essential element in the function of insulin in the body.

When used for the treatment of elevated blood sugar, such as in diabetes mellitus, insulin may be obtained in several forms. Regular crystalline insulin is a quick-acting, clear solution which has its greatest effect from 2 to 6 hours after injection. NPH insulin is a long-acting preparation which lasts about 24 hours and has its peak effect from 7 to 11 hours after injection. Lente insulin is about the same in effect as NPH. Lente, itself, is a mixture of 70% Ultralente and 30% Semilente, but it can be mixed with additional Ultra or Semi to obtain varying effects.

Commercial insulin products are prepared from the pancreas of cows and pigs. This insulin is slightly different in its chemical make-up from human insulin. Since the human body may not accept foreign material, it is understandable that at times patients develop an allergy (antibodies) to the insulin of animal origin. Biochemical research may soon devise a way to alter animal insulin so it is identical with the human product.

A promising lead is the recent capability for producing insulin in bacteria. The gene for insulin production in the rat has been successfully transplanted into harmless bacteria that then produce rat insulin. The gene for the production of human insulin has also been transplanted into bacteria and human insulin has been harvested from the bacteria. Further speculation stemming from such recombinant DNA research can stagger the imagination. One can suppose, for example, the situation in which a defective gene for diabetes is replaced by a gene for normal carbohydrate metabolism.

(A fascinating insight into the early clinical use of insulin is given in Appendix C of this book, where Dr. Ross Cameron (M.D.) tells of his participation in early trials with this then unknown substance.)

All insulins are given by injection — that is, under the skin, in the muscle, or in the veins. Taken orally, this protein hormone is digested like any other protein and hence becomes inaccessible to the body. Insulin for treatment of diabetes has been available as 40 units per cc. or 80 units per cc., but it is being replaced with one standard insulin concentration, 100 units per cc. (U-100: Eli Lilly & Company). The use of this preparation facilitates the calculation and administration of insulin dosage. Furthermore, the U-100 insulin is essentially nonantigenic in the body. This means that the problem of allergy to insulin may be eliminated.

Mention has already been made of proinsulin, the chemical precursor of insulin. C-peptide has also been isolated. It is the fragment which connects the 2 insulin chains together. The availability of these substances permits scientists to more accurately determine the state of beta-cell pancreatic function in a patient. For example, injected insulin does not contain C-peptide, whereas insulin manufactured in the body will be accompanied by the production of C-peptide. Hence, a diabetic patient receiving insulin injections may have a determination made of the amount of insulin that his own pancreas is producing.

Proinsulin can be found in the blood along with insulin that is formed in the body and accounts for about 5 to 15 percent of total blood insulin.

The beta cells initially synthesize proinsulin, the 80 amino acid polypeptide precursor of insulin. Proinsulin is then split by enzymatic action to produce insulin, the 51 amino acid protein in the form of 2 polypeptide chains bound together by disulfide bridges. The small fragment which is split off in the cleavage of proinsulin is the connecting C-peptide. For every molecule of insulin, a corresponding amount of C-peptide is produced.

The normal insulin response to an elevation of blood sugar is two-pronged or biphasic.

Whenever the blood sugar becomes elevated, the normal individual produces a rapid spike of insulin release into the blood. This reaches a peak in 5 or 6 minutes and lasts only 15 to 20 minutes. This sudden insulin increase is called Phase I and probably represents discharge of stored insulin from a highly mobile pool.

Under the stimulation of the rise in glucose, a delayed Phase II secretion of insulin also occurs in the normal individual. This insulin release occurs 30 to 60 minutes after the initial rise in blood sugar. Phase II probably results from a more slowly responsive pool of stored insulin, newly formed insulin, and small amounts of proinsulin.

In persons with diabetes mellitus, the first phase of insulin secretion is characteristically absent. The diabetic may, however, show a total insulin secretion which is normal or greater than normal in amount.

According to the papers of the Jarvis Correspondence Group,[8] there are many other uses for insulin besides the treatment of diabetes. According to this source, insulin may be used to treat an overactive thyroid condition, to promote general healing, and for its possible anticancer activity. These uses of insulin are not proven and are not generally accepted; but neither, to my knowledge, have they been shown to be ineffective for these purposes.

Nonsuppressible insulinlike activity (NSILA) refers to a substance in the blood which has 60,000 times the activity of insulin. Blood insulin and proinsulin levels, determined as immunoreactive insulin, probably represent the tip of the iceberg as far as the total occurrence of insulinlike activity is concerned. Nonsuppressible insulinlike activity (NSILA), of either the soluble or protein-bound type, probably accounts for a huge portion of general insulin activity in the body. Nothing more can be said about its function at this time. Much is yet to be learned about the entire spectrum of insulin activity.

It has been postulated that insulin exerts its various effects by facilitating a rise in free calcium ions within the cell. If this is so, then calcium acts as a second messenger substance within the cell, responsible for insulin effect.

Recently, the existence of insulin-sensitive chemical receptors in the brain has been shown.[9] These are thought to lie in the hypothalamus and to be a part of the parasympathetic nervous system. In rats, a small amount of insulin given into the main artery leading to the brain is capable of producing a definite blood-sugar-lowering effect. The same amount of insulin given in a peripheral vein has no effect on blood sugar. This regulatory system in the central nervous system acts rapidly and is of short duration. It is thought to be separate from the hormonal regulatory system which is slow and of fairly prolonged duration. The neural center is thought to inhibit glucose production by the liver in some way not yet understood.

---

[8]Courtesy of Dr. Melvin E. Page, D.D.S., St. Petersburg Beach, Fla. Also see "Non-Diabetic Uses for Insulin" by Hal A. Huggins, D.D.S. Published in *Journal of the International Academy of Preventive Medicine*, Vol. 4, No. 1, July 1977, pages 100 - 107.

[9]"New Evidence of Insulin-Sensitive CNS Sites." A report in *Medical Tribune*, Vol. 16, No. 22, June 11, 1975.

HYPERINSULINISM

Hyperinsulinism, in the broad definition, includes a number of varying conditions. They range from the insulin-secreting tumor of the pancreas, to the overgrowth of the pancreatic-islet tissue (hyperplasia), to the normal appearing pancreatic isles whose function is excessive, to the condition of the body in which there is hypersensitive response to normal amounts of secreted insulin. Many times, this latter condition is associated with functional deficiency of adrenal-corticoid hormones.

Classical hyperinsulinism is a relatively uncommon disease of the pancreatic isles in which insulin and sometimes proinsulin are secreted in excess. This disorder is due to an insulin-secreting tumor of the pancreas (benign or malignant) or hyperplasia. Originally the disorder was suspected when the Whipple-Frantz Triad[10] was noted. It is now diagnosed by finding high levels of serum insulin or proinsulin with low blood-glucose levels.

The hypoglycemia is usually profound in classical hyperinsulinism of this type. Neuropsychiatric symptoms and obesity are found in adults with this problem. Children often have abdominal pain. Infants may have seizures, collapse, altered circulatory status, respiratory arrest, etc. Severe brain damage or death may result if the diagnosis is not quickly established.

As soon as the diagnosis of classical hyperinsulinism due to tumor or hyperplasia is made, surgery on the pancreas is indicated. Removal of the tumor or resection of part of the pancreas is performed.

Much more commonly encountered is a different kind of hyperinsulinism. In these individuals, the blood-sugar values behave as if a high level of blood insulin is present, even though blood-insulin levels may or may not be elevated from the standpoint of accepted "normal" values. The term *hyperinsulinism*, thus, may refer to the situation in which elevated levels of insulin have been demonstrated, merely presumed, or in which the blood-sugar levels are similar to those in high insulin states. It would be desirable to establish more precise terminology, and un-

[10]The Whipple-Frantz Triad includes
1. symptoms of hypoglycemia,
2. an abnormally low level of blood glucose, and
3. relief of symptoms by administration of glucose or other carbohydrates.

doubtedly this will come about when levels of insulin and adrenocorticoid hormones in the blood are more readily available to and utilized by the clinician.

Hyperinsulinism is not a desirable state. When it occurs in utero, the fetus is subject to death or overgrowth in size, or the subsequent newborn may have hypoglycemia (see Chapter 10, sections entitled "Effect of Pregnancy on Blood Sugar," page 205 and "Infant of the Diabetic Mother," pages 205 - 207.

Blood-sugar levels have been determined in the laboratory for many years. Every medical laboratory has a well-worked-out procedure for determining blood glucose. The procedure can be carried out with relative ease and simplicity and without great expense. For these reasons, blood-sugar levels are studied most often rather than blood-insulin levels and adrenal-corticoid levels. Hormone determinations have been available at many fewer laboratories and have been primarily a research tool.

The pancreatic beta cells are variable in number and have different degrees of sensitivity. The "set" of the beta cells for insulin production may have a hereditary basis but appears to be quite responsive to environmental variables — especially diet and psychological stress. The beta cells can become so sensitive that they are known as "trigger happy." In this state, an excessive amount of insulin is released. This hyperinsulinism is considered by some to be the opposite of diabetes, and by others to be a forerunner of diabetes.

Caffeine appears to sensitize the beta cells. This may occur because of adrenocortical stimulation which in turn induces the liver to break down glycogen into glucose. The net effect of caffeine ingestion appears to be as though the patient has eaten a load of sugar.

Most individuals do not appreciate the caffeine exposure that abounds in civilized cultures. The average reader may not realize that beverages commonly consumed in our culture contain significant amounts of caffeine.[11] An 8-ounce cup of coffee or strong tea usually contains between 100 and 150 mg. of caffeine (up to 18 mg. per ounce). A commonly available pain-relieving tablet contains 30 mg. of caffeine. Tea of average strength con-

[11]Committee Statement by the Committee on Drugs of the American Academy of Pediatrics. Published in the *Newsletter* of the American Academy of Pediatrics, Vol. 22, No. 18, December 1, 1971.

tains approximately 12 to 15 mg. of caffeine per ounce. Eight ounces of cocoa may have as much as 50 mg. The maximum caffeine content of cola drinks is 50 mg. per 8-ounce bottle. "Decaffeinated" coffees that the author is familiar with still contain some caffeine.

The question might well be asked, "Why do we need caffeine in our beverages at all?" The answer is that caffeine gives a lift, an exhilaration that urges one to buy again. As far as I 'm aware, no study has ever been done to show that caffeine in our beverages is a health-promoting feature. It is amazing how tacitly and without question we accept practices that are ongoing around us.

There is no question that we readily become accustomed and/or addicted to our "caffeine fix." It is so much simpler than problem solving, working out a long-term graduated program of exercise-conditioning, optimalizing our diet, reducing mental stress, and developing a lifestyle in harmony with natural laws.

Overindulgence in caffeine and sweets tends to sensitize the beta cells by subjecting them to repeated stimulation and discharge. After repeated stimulation, the isles may become so sensitive that they overrespond to a normal or minimal stimulus. In the case of the beta cells, at least in some persons, overuse can lead to overresponse.

Insulin encourages growth in the body. Growth occurs by the accumulation of protein in tissues. Protein is formed when amino acids from the diet are cleared from the blood and made into protein. Insulin depletes the blood of all amino acids except one, tryptophane. Tryptophane in the blood is carried to the brain and is transformed there into serotonin. Serotonin is one of the major neurochemicals involved in brain function.

Because of this relationship between insulin and brain neurochemistry, it is understandable that persons with hyperinsulinism might have some rather significant deviations in neuromental behavior. Such, indeed, is the case. The case studies in Chapters 12, 13, and 14, and in Appendix B illustrate the wide range of behavioral disturbances that are often noted in association with hyperinsulinism.

Let us also remember that insulin promotes the buildup of fat. Unrestrained hyperinsulinism may be associated with obesity and disturbed balance in blood fats.

Scotland is the unenviable leader in deaths from coronary artery disease. The rate of death from coronary artery disease is 94 per 100,000 in Scotland, the highest in the world. Compared with other groups, an investigation of healthy 40-year-old males showed that the Scots produced more insulin and produced it earlier.[12] Disordered excessive insulin secretion may be the common metabolic denominator of obesity, maturity-onset diabetes, elevated blood fats, and coronary artery disease. A most likely and most common origin of hyperinsulinism is the widespread dietary use of refined sugar.

Dr. L. J. Kryston (M.D.) of the Endocrinology Department at Hahneman Hospital in Philadelphia made a detailed study of potential diabetic patients.[13] He obtained blood sugars and blood insulins on individuals with a history of diabetes on both the mother's and father's sides of the family. A high percentage of atypical 5-hour glucose tolerance tests were found. These atypical tests included the following: alimentary (abnormally high early peak with rest of curve normal), flat, saw-tooth, slow response (delayed), and hypoglycemic (low blood sugar at 3, 4, or 5 hours).[14]

Of even more interest in this study was the finding of an elevated blood-insulin level in all of these potential diabetics. The elevated blood insulin was present regardless of which atypical GTT pattern was recorded. A delayed release of insulin occurred in 50% of these patients. Hypoglycemic symptoms were better correlated with insulin levels than with blood-sugar levels.

As I will point out later in this book, diabetic states may not always be associated with an absolute deficiency of insulin. As I have already indicated, what we call hyperinsulinism may not always be associated with absolutely elevated levels of insulin. It may be that hyperinsulinism is a state, reflecting a number of biochemical situations. A defective type of insulin may, for example, be produced by some individuals due to nutritive deficiency or genetic trait. In such cases, the pancreas may be overactive but its insulin product ineffective. Until further progress is made in

[12]"Investigators Puzzled Over Insulin Role in CHD Deaths." Report of an international study in *Medical Tribune*, Wednesday, November 16, 1977.

[13]Paper presented at the Fourth Annual Meeting of the Academy of Orthomolecular Psychiatry, Detroit, Mich., May 4, 1974.

[14]Examples of varieties of glucose tolerance tests may be found in Appendix A and Appendix B of this book.

understanding basic abnormalities, the terms *diabetes* and *hyperinsulinism* serve useful descriptive purposes. No doubt, a more accurate terminology for these conditions will emerge as scientific progress comes about.

Dr. William H. Philpott (M.D.) believes that hyperinsulinism is one stage on the continuum of carbohydrate dysfunction produced by food allergy.[15] I suspect that he is largely correct. Dr. Philpott has also pointed out that dysfunction of the exocrine (digestive) part of the pancreas may be associated with states of disordered blood sugar.[16]

Figures 34 through 45 show blood-sugar levels and other characteristics of patients with probable hyperinsulinism. This hyperinsulinism has been inferred because of the finding of a *lowered* blood glucose after an oral-loading dose of carbohydrate. These patients may have hypoadrenocorticism, accounting for relative insulin excess. Other disorders such as low thyroid function and faulty gastrointestinal function may account for such findings.

In Figures 34 through 38 and Figures 40 and 41, the *decline* in blood sugar occurs following a standard-loading dose of glucose (a glucose tolerance test).[17] In Figures 42 through 45, the decline in blood sugar has occurred following the patient's usual, normal breakfast (a "natural" tolerance test).[18] For further discussion of hyperinsulinism see Chapter 14.

Glucose Tolerance Test Cases

These glucose tolerance tests are of a variety in which the

[15]"The Significance of Selective Food and Chemical Stressors in Ecologic Hypoglycemia and Hyperglycemia as Demonstrated by Induction Testing Techniques" by William H. Philpott, M.D. Published in *Journal of International Academy of Metabology*, Vol. V, No. 1, March 1976, pages 80 - 89.

[16]"Proteolytic Enzyme and Amino Acid Therapy in Degenerative Disease," the thinking of William H. Philpott, M.D. Thoughts derived from two articles: "The Significance of Reduced Proteolytic Enzymes in the Diabetes Mellitus Disease Process and in the Schizophrenia Syndrome Variable" and "Ecologic Stimulus Evoked Pancreatic Insufficiency in Chronic Degenerative Disease in General and Cardiovascular Disease in Particular," edited, condensed, and distributed by Dwight A. Kalita, Ph.D., June 1977.

[17]See Chapter 11 for a discussion of the glucose tolerance test.

[18]See Chapter 13 for a discussion of the natural tolerance test.

blood sugar declines following an oral glucose load (Figures 34 through 38 and Figures 40 and 41).

Those readers who are not familiar with the glucose tolerance test may wish to first read Chapter 11 and then return to these cases.

Those readers who are familiar with the glucose tolerance test will find these cases instructive. They show a *decline* in blood sugar in response to an oral glucose load. This decline is, in all probability, due to excessive insulin secretion or excessive insulin effect in the body, triggered by the presence of glucose in the gut and its absorption into the blood.

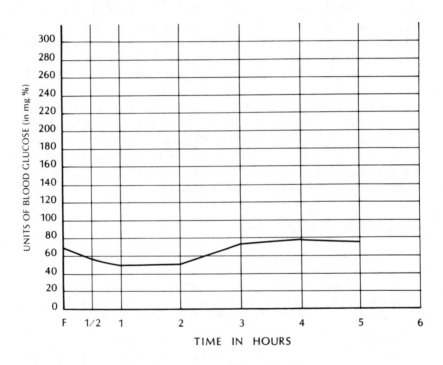

Figure 34. A glucose tolerance test curve

A 16-year-old girl who is a slow learner.

Blood-sugar level lowered in response to oral glucose load and stays low for at least 2 hours. Drop from 68 to 49 mg.% at 1 - 2 hours. Fasting level may be somewhat lower than desirable.

In my experience, this kind of glucose tolerance curve is not often seen in well-nourished, achieving persons without symptoms.

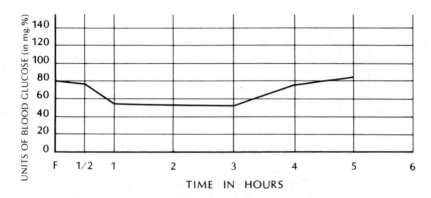

Figure 35. A glucose tolerance test curve

A 14-year-old girl with mild mental retardation.

Fasting blood glucose is at a fairly good level of 80 mg.%. After a loading dose of oral glucose, however, her blood sugars decline within 1 hour to a low of 55 mg.% and remain at that level through the 3-hour period. The effect of the glucose loading is over by 4 hours, when the blood sugar returns to a normal level. Such is the profound effect of sugar in some individuals.

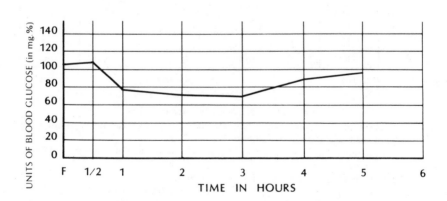

Figure 36. A glucose tolerance test curve

A 12-year-old boy with disruptive school behavior and delinquency. There was an alcoholic father.

This glucose tolerance curve descends instead of rising in response to an oral glucose load. This is the type of curve frequently seen in the relatives of alcoholics as well as in alcoholics themselves when they are not drinking. The resemblance of this curve to that in Figure 35 is striking. In my experience, this is not a desirable type of glucose tolerance curve.

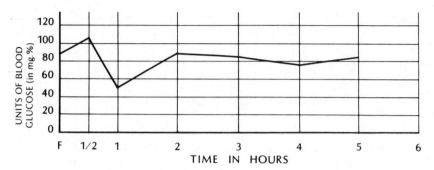

Figure 37. A glucose tolerance test curve

A 14-year-old boy with fatigue and obesity and with 11% eosinophil cells ("allergy cells") in the blood.

In this glucose tolerance test, after a slight initial rise, the blood sugar plummets from 108 to 50, a change of 58 mg.% in a half-hour period. This hyperinsulinism is brief but rather pronounced. The effect of the glucose load has disappeared by 2 hours.

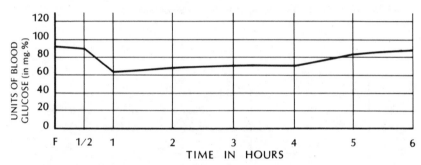

Figure 38. A glucose tolerance test curve

A 14½-year-old boy who was evaluated for hyperactivity, diminished attention span, and chronic argumentative behavior.

Eosinophils ("allergy cells") in the blood count ranged from 8 - 11% on multiple occasions.

The glucose tolerance test showed that his blood sugar did not rise after a loading dose of glucose. In fact, it dropped from a fasting level of 92 mg.% to a low of 63 mg.% at 1 hour. The fasting level was almost, but not quite, restored by 6 hours, when a level of 87 mg.% was reached.

The blood-sugar-lowering effect of a glucose load (hyperinsulinism) is more commonly terminated by 2 or 3 hours when counterregulatory factors such as glucagon, liver function, adrenalin and noradrenalin, and adrenocortical secretions come into play.

Increased blood levels of eosinophils (normal is reported to be up to 2 or 3%) often indicate the presence of respiratory or food allergy. The relationship of allergy to disturbances in carbohydrate metabolism is not known with certainty. The conditions are frequently found together.

Figure 39. Drawing made by 14½-year-old boy referred to in Figure 38

This drawing of a boy, said to be 10 years old, shows a simple stick-figure approach. This austere figure with its stark representation of fingers and toes, absence of hands and feet, and the lack of other details, indicates a large degree of developmental delay, emotional constriction, or body chemical imbalance. Hyperinsulinism with less than optimal blood-sugar levels may contribute to this boy's hyperactivity, diminished attention span, argumentative behavior, developmental delay, or emotional constriction.

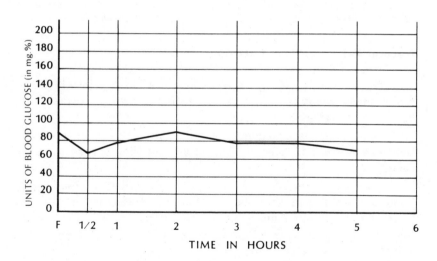

Figure 40. A glucose tolerance test curve

An 8-year-old boy with recurrent respiratory infections, hyperactivity, and allergic rhinitis. Eosinophils 6% in the blood.

Notice the prompt decline in blood-sugar levels at one-half hour after drinking glucose. Such glucose tolerance curves may be associated with allergic or maladaptive reactions to foods.

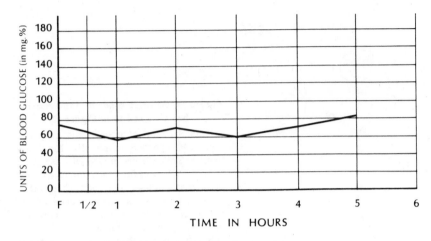

Figure 41. A glucose tolerance test curve

A 7-year-old, asthmatic, undergrown child with chronic allergic rhinitis. Seven percent eosinophils found in the differential white blood cell count.

The glucose tolerance test shows a decline in blood-sugar values after administration of the oral-glucose load. The fasting level is not restored until the 4th to 5th hour. Although these changes are relatively slight, their importance should not be overlooked. Even small blood-sugar changes below 75 mgm.% may be of more significance than similar changes occurring at higher blood-sugar levels.

Natural Tolerance Test Cases

These natural tolerance tests are of a variety in which the blood sugar declines following the patient's own breakfast (Figures 42 through 45).

Those readers who are not familiar with a natural tolerance test may wish to first read Chapter 13 and then return to these cases.

The following cases are those of children whose blood-sugar levels were obtained after the children ate their usual breakfasts. In most cases in our American culture, breakfast includes a large quantity of refined carbohydrate, and often a high amount of sugar itself is eaten.

Such a "natural" tolerance test measures the effect of food on blood-sugar levels, rather than the effect of a standard glucose load as is done in the glucose tolerance test.

The hyperinsulinism shown in these cases is usually eliminated when the diet is altered (see Chapter 15).

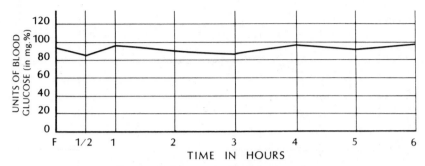

Figure 42. A natural tolerance test curve

An 8-year-old boy with chronic nasal allergy and reading disability.

This natural tolerance test shows a decline of 10 mg.% in blood glucose after a breakfast which consisted of a boxed cereal with milk. Although this blood-sugar decline is not great in amount, it is probably quite significant. It occurs in response to the release of insulin in response to food. It is usually expected that the process of food digestion and absorption into the body will be followed by an elevation of blood sugar. This is often tacitly assumed; but as this case shows, it may not be so.

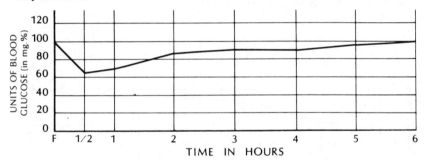

Figure 43. A natural tolerance test curve

A 13-year-old boy with a long history of difficulty achieving in school due to poor reading and math ability.

A usual breakfast was eaten by this boy after the fasting blood specimen was taken. The breakfast consisted of 1 tangerine, 6 oz. of milk, 1 egg, 2 pieces of cornbread with white sauce and honey, and a multiple vitamin tablet.

Notice that the blood sugar changes from a fasting level of 99 mg.% to a low of 64 mg.% at ½ hour, and 67 mg.% at 1 hour. This curve is consistent with a "hair-trigger" hyperinsulin release in response to food or a heightened sensitivity to normal amounts of insulin.

In this boy at this time, the result of eating is to bring about a lowering of blood sugar. Is this of any significance? In those patients who alter their dietary habits in accord with the recommendations made in this book (see Chapter 15), this kind of glucose curve often changes to one which rises after eating. In addition, learning ability may improve, and a more favorable personality may come about. In some people, specific food sensitivities must be discovered and treated before favorable change occurs.

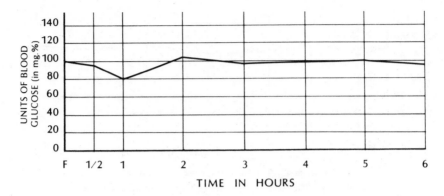

Figure 44. A natural tolerance test curve

A 12-year-old boy with learning disability and hyperactivity.

After a breakfast consisting of cinnamon toast with sugar and butter, a chocolate form of instant breakfast, and 8 oz. of milk, this natural tolerance test was recorded.

Note the decline of blood sugar of 20 mg.% (100 - 80) by 1 hour. The fasting level of 100 mg.% may be higher than desirable.

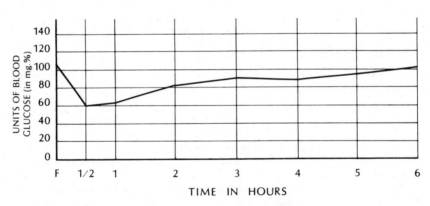

Figure 45. A natural tolerance test curve

A 15-year-old boy who was involved with the law because of stealing and other delinquent behavior.

Breakfast consisted of fruit, chocolate milk, 1 egg, and 2 pieces of toast with jelly.

Note the rather severe drop in blood sugar from 105 to 60 at ½ hour. After recovery from this drop, the blood sugar remains fairly steady.

The effect of food on the blood sugar has usually disappeared by the 2nd or 3rd hour after eating.

Repetitive bouts of hyperinsulinism such as this in response to food undoubtedly influence behavior and well-being. Removal of sugar from the diet very often eliminates this pattern of blood-sugar drop in response to food.

CHAPTER 10

DIABETES MELLITUS

Diabetes mellitus is a disorder of sugar metabolism. It is a syndrome — a condition which has a cluster of characteristic symptoms and laboratory findings. Although the block in sugar metabolism is most evident, other metabolic disturbances also are present.

Diabetes mellitus is often referred to as "sugar" by the laity. A person may say, for example, "I have sugar." The disease is also characterized by the phrase "sugar in the urine."

The word *diabetes* means "to run through." In a diabetic, the sugar "runs through" the body when it is excreted as sugar in the urine. Elevated levels of sugar in the blood are responsible for the overflow of sugar into the urine. *Mellitus* is a word which means "honeyed," referring to the sweetness of the urine.

Another type of diabetes, much less well-known, is diabetes insipidus. It has nothing to do with sugar. In this book, the word *diabetes* is used exclusively to mean diabetes mellitus.

Wherever we find civilized populations, we find a high prevalence of diabetes. It is one of the many disorders found with high frequency in developed or modernized cultures as opposed to primitive societies. Many consider diabetes to be our country's number 1 health problem.

Figures for the prevalence of diabetes range from 1 to 10 percent of the general population. Criteria for the diagnosis of the disorder vary somewhat, probably accounting for this wide range. Ninety percent of cases are found in adults. Medical experts agree that diabetes is one of the most common chronic diseases. It is found more often in those who are older, obese, female, black, Jewish, American Indian, and poor. It occurs frequently amongst physicians. The chance of developing diabetes doubles with every 20% of an individual's excess weight and also doubles with every decade of increasing age.

It has been found in 1.6 to 3.5 persons out of 1000 who are less than 25 or 26 years of age. One out of 1000 to 2000 children less than 15 years of age has diabetes.

Diabetes is the third leading cause of death in the U.S.A., and it is among the commonest causes of blindness. It is the commonest pediatric endocrine disease. The incidence increases around the time of puberty.

There are some ten million diabetics in the United States. Even more alarming is the fact that the prevalence of diabetes increased by more than 50% between 1965 and 1975. There is reason to believe that the occurrence of the syndrome is rising at the very rapid rate of 6% per year. Diabetes is expected to double every 15 years. My experience, as well as that of others, indicates that there are many individuals with "chemical diabetes" who have diabetic glucose tolerance tests, but who do not have clinical diabetes. This situation is not uncommon in children.

There is a strong hereditary factor in diabetes (especially in that of the adult-onset type); but in many cases, it is probable that a person's environment — what he eats, drinks, breathes, thinks, and does — is of major importance in determining the presence or absence of the syndrome. One of every four persons is a carrier of diabetes. The inheritance is thought to be multifactorial (polygenic), or the frequency of the diabetic gene is low enough to be inapparent in the blood-sugar test (which itself is influenced by many other factors). When both parents have diabetes, clinical diabetes is found in about 8% of their children. Many more, perhaps 30%, will probably develop maturity-onset diabetes unless stringent preventive measures are taken.

SYMPTOMS OF DIABETES

The outstanding symptoms of diabetes in children are excessive hunger, thirst, and urination. In technical terms, these symptoms are known as the 3 P's: *polyphagia, polydipsia,* and *polyuria.* They are the classic acute symptoms of the onset of diabetes.

The hunger reflects a block in metabolism which prevents the normal production of energy. One could say that the diabetic body, sensing its need for energy, becomes hungry in an attempt to provide the body with adequate fuel for metabolism.

Excessive thirst develops as a compensation for the large amounts of fluid lost in the urine. This large quantity of fluid is excreted as a necessity, because sugar carries out much fluid with it in the urine (osmotic effect).

A common symptom seen in new diabetics is the recent onset of bedwetting when it had not been previously present. Weakness is also a prominent symptom stemming from the block in glucose metabolism. Very often a new diabetic, while eating voraciously, will lose weight because of this metabolic defect. ("His food does him no good.")

Many clinicians believe that the onset of diabetes in children is more gradual in recent years than it was in the past. Thus the classic acute symptoms of onset, the 3 P's, may not be readily apparent.

In adults, the symptoms of diabetes are not usually as acute or dramatic as they are in children. The patient may feel sluggish, sleepy, weak, or dizzy; his weight accumulates; the blood pressure may rise; and he may be plagued by recurrent skin infections such as boils. Diabetes in women may first be recognized by the gynecologist because of the fungi (Monilia) that overgrow in sugar-containing urine and cause a distressing vaginitis or vulvitis. Genital itch may be troublesome. The first sign of diabetes in middle-aged men may be impotence.

Diabetes may also become evident because of fluctuations in a person's vision. Elevated and changing levels of blood sugar give rise to alterations in the refractive index of the eyes. Blurred vision may be noted. Vision specialists are usually quite aware of the importance of blood-sugar deviations in relation to vision.

Children with infectious illness who fail to recover fully

may cause their physician to order a urinalysis or blood-sugar test that may lead to a diagnosis of diabetes.

Celiac disease and thyroid problems are not uncommonly associated with diabetes. An underactive thyroid is not infrequently encountered.

Many diabetics may harbor the disorder without experiencing symptoms that they recognize as illness. Such "hidden diabetics" may only be identified by blood-glucose testing.

DIAGNOSIS

Diabetes is diagnosed by performing a glucose tolerance test (GTT) and finding elevated levels of blood glucose which do not return to normal within the usual time period (see Chapter 11, "The Glucose Tolerance Test (GTT)" and Chapter 12, "Glucose Tolerance Curves," for diagnostic criteria and examples). A GTT may not be necessary for the diagnosis of diabetes if the clinical picture is characteristic. This would include elevated blood-sugar values and the presence of sugar and perhaps acetone in the urine. A GTT may be indicated, however, even in cases with a characteristic clinical picture in order to exclude low blood sugar, which may occur at some time during the GTT which is otherwise diabetic.

For all practical purposes, the GTT that shows higher than normal blood-sugar levels and a delayed return to a normal sugar level indicates diabetes. An overactive adrenal gland or an overactive thyroid gland, however, may also produce these findings on GTT. Elevated blood sodium, brain tumors, meningitis, strokes, or aspirin poisoning can mimic diabetes but usually pose no diagnostic problem.

Glucose values are higher in obese individuals than they are in thin persons. The diagnosis of diabetes in the obese may thus be more difficult.

Some physicians are content to use 1 or 2 blood levels of glucose in the diagnosis of diabetes. Minimum plasma glucose levels indicative of diabetes have been given as:[1]

[1]"Evaluation of a Positive Urinary Sugar Test" by Harvey C. Knowles Jr., M.D. Published in *Journal of the American Medical Association*, Vol. 234, No. 9, pages 961 - 963.

	Age less than 50 years	Age more than 50 years
fasting	150 mg.%	150 mg.%
1 - 2 hours after fasting	230 mg.%	200 mg.%

Another investigator indicates that a fasting plasma glucose over 130 or a random blood sugar above 195 gives a clearly evident diagnosis of diabetes.[2]

Dr. Robert L. Jackson (M.D.) has suggested the following screening procedure for diagnostic detection of diabetes in children:[3] When the child is ill with an infectious disease, have him drink orange juice or grape juice (a child less than 6 years old, 4 oz.; more than 6 years old, 8 oz.) to which 1 tablespoon of sugar has been added. Twenty minutes later have him drink Coca Cola (a child less than 6 years old, 3 oz.; more than 6 years old, 6 oz.). After 2 - 3 hours, test the child's urine for sugar. If the test is positive during this period of infectious stress, it is likely that the child has diabetes, and a glucose tolerance test should be done as soon as practical. (I do not recommend sugar or Coca Cola for any person.)

In the practice of pediatrics, one frequently sees children who are ill and who have been drinking sugar-containing fluids such as cola drinks, Gatorade, etc. If these children have sugar in their urine, it is best to investigate them with a glucose tolerance test when they are well. In my experience, even a trace of sugar in the urine indicates that that individual has a faulty diet (see Chapter 15).

More important than making a diagnosis is altering the diet toward one that leads to optimal health and freedom from illness. Administering sugar to children during illness or for testing purposes is not usually necessary. Sugar tends to make well children sick and sick children sicker. It may lengthen illness and increase the chances of infection with opportunistic bacterial invaders. When glucose tolerance testing is needed, a food high in carbohydrates may be used as a test meal (see Chapter 13, "The Natural Tolerance Test [NTT]").

[2]By Robert N. Alsever, M.D., quoted in the report "Glucose Tolerance Testing Under Fire." Published in *Medical Group News*, Vol. 9, No. 1, January 1976.

[3]"The Child with Diabetes" by Robert L. Jackson, M.D. Published in *Nutrition Today*, Vol. 6, No. 2, March/April 1971.

In the normal individual receiving a glucose load, there is some insulin release into the bloodstream within the first 30 minutes after the administration of the glucose. In the diabetic, this early insulin discharge is characteristically absent and appears to be a hallmark of the diabetic state.

According to Dr. R. M. Ehrlich (M.D.), diabetes in childhood is indicated by a random blood sugar over 300 mg.%. He indicates that the fasting blood sugar may be normal in mild cases.[4]

In children who have no insulin response to a glucose load, an intravenous glucagon test may be done. This may bring about an insulin response, indicating that some beta-cell function is present.

A preliminary report has recently indicated that measurement of glucose metabolism in the red blood cells of children may be an early indicator of glucose intolerance.[5] More research is needed to see if this probe of cellular function can truly identify the child at high risk for the development of diabetes. Measuring the concentration of a specific kind of hemoglobin (AIc) in the blood may aid in the assessment of blood glucose.

Recent research with blood-insulin levels has confirmed that diabetics may be deficient in fasting levels of insulin (the juvenile pattern). They may also possess insulin levels which rise in response to glucose but which have a delayed return to normal. The insulin peak may also come about later than normal (at 2 - 3 hours instead of within the first hour), or the fasting insulin level may be excessive.

Although hair analysis of trace minerals is not yet utilized by the average physician, this diagnostic technique may well become commonplace within the next decade. According to Dr. Lloyd Horton (D.C.) and Dr. Paul Eck (D.N.), a diabetic state is indicated by a relative deficiency of calcium in relation to magnesium on hair analysis.[6] A ratio of magnesium to calcium of 2 to 1 indicates diabetes. In addition, high calcium and magnesium in relation to zinc (relative zinc deficiency) is also seen, as well as a relative deficiency of manganese (calcium to manganese ratio of 226

[4]"Diabetes Mellitus in Childhood" by R. M. Ehrlich, M.D. Published in *Pediatric Clinics of North America*, Vol. 21, No. 4, November 1974, pages 871 - 884.

[5]Report of the Annual Meeting of the American Diabetes Association, reported in *Medical Tribune*, Wednesday, July 24, 1974.

[6]"The Total Metabolic Approach," audio-tape featuring Lloyd Horton, D.C., and Paul Eck, D.N., January 1977. Supplied by Western Academy of Biological Sciences, 115 South 38th Street, Tacoma, Wash. 98408.

to 1 or greater; normal is 73 to 1). The diabetic also usually has a low sodium and potassium in hair, suggesting adrenal insufficiency. Chromium disturbance is usual.

TYPES OF DIABETES

There are two main types of diabetes:

1. that in which the pancreas secretes insulin and
2. that in which the pancreas secretes little or no insulin.

Diabetes in adults is usually the first type, the insulin-secreting variety. This type is known as *adult diabetes*, but is more accurately described as *maturity-onset diabetes*. Two out of three of such cases are women, and many of them have a history of delivering particularly large (oversized) babies. In maturity-onset diabetes, the isles of Langerhans in the pancreas are about 40% reduced in number, but blood-insulin levels are often normal or higher than normal. This type of diabetes is best known as *non-insulin-dependent diabetes*.

Adult-onset diabetics are usually overweight. They can often be managed successfully by diet, weight reduction, graduated programs of physical conditioning, allergy management, or the use of oral medication to lower blood sugar (see Chapter 15).

Diabetes in children is usually the insulin-deficient or insulin-absent variety. Such juvenile diabetes is known as *growth-onset diabetes* or *insulin-dependent diabetes* and has its beginning before 20 years of age. It is most common in preadolescents of 10 to 12 years of age. In juvenile cases, which amount to some 10% of total diabetics, the isles of Langerhans contain practically no beta cells.

Juvenile diabetics almost invariably have a severe form of the disease and require injectable insulin to control blood sugar, to provide proper growth, and to prevent complications.

Juvenile diabetes showed an alarming increase between 1959 and 1972.[7] Whether this apparent epidemic is continuing or not is not actually known at this point. A shift from breast feeding to formula and the use of the mumps vaccine are factors to be

[7]"Juvenile Diabetes, Researchers Worried About an 'Epidemic,'" a report of research conducted by A. Frederick North Jr., M.D., St. Petersburg *Independent*, Wednesday, November 9, 1977.

considered in the origin of juvenile diabetes. Several patients that I have cared for were nutritionally disadvantaged and then suffered a severe emotional shock. There is a distinct seasonal variation in the onset of diabetes in children, with peak incidence in September and the winter months.[8]

Despite proper treatment, juvenile diabetics have been reported to have an abbreviated life span averaging 27 years after diagnosis. The life expectancy of the diabetic is estimated to be shorter by one-third than that of the nondiabetic. Long-term complications primarily affect the blood vessels in the kidneys, nervous system, and eyes. Blindness, heart attacks, strokes, kidney failure, and gangrene in the limbs may occur. The diabetic's risk of blindness is 25 times that of the individual without the disorder.

Vascular diseases are directly responsible for 70% of deaths in diabetics. This is a rate 2½ times that of persons without diabetes. Most diabetes specialists agree that the blood-vessel complications of the disease increase with the duration of the disease, but some individuals who have had juvenile diabetes for 40 years have little or no disability. It is strongly suspected that rigid medical control of the diabetes is an important factor in minimizing complications. This is especially the case in regard to cataracts and diabetic neuropathy. The life expectancy after the onset of diabetes has increased from 4.9 years in 1914 to 18.1 years in 1970.

Thickening of the basement membrane in small blood vessels is a characteristic feature of diabetes. Moreover, the more severe the diabetes, the thicker is the abnormal basement membrane. The membrane contains considerable carbohydrate material.

Diabetics have been reported to have a three-fold higher incidence of cancer than nondiabetics. It is interesting that cancer patients commonly have blood-sugar values which are higher than noncancer patients. Tissue culture techniques used for growing cancer cells require considerable amounts of sugar to sustain the growth of the cancer cells. Although unproven, it is

[8]"Seasonal Variation in the Onset of Diabetes in Children," a report of Duncan R. MacMillan, M.D., F.R.C.P. (C), Marcos Kotoyan, M.D., Daniel Zeidner, M.D., and Bijan Hafezi, M.D. Published in *Pediatrics*, Vol. 59, No. 1, January 1977, pages 113 - 115.

quite possible that avoidance of chronic or recurrent high levels of blood sugar may decrease the likelihood of developing cancer.

Evidence of a link between sugar levels in the body and cancer may be contained in the observation that tiny doses of long-acting insulin may exert a cancer-suppressing effect.

A further suggestion of a link between sugar and cancer is the effect of the experimental anticancer drug hydrazine sulfate. Information from the Syracuse Cancer Research Institute, Inc.[9] indicates an overall high degree of favorable results with this drug to date. Hydrazine sulfate inhibits gluconeogenesis in the liver and kidney, thus tending to bring about lowered blood-sugar levels.

THE BASIC PROBLEM

Now let us look more closely at diabetes. What is the problem in this metabolic disorder? It is likely that multiple causes will be found to explain the condition that we presently know as diabetes.

In diabetes, there is an impairment of insulin activity. Either the body doesn't produce enough insulin or that which is available is prohibited from proper function by other substances or circumstances in the body.

As far as is known at the present time, there is a block in the metabolism of glucose in diabetes. Cells starved for energy call upon reserves of fat in the body, but these fatty substances (an alternate fuel source) are not able to entirely substitute for glucose. Profound chemical imbalances occur with elevated blood sugar and the appearance of fatty chemicals (ketones) in the blood and urine. The liver, working overtime to produce ketones from mobilized fat, often becomes enlarged and tender.

Why is there a block in the processing of glucose in diabetes? Certainly, all the answers are not known. Nevertheless, an absolute deficiency in insulin is certainly one explanation. Insulin deficiency is characteristic of juvenile diabetics. In fact, there is a

[9]Syracuse Cancer Research Institute, Inc., Presidential Plaza, 600 East Genesee Street, Syracuse, N.Y. 13202.

correlation between the severity of diabetes and the condition and number of the islet cells in the pancreas. In diabetes, they are fewer in number, and the beta cells are less granulated than in normals. We can certainly say that a blunted insulin response to a rise in blood sugar is a hallmark of juvenile diabetes.

Individuals with insulin deficiency behave as though they are in a superfasting state. Lack of insulin disallows the normal processing of ingested food, and profound metabolic disturbances occur.

There is evidence that the alpha cells as well as the beta cells in the pancreas are not working correctly in diabetes. Glucagon and insulin, the hormone products of these cells, appear to be imbalanced.

More basic cellular abnormalities in diabetes may be discovered as research techniques improve. Already, a deviation in red blood cell metabolism of glucose has been reported. The basement membrane abnormality in blood vessels has already been noted.

Unresponsiveness to insulin may be another explanation for diabetes. Indeed, there is evidence to show that insulin resistance at a tissue level is involved. This appears to be the case particularly in maturity-onset diabetes. A model for this sort of problem is already known. It is a clinical disorder which involves lack of response to a hormone in the body. In the disorder known as *pseudohypoparathyroidism*, normal levels of parathyroid secretion fail to elicit calcium and phosphorous regulation in the kidney. The result of this failure to respond to the hormone is an accumulation of phosphorous and a low calcium level in the blood. Growth failure, bony abnormalities, and brain dysfunction occur. Then, too, in one type of familial dwarfism, growth hormone levels are normal, but the body cells appear to be insensitive to this hormone. In the case of thyroid, also, it is likely that tissues in some individuals may be underresponsive to normal levels of thyroid hormones. At least some diabetics may possess cells which are unresponsive to whatever insulin there is in their bodies.

Unresponsiveness of cells to insulin leads to elevated blood sugar, burning of fat for energy, and the entire complex of chemical and clinical disabilities which characterize diabetes.

One reason for insulin unresponsiveness is a deficiency

or unavailability of insulin-binding sites on cells. Dr. Melvin Blecher (Ph.D.) has reported that white blood cells of adult-onset diabetics have fewer binding sites than nondiabetic controls.[10]

Certain genetic determinants known as *HLA (histocompatability locus antigen)* factors bear a statistically certain relationship with the presence of juvenile diabetes mellitus. The frequencies of HL-A8 and W15 histocompatability antigens are increased in juvenile diabetics. These factors may influence binding sites, pancreatic makeup, and immune function in diabetics.

The adrenocortical (glucocorticoid) hormones as well as estrogen appear to bring about a state of insulin resistance and can be associated with the diabetic condition. Perhaps subtle imbalances in these hormones may be responsible for some diabetic states.

There may even be chemicals released in the body which "coat" cells or otherwise alter them, so insulin may not be able to exert its usual effect. This concept, however, is speculative at this point.

Closely allied to this concept is the idea that prostaglandin in the body may inhibit insulin release from the pancreas. The presence of prostaglandin E-2 may influence pancreatic beta cells so they fail to recognize elevated levels of circulating blood glucose. Prostaglandin E-2 is believed to play a role in "downmodulating" insulin secretion in the body. Support for this concept has been obtained by experiments on maturity-onset diabetics.[11] Intravenous sodium salicylate equivalent to about 8 aspirin tablets given to these patients increased blood-insulin levels, and a desirable fall in blood glucose took place. Observations have also been made through the years, indicating that aspirin has at least partial effectiveness in treating diabetics. Aspirin and sodium salicylate are inhibitors of prostaglandin chemicals in the body.

Before anyone begins to take aspirin, let me point out that these results are preliminary and that dietary improvement, exer-

[10]Dr. Blecher is Professor of Biochemistry at Georgetown University School of Medicine, Washington, D.C. A report of his work was included in the "Medical News" section of the *Journal of the American Medical Association,* Vol. 235, No. 9, March 1, 1976, pages 893 and 894.
[11]"Insulin Defect Reversed by IV Aspirin," a report in *Medical Tribune,* Vol. 18, No. 24, Wednesday, July 20, 1977; investigative work of Dr. Mei Chen, Assistant Professor of Medicine, University of Washington, Seattle, Wash.

cise, weight loss, and lifestyle change constitute the preferred management of maturity-onset diabetes.

Anti-insulin antibodies may be a cause of insulin insufficiency or insulin unresponsiveness. In a survey of 105 juvenile diabetics, antibodies against pancreatic islet cells were found in 51 (48.6%).[12] Infiltration of the isles with lymphocytes (immune cells) and evidence of cell-mediated immune reactions to pancreas tissue also occur in juvenile diabetics. It is increasingly evident that autoimmune reactions[13] are associated with the onset of juvenile diabetes mellitus.

I believe that it will eventually be shown that there is a strong relationship between lack of physical activity and the development of diabetes. The beneficial effect of exercise on sugar metabolism is already known. The person who exercises can be shown to have increased utilization of glucose compared with the same person who is sedentary. Individuals confined to bed develop an impairment in glucose utilization within a few days. Elderly persons who are less physically active develop higher blood-sugar levels than younger, more active individuals.

Diabetics who take insulin become quite aware of the effect of exercise in lowering their blood glucose and thus lowering their insulin requirement. Diabetes appears to be a disorder of civilized peoples who are, by and large, sedentary in their habits and who consume large amounts of sugar and other refined carbohydrates.

In maturity-onset diabetes, there appears to be some defect of the pancreatic beta cells in their response to glucose. Many times it appears that the timing is off with the pancreas failing to respond quickly enough to glucose input.

Then, too, it appears that many overweight adult diabetics actually become diabetic because of the presence of excess fat. In these persons, excess fat cells seem to bind or "tieup" insulin, preventing it from exerting its usual blood-sugar lowering effect in the body. The fatter the person, the more insulin resistance he has. Excessive insulin secretion fosters fat formation, and fat formation appears to hinder insulin action.

[12]"Islet Cell Antibodies in Juvenile Diabetes Mellitus of Recent Onset" by R. Lendrum, G. Walker, and D. R. Gamble. Published in *Lancet,* Vol. 1, 1975, pages 880 - 883.

[13]Autoimmune reactions are disturbances of the immunity system in which one's own tissue is attacked as though it were a foreign substance. It is self-allergy.

Lack of adequate amounts of vitamin C may reduce the sensitivity of body tissues to insulin, leading to hyperinsulinism and eventual diabetes. It has also been shown that insulin enhances the activity of vitamin C in the peripheral body tissues, but there is less of this response in the diabetic patient. In some diabetics, the administration of vitamin C may lower their daily insulin requirement.[14]

An increasing number of diabetics in a society which has a galloping consumption of sugar, leads one to consider sugar intake as a cause of diabetes. Lay persons have long suspected such a relationship. It was not rare in the author's generation to hear parents or grandparents warn children to not take too much sugar, because it might lead to diabetes. In the present era of diminished family influence on children, such admonitions are probably less frequently given.

Laboratory rats can be made diabetic by depeleting their beta cells of insulin by long-term feeding of sugar. Dr. A. M. Cohen in Israel has presented considerable data, linking high sugar intake and the development of diabetes.

From a statistical standpoint, eating sugar is indeed associated with diabetes. Any country which has a sugar consumption of roughly 70 pounds per person per year, or greater, has a high incidence of diabetes roughly approximating that of westernized nations. The average sugar consumption in countries where diabetes is common is three times that where it is uncommon. Furthermore, there is virtual absence of diabetes in primitive areas which do not have access to refined carbohydrate in the diet. Sugar is the most commonly eaten refined carbohydrate.

It is quite conceivable that repetitive use and overuse of the insulin-secreting apparatus in humans might lead to exhaustion of the gland and the production of diabetes. This may have something to do with the fact that diabetes is linked with an increase in the rate of glucose absorption. Further discussion of these matters will be found in the chapter on low blood sugar (Chapter 14).

T. L. Cleave (M.R.C.P.) and his associates (see "References," page 489) point out that the pancreas has evolved as an

[14]It should be noted that vitamin C intake may alter the usual urine tests used to detect sugar in the urine. Close planning with one's physician is indicated so this does not create difficulties in management for patient or physician.

endocrine organ to deal with *dilute*, unrefined carbohydrate as found in nature. The diet of developed nations worldwide contains large amounts of refined, *concentrated* carbohydrate (white bread, sugar, boxed cereals, etc.). A situation results, according to T. L. Cleave, in which chronic pancreatic strain over a period of some 20 years or so leads to diabetes.

When we consider the rapidity and magnitude of blood-sugar increase in some persons following the intake of sugar, we can realize the degree of homeostatic disruption that occurs. Figure 85 in Chapter 12 (page 256) shows the graph of a patient with a blood-sugar increase of 64 mg.% within 15 minutes after drinking sugar solution! Cleave's thesis of pancreatic strain (more likely, multifold metabolic strain) seems likely when such glucose tolerance tests are reviewed (see Chapter 12, "Glucose Tolerance Curves").

It is believed that low blood sugar precedes the diabetic state, at least in some cases. The degree to which this is true, remains to be seen. One problem is the definition of low blood sugar (see Chapter 14). It is likely that some form of carbohydrate metabolic dysfunction precedes the diabetic state. Some physicians believe that the patient who has a low, flat glucose tolerance curve (see Chapters 11 and 12) in early adulthood, is prone to ultimately develop diabetes in later life.

Diabetes itself is known to occur in some cases simultaneously with one form of hypoglycemia. When this occurs, the diabetes is usually mild (see Chapter 14).

Overproduction of growth hormone or excessive sensitivity to it may be a factor in the origin of diabetes. Diabetic patients have elevated levels of growth hormone after many stimuli such as exercise; but with better medical management of their disease, the elevated growth hormone levels return toward normal.

In dwarfs who have a hereditary growth hormone deficiency, diabetes, when it occurs, is extremely mild; and there is no elevation of the fasting blood-sugar level. There is also some evidence that diabetic patients with blood vessel disease in the eyes have higher levels of growth hormone than those patients who do not have such eye complications.

Dr. Melvin Page (D.D.S.) indicates that diabetes is more often due to an overactive anterior pituitary gland than to insulin deficiency. He points out that large amounts of insulin, far in excess of the insulin output of the pancreas, may be needed for treatment of the diabetic. It may be, according to Dr. Page, that this insulin is acting as an antagonist to an overactive anterior pituitary gland. Dr. Page speaks in general of the anterior pituitary, but modern research is closing in on growth hormone as a most important diabetes-producing substance. It must also be remembered that anterior pituitary overactivity is likely to result in adrenocortical excess, with relative insulin deficiency or ineffectiveness.

Patients with acromegaly due to excess anterior pituitary secretion in adulthood have diabetes or decreased glucose tolerance in 50% of cases.

Experimentally, the use of growth hormone, adrenocortical hormones, thyroid hormone, or glucagon can produce a diabetic-like syndrome. Producing permanent diabetes in animals, however, usually requires long-term use of a hormone plus removal of half or more of the pancreas. The chemical alloxan, however, selectively destroys the isles of Langerhans, producing permanent diabetes in experimental animals.

Cushing's syndrome is due to an overactivity of the adrenal cortex. Diabetes or impaired glucose tolerance is found in 75% of cases.

Somatostatin has been discussed in Chapter 8. It inhibits growth hormone of the anterior pituitary and directly suppresses insulin from the beta cells of the isles of Langerhans, as well as glucagon from the alpha cells. In times past, diabetic patients with severe eye complications were treated by removal of the pituitary gland. Treatment with somatostatin may in effect perform a "medical" pituitary removal, at least in part. The full significance of this hormone in diabetes is not known at this time, but treatment with it may become a reality. The kind of somatostatin presently available is extremely short-acting in the bloodstream. Treatment with this material awaits the availability of a longer-acting preparation. Already, however, insulin and somatostatin together have proved more effective than insulin alone in treating elevated blood-sugar levels of diabetics.

Liver regulation of glucose may be a key area in determining whether a patient is diabetic or not. The diabetic may be a person whose liver is unable to shut off the production of glucose from glycogen.

Does diabetes involve an impairment of liver function due to protracted mineral and B-vitamin deficiency? Could it be that the liver responds differently to insulin in the presence of such deficiencies? What is the role of dietary amino acids? What about liver function in regard to glucose metabolism in the presence of obesity? These questions cannot be answered with certainty at this time.

The complete role of the hormone glucagon has not been unravelled in diabetics or in normal individuals. Glucagon, as the reader will remember, is the "anti-insulin" hormone produced in the alpha cells of the pancreas. It is said to suppress the secretion of insulin and decrease the uptake of glucose by tissues. Nevertheless, intravenously administered glucagon has been shown by Dr. ZVI Laron (M.D.) to stimulate a rise in blood insulin.[15]

Glucagon has been found to be the first hormone to function in the embryo. Also, levels of glucagon in the rat embryo are a hundred times greater than the first secretions of insulin. Under normal conditions, elevation of blood sugar suppresses glucagon secretion, and hypoglycemia increases it. In diabetic patients, excessive amounts of glucagon are found. Furthermore, the more out of control they are, the more glucagon they produce. In addition, diabetic patients, unlike normals, are unable to shut off the production of glucagon when glucose is taken into the body.

In other research, diabetic patients have been shown to have a defective glucagon response to insulin-induced low blood sugar. The glucagon-secreting alpha cells, like the insulin-producing beta cells, seem not to be functioning properly in the juvenile diabetic and perhaps in the maturity-onset diabetic as well. One can say that glucagon secretion is inappropriately controlled in the diabetic.

Dr. A. K. Khachadurian (M.D.) and Dr. M. Kletzkin (Ph.D.) discussed the matter this way in a recent article:[16]

[15]"Treatment of Diabetes in Children: Revised" by ZVI Laron, M.D. Published in *Pediatric Annals*, Vol. 3, No. 7, July 1974, pages 63 - 77.

[16]"Diabetes in Concept and Fact" by A. K. Khachadurian, M.D., and M. Kletzkin, Ph.D. Published in *Drug Therapy*, March 1975. This quotation is used with the permission of the authors and the publishers of *Drug Therapy*.

Although the central role of insulin deficiency in diabetes cannot be deemphasized, there is growing awareness that it is the interplay of insulin with other factors, particularly glucagon, which determines the resultant blood-glucose level. The relative concentrations of these hormones determine whether the prevailing metabolic state is anabolic (with high insulin levels) or catabolic (with high glucagon levels). The opposing actions of insulin and glucagon provide a metabolic regulatory mechanism which depends upon proper functioning of both alpha (glucagon-secreting) and beta (insulin-secreting) cells in the pancreas. In diabetes, this regulatory mechanism is disturbed because both types of cells are not functioning properly. Thus, abnormal glucose tolerance is a consequence of a dihormonal disturbance in which the proportion of insulin to glucagon, rather than the absolute concentration of each hormone, is critical.

The low insulin-to-glucagon ratios reported in diabetes result from a combination of reduced insulin secretion and increased glucagon secretion. In diabetes, glucose does not seem to inhibit glucagon secretion to the extent that it does in normals, and this failure appears to be the result of a genetic defect in the alpha cell. As a consequence, glucagon levels are inappropriately high for existing blood glucose levels.

More recently, Dr. Rachmiel Levine (M.D.) has concluded that glucagon is "a potent diabetogenic factor in the absence of insulin but that physiologic amounts of insulin can overcome or prevent the effects of appreciably increased glucagon levels, at least in man."[17]

At least one case has been described in which diabetes has been cured by surgical removal of a glucagon-secreting pancreatic tumor. There is some evidence that a specific glucagonoma syndrome may exist in which diabetes, anemia, and a typical skin rash occur in postmenopausal women.[18]

When one ponders hypoglycemia as well as diabetes, he sees the distinct possibility of hormonal imbalance in each

[17]"Glucagon and the Regulation of Blood Sugar" by R. Levine. Published in *New England Journal of Medicine*, Vol. 294, 1976, pages 494 - 495.

[18]"Aglucagonemia Syndrome," a condensation of "Aglucagonoma Syndrome" by C. N. Mallinson, S. R. Bloom, A. P. Warin, P. R. Salmon, and B. Cox. Condensation published in the *Lancet*, Vol. II, 1974, page 1. Original article published in *Current Medical Digest*, Vol. 42, No. 2, February 1975.

condition. The main thread of hormonal disturbance in hypoglycemia appears to be adrenocorticoid-insulin imbalance. The main thread of hormonal disturbance in diabetes may be glucagon-insulin imbalance. As one progresses from hypoglycemia toward and into diabetes, the symphony of endocrine function may become increasingly discordant. Many factors contribute to the dissonance. Pituitary, thyroid, gut, liver, pancreatic, exercise, mental, and mineral factors may all intertwine and play varying roles at varying times.

Although incompletely understood, there is a relationship of glucose metabolism to body potassium. Potassium is synergistic with insulin. Patients who have deficiencies of potassium in their body often have impaired glucose tolerance with elevated blood-sugar values. When potassium is given to these patients, the sugar levels return to normal. The use of thiazide diuretics may bring on elevated blood sugars, a diabeticlike GTT, and a decreased potassium level in the blood. Giving potassium to these patients raises the blood sugar and GTT to normal. Insulin from the beta cells of the pancreas may be dependent upon potassium.

One wonders about the long-term effects of caffeine on sugar and potassium balance in the body. Caffeine is a mild diuretic and acts to increase the blood sugar.

It is well known that the administration of insulin brings about a transfer of potassium into cells and a lowering of blood potassium. In fact, clinical states of potassium excess in the blood may be managed by treatment with insulin.

When an analysis of minerals in the hair is made, the finding of high potassium in relationship to zinc often indicates the presence of a diabetic condition, according to the late Dr. John Miller (Ph.D.) of the Bio-Medical Laboratories in Chicago.[19] Figure 46 illustrates this relationship. Figures 47 and 48 show the glucose tolerance test and natural tolerance test of the patient whose hair minerals are displayed in Figure 46.

Release of insulin from the beta cells of the pancreas may be dependent upon potassium. It is obvious that the final chapter has not been written on diabetes, potassium, or the relationship between them.

[19]Doctor's Data, Inc., P. O. Box 111, 245 Roosevelt Road, West Chicago, Ill. 60185.

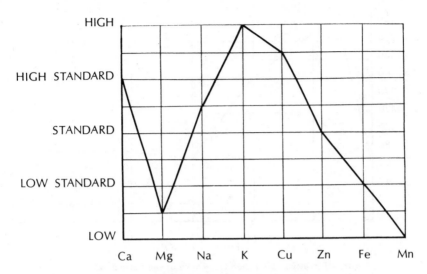

**Figure 46. Graph of mineral elements in hair showing
the relationship of one mineral to another**

This patient has a high potassium to zinc ratio which suggests a diabetic tendency. Administration of manganese, chromium, and zinc to persons with carbohydrate dysmetabolism can significantly improve their conditions. Hair analysis is a valuable tool. The therapist must be sure that all minerals are in correct balance, with deficiencies and excesses corrected.

Ca:	calcium	Cu:	copper
Mg:	magnesium	Zn:	zinc
Na:	sodium	Fe:	iron
K:	potassium	Mn:	manganese

According to Dr. Charles J. Rudolph Jr. (D.O., Ph.D.),[20] juvenile diabetes is characterized by a deficiency in potassium on hair analysis. In adult-onset diabetes, however, low levels of magnesium, zinc, manganese, and chromium are characteristically noted.

There is a condition known as hypernatremic dehydration, which is seen in young children as a result of vomiting and/or diarrhea. This form of fluid deficit is characterized by elevated

[20]"Trace Element Patterning in Degenerative Diseases" by Charles J. Rudolph Jr., D.O., Ph.D. Published in the *Journal of the International Academy of Preventive Medicine*, Vol. IV, No. 1, July 1977, pages 9 - 31.

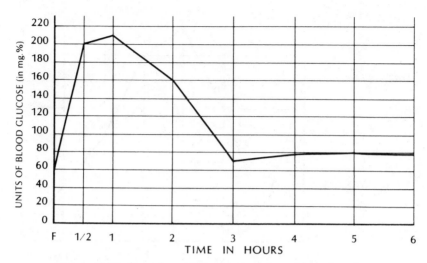

**Figure 47. Glucose tolerance test curve of the patient whose hair
mineral pattern is displayed in Figure 46**

This glucose tolerance test is a diabetic curve — that is, the blood sugar
rises excessively high and remains elevated longer than it should. (See Chapter 11
for a discussion of the glucose tolerance test.)

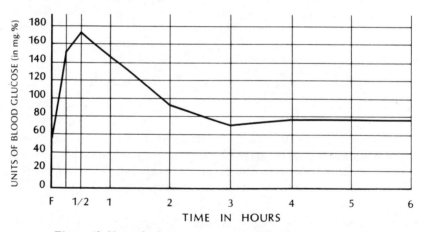

**Figure 48. Natural tolerance test curve of the patient whose hair
minerals and glucose tolerance test are shown in Figures 46 and 47,
respectively**

When the patient ate his normal breakfast, this diabetic type of glucose
response was obtained. Eating food should ordinarily not elevate the blood sugar
this high for so prolonged a time. (See Chapter 13 for a discussion of the natural
tolerance test.)

levels of sodium in the blood. Very often, elevated levels of blood sugar are also found in this condition. Again we see the interlocking relationship of body chemicals to blood sugar.

In diabetic patients, blood copper and zinc may be high, whereas manganese is low, especially during the first three years of illness. Furthermore, the lower the manganese levels, the worse the diabetic state.

Proper sodium levels in the body are intimately related to proper cellular functions in the body, one of which is the absorption of sugar from the gut. It is possible that the restriction of sodium in the diet could bring about low blood sugar.

There is a strong relationship between chromium deficiency and diabetes. Chromium is necessary in the body for proper insulin function and chromium deficiency, at least in rats, is associated with high levels of atherosclerosis, a disorder that occurs with increased frequency in diabetes. There is also evidence that chromium is an integral part of the binding process in which insulin attaches itself to cell-receptor sites. Chromium may be thought of as a cofactor with insulin to permit glucose entry into cells.

Chromium deficiency on a nutritional basis is reported to be widespread in our country. The reason for this may be the common use of chromium-poor, refined foods in our diets. Judging by animal experiments, the diet in the United States should supply ten times the amount of biologically available chromium that it now provides, to insure proper insulin function. Whole grains, seeds, and brewer's yeast are rich in chromium. Blackstrap molasses is another chromium-rich food, although its sugar content may not be desirable for some diabetics. Beer is generally not desirable from the standpoint of alcohol and additives, but it contains considerable chromium.

The matter of chromium excess in the body is just now being recognized. In my experience, high chromium levels in hair have been associated with disturbed carbohydrate metabolism. Much more investigation is needed.

Zinc is another mineral of central importance in carbohydrate metabolism. Pancreatic zinc has been reported to be reduced in amount in diabetic patients.[21] Zinc is an essential

[21] By D. A. Scott and A. M. Fisher in *Journal of Biochemistry*, Vol. 29, page 1043. Also, I. Kodata in *Journal of Laboratory and Clinical Medicine*, Vol. 35, 1950, page 568.

ingredient of the insulin molecule, and it has been said that zinc holds insulin in the granules of the beta cells in the pancreatic islets.[22] Increased urinary zinc loss and low serum zinc levels have been reported in diabetics. In addition, zinc concentrations have been shown to be low in the hair of juvenile diabetics.[23]

In a considerable number of patients with diabetic glucose tolerance tests in my practice, provision of an optimal diet with appropriate nutritional supplements has been followed by return of the glucose tolerance curve to or toward normal (see Chapter 15).

Vitamin B_6 (pyridoxine) deficiency in humans can induce a diabetic GTT (glucose tolerance test). In pregnancy there is an increased need for vitamin B_6. Women who develop diabetic glucose intolerance during pregnancy improve when supplementary B_6 is given. An association between B_6 deficiency and deterioration in glucose tolerance has been found in women who take birth control pills.[24] Treatment with vitamin B_6 reverses the glucose intolerance (and often makes the women feel a lot better!) Some cases of diabetes in the general population may be associated with lower than optimal intake of vitamin B_6.

Werner's syndrome is a condition of premature aging in which diabetes occurs in 50% of the cases.

In psoriasis, alterations in glucose metabolism are very frequent with borderline or frank diabetes occurring in the majority of patients.

There is a chronic life-shortening disease known as *cystic fibrosis of the pancreas.* In this disorder, as well as in chronic inflammation of the pancreas, elevated levels of blood glucose in response to a glucose load are not rare. Such glucose intolerance with symptoms is known to occur in 11 out of 1000 patients with cystic fibrosis. An even greater number have low insulin in the blood and glucose intolerance without symptoms. The glucose disorder of cystic fibrosis is milder than in chemical

[22]By M. Aksoy in *American Journal of Clinical Nutrition,* Vol. 25, 1972, page 262.

[23]"Hair Zinc Concentration in Diabetic Children" by M. Amador, M. Hermelo, P. Flores, and A. Gonzalez. Published in *Lancet,* December 6, 1975, page 1146.

[24]By D. P. Rose and others in *American Journal of Clinical Nutrition,* Vol. 28, 1975, pages 872 - 878.

diabetes and is not associated with ketoacidosis.

B. M. Frier[25] and Dr. William H. Philpott (M.D.)[26] indicate that diabetes is a state of generalized pancreatic insufficiency with disturbed production of bicarbonate, digestive enzymes, and insulin.

In explaining why one person becomes diabetic and another not, we must always keep in mind the important factor of hereditary variability. We know that diabetes runs in families and that the disease has a strong genetic background. It may be, for example, that diabetics are those persons who have an insufficient number of beta cells or an excess number of alpha cells in their pancreatic islets. Or, it may be that diabetics are those who have stronger function of alpha cells than beta cells; or, alpha and beta cells which are just nonoperative.

The pancreas may be injured by viral infections, allergies, metal toxicities, or chronic nutritional deficiency. The pancreas may produce an incomplete or altered form of insulin that is ineffective.

Iron deposits are one example of pancreatic injury. In the disorder known as *hemochromatosis*, iron deposits are characteristically found in the pancreas, liver, and skin. The deposits occur because of an overload of iron in the body. Diabetes is characteristic of this disorder, occurring in 63% of cases.

Infection before or after birth has been implicated in the development of diabetes. Infection with rubella virus (German measles) seems to be particularly associated with later diabetes. Studies of patients who have had prenatal rubella indicate that up to 40% of them develop diabetes. It may be that the pancreas is involved in the infection.

In addition to rubella, mumps, coxsackie, mice encephalomyocarditis virus, and Venezuelan equine encephalitis virus in hamsters have been implicated in the production of diabetes. It is notable that animals in the acute phase of infection may be hypoglycemic.

There is some evidence suggesting a coincidence in

[25]"Exocrine Pancreatic Function in Juvenile-Onset Diabetes Mellitus" by B. M. Frier. Published in *Gut*, Vol. 17, 1976, pages 685 - 691m.

[26]"Proteolytic Enzyme and Amino Acid Therapy in Degenerative Disease," an abstract of two articles by William H. Philpott, M.D. Edited, condensed, and distributed by Dwight Kalita, Ph.D., June 1977.

coxsackie virus B_4 infection and the seasonal incidence of onset in juvenile diabetics.

Since the mumps virus is known to affect the pancreas in a considerable percentage of mumps cases, this virus has been considered suspect as a likely cause of pancreatic dysfunction and subsequent diabetes. One investigator reported epidemiologic data supporting a link between mumps infection and diabetes.[27]

Mice genetically predisposed to diabetes are available for research, and inbreeding of low sugar-tolerant rats has produced a strain with spontaneous diabetes.

The finding that Chinese hamsters spontaneously develop diabetes has provided an animal model of this disease which closely resembles the disorder in man.[28] Hamster diabetes, like that in man, comes in the juvenile as well as the maturity-onset variety. The hamster condition is also characterized by degenerative complications in the eyes, kidneys, and nerves and by death at a premature age. Most instructive of all is the observation that hamsters with diabetic genes *overeat* before they develop their diabetes. Furthermore, restriction of food can delay the onset of the disease, lessen its severity, and ensure a normal life span.

In humans, overeating is a considerable problem. Perhaps of greater importance is the ready availability of "junk" foods in our food bins. What we do to our bodies through our lifetimes is just as important as our heritage. Moreover, what our parents, grandparents, and more distant relatives have done to their bodies greatly influences our abilities and disabilities.

An allergic (hypersensitive) origin of diabetes has been postulated. Circulating antibodies to pancreatic islet cells have been identified in some diabetics.[29] These findings provide evi-

[27]"Study of Childhood Diabetes Supports Link with Mumps" by Harry A. Sultz, D.D.S. A report to American Public Health Association, reported in *Medical Tribune*, December 11, 1974.

[28]Even though he may overeat, the Chinese hamster is an animal that becomes diabetic without obesity. The Egyptian sand rat develops diabetes on a high calorie diet. In the KK mouse, the pancreas is normal, but the peripheral body tissues are apparently insensitive to insulin.

[29]"Antibodies to Pancreatic Islet Cells in Insulin-Dependent Diabetics with Coexistent Autoimmune Disease" by A. C. MacGuish, W. J. Irvine, E. W. Barnes, and L. J. P. Duncan, Royal Infirmary, Edinburgh, Scotland, in the *Lancet*, December 28, 1974.

dence for an autoimmune form of diabetes.

According to Dr. William Philpott (M.D.), an orthomolecular psychiatrist, allergic and maladaptive reactions to foods can be responsible for extreme deviations in blood-sugar levels.[30] At least some cases of diabetes are viewed by Dr. Philpott as hypersensitive (allergic, allergylike, and addictive) reactions to foods, chemicals, and inhalents. He emphasizes that sugar derived from corn syrup is just one of the many foods that can be responsible for clinically significant deviations in blood-sugar levels.

Patients who are mentally ill have a high frequency of carbohydrate intolerance. Schizophrenics usually are found to have chemical diabetes when they are subjected to glucose tolerance testing. In one paper, 18 schizophrenics were examined and all had chemical diabetes.[31]

In the newborn infant, a condition of temporary diabetes may occur in association with intrauterine growth retardation and a tongue size larger than normal.

What is the basic problem in diabetes? The reader can see from the discussion in this section that many factors are involved. It is likely that there are many basic problems in diabetes rather than a single disturbance to explain all cases. In regard to insulin, however, these statements can be made: Altered insulin secretion or effectiveness is characteristic of all diabetic stages. In early chemical and prediabetes, insulin secreted in excess may produce symptoms of reactive hypoglycemia. Excess insulin also promotes fat buildup, overweight occurs, and this encourages the development of diabetes. Ineffective insulin or resistance to insulin often occurs. Beta cell function may "wear out," insulin may no longer be secreted in sufficient amounts, and severe overt clinical

[30]"Orthomolecular and Other Biologic Factors" by William Philpott, M.D. An audiotape presentation to Clinical Ecology meeting November 1972. Tape produced by Insta-Tape, Inc., 1139 South Fair Oaks Avenue, Pasadena, Calif. 91105. Also,, "The Significance of Selective Food and Chemical Stressors in Ecologic Hypoglycemia and Hyperglycemia as Demonstrated by Induction Testing Techniques" by William Philpott, M.D. Published in *Journal of the International Academy of Metabology, Inc.,* Vol. V, No. 1, March 1976, pages 80 - 89.

[31]"Glucose-Insulin Metabolism in Chronic Schizophrenia" by F. Brambilla, A. Guastella, A. Guerrini, F. Riggi, C. Rovere, A. Zanoboni, and W. Zanoboni-Muciaccia. Published in *Diseases of the Nervous System,* Vol. 37, February 1976, pages 98 - 103.

diabetes may come about.

Dr. Leonard Kryston (M.D.) and Dr. Ralph Shaw (M.D., Ph.D.) have published a most informative and practical review of diabetes and its treatment.[32] This article effectively integrates information on insulin levels in diabetes and therapeutic management.

Dr. Samuel Bessman (M.D.) has given us this overview of the situation: "The diabetic is a battleground of organized catabolic forces resulting from stresses of living. These stresses occur intermittently with little anabolic insulin response at the appropriate times."[33]

COMPLICATIONS OF DIABETES

The most severe complication of diabetes is death. The mean life expectancy of the diabetic is at least 25% shorter than that of the nondiabetic. If this statistic seems harsh, perhaps the reader with diabetes will use it as a motivating factor to help him improve his lifestyle.

Doctors generally have some difficulty in gaining sufficient cooperation from diabetic patients regarding diet and weight control. In our "civilized" life with its hectic pace, convenience foods, and high sugar consumption, it does, indeed, require considerable effort and dedication to obtain and consume a proper diet. Nevertheless, it can be done and should be done. More and more physicians are learning the importance of nutrition in treating and preventing the diabetic state.

A rigid diet is not necessarily a primary aim in juvenile diabetes. In maturity-onset disease, however, weight reduction accomplished through diet and exercise is apt to be the single most important tool of treatment. Feeding the lean body mass (muscles, organs, and supporting tissues) with appropriate protein, fat, vitamins, minerals, and enzymes is important so that metabolic activity may "burn" away excess fat deposits.

[32]"Lessening the Swings in Diabetes" by Leonard J. Kryston, M.D., and Ralph A. Shaw, M.D., Ph.D. Published in *Drug Therapy*, March 1975.

[33]"Status of the Artificial Beta Cell" by Samuel P. Bessman, M.D. Published in *Guidelines to Metabolic Therapy*, Vol. 3, No. 4, Winter 1974. Published by Upjohn Company.

Diabetics have a way of getting their body chemistry "out of whack." Emotional upsets, infections, exercise variations, biorhythmic patterns, and diet can all be responsible for chemistry problems within the body of the diabetic. It becomes very important, therefore, for the diabetic patient to reach a state of homeostasis in which school, job, hobbies, diet, exercise, and emotions are integrated into a state of balanced equilibrium.

Of great danger to the diabetic patient whose disease is out of control is the development of dehydration, thickening of the blood (hyperosmolarity), and coma. In typical diabetic acidosis, the patient vomits, has a sweet acetone (ketone) breath, and has high levels of blood sugar and acetone, as well as acid blood and thickening (hyperosmolarity) of the blood. In nonketotic coma, the patient has dehydration, highly elevated blood sugar, and thickened blood (hyperosmolarity), but no ketosis. Convulsions, speech difficulty, or signs of stroke may progress to coma and death. It is vitally important that both conditions, typical diabetic acidosis as well as nonketotic coma, receive prompt medical treatment.

Failure of function of the peripheral nerves (neuropathy) is a complication of diabetes which is closely correlated with lack of control of the diabetic condition.

Blindness has already been mentioned as a prominent complication of the diabetic condition. This is largely due to small blood vessel disease in the retinas of the eye. Cataracts may also be a problem. Diabetics have poor color discrimination when tested by sophisticated techniques. These color vision changes are similar to those which occur with aging.

By the use of a special photographic technique, Dr. John Malone (M.D.) and colleagues at the University of South Florida Medical School have found that 77% of juvenile diabetics have microaneurysms[34] in their retinas. It is not known for sure whether blood vessel abnormalities occur before the onset of diabetes, accompanying the disorder, or as a result of the condition. The high incidence of retinal microaneurysms found in juvenile diabetes is evidence that a disturbance in blood-vessel integrity is basic to the disease.

The thickening of the basement membrane of blood vessels in diabetes is a hallmark of the condition. Biochemical imbalance

[34]Microaneurysms are tiny ballooned-out sections of blood vessels.

over many years undoubtedly provides additional insults (damages) to blood vessels leading to premature arteriosclerosis. More than half of all heart attacks are diabetes-related. Three-quarters of all strokes are attributable to diabetes. Out of every 6 amputations necessitated by gangrene, 5 are a result of diabetes.

It is not known with certainty whether low blood sugar regularly precedes diabetes or not. This may well be so. Certainly, one type of mild diabetic often displays a form of low blood sugar. This matter is further discussed in the chapter on low blood sugar (Chapter 14).

The individual who has low blood sugar as the result of excessive insulin treatment of diabetes may have dizzy, weak, or fainting spells, as well as insulin shock. In some cases, the insulin-induced low blood sugar leads to a compensatory rise in blood sugar (Somogyi effect) which is then treated with a further increase in insulin, thus compounding the patient's problems. Physicians experienced in the care of diabetics are alert for this problem.

The diabetic patient is more susceptible to infections than a nondiabetic person. It has been shown that elevated blood-sugar levels as well as ketosis decrease the ability of the white blood cells to ingest and kill bacteria (phagocytosis). The well-controlled diabetic, however, is thought to be no more susceptible to infection than the nondiabetic.

The diabetic who becomes pregnant runs a higher than average risk of fetal complications.[35] Patients with severe diabetes who have vascular disease as a complication of their diabetes have an 11% to 12% incidence of fetal abnormalities. Patients with less severe diabetes have a congenital abnormality rate of 5%. The population at large has a 1% rate.

The outcome of pregnancy in a diabetic patient depends on how well the blood glucose has been controlled during the pregnancy. The more normal the blood sugar, the more nearly normal is the outcome. The presence of ketosis or acidosis during pregnancy is a severe factor that threatens the life and well-being of the fetus.

[35]"Normoglycemia: essential in pregnant diabetic women" by David J. S. Hunter, M.D. Published in *Consultant*, November 1977, pages 162 - 172.

SUGAR IN THE URINE

One of the hallmarks of diabetes is sugar in the urine (glycosuria, mellituria). Diabetes became associated with sugar in the urine when some enterprising researcher stuck his finger in the urine of a diabetic and noted the sweet taste. In order to understand how sugar gets in the urine, I must first briefly explain the process of urine formation.

Blood, pumped by the heart, circulates through the kidney and back to the heart for recirculation (see Figure 49).

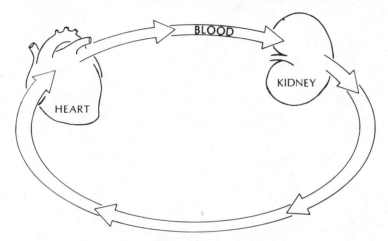

Figure 49. Blood circulation to the kidney
See text, this page, for explanation.

As the blood flows through the kidneys, it is filtered through the walls of small capillaries. As a result, the liquid portion of the blood is separated from the solids (cells) in the blood (see Figure 50).

The liquid filtrate from millions of those blood vessels collects together and drains along tubules in the kidney.

Certain chemicals in the liquid are absorbed back into the blood by the cells of the kidney tubules; other chemicals (wastes) are not reabsorbed and pass out of the body in the urine.

Normally, urine contains no sugar. When sugar is present in the urine, it may be there because 1) the blood-sugar level is abnormally high or 2) because the kidney tubule cells have

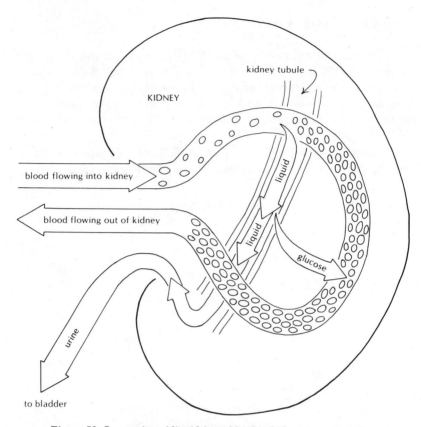

Figure 50. Separation of liquid from blood in kidney to make urine
Glucose dissolved in the liquid part of the blood is filtered from the blood into the kidney tubule and then reabsorbed into the blood. If the blood glucose is too high (usually more than 170 mg.%), the reabsorptive capacity is surpassed, and sugar appears in the urine.

failed to take back the sugar from the urine. The first type is known as hyperglycemic glycosuria; and the second type, renal (kidney) glycosuria.

Phlorizin is a chemical compound which selectively inhibits the kidney tubules, preventing them from reabsorbing glucose. In phlorizin-induced glycosuria, glucose is lost in the urine, but the blood glucose concentration is normal or lowered. This artificial condition mimics that of the natural disorder, renal glycosuria, in which the kidney tubules do not function properly to reabsorb sugar. Interestingly enough, phlorizin also interferes with the cellular absorption of glucose from the intestines.

From a practical standpoint, sugar in the urine usually means diabetes mellitus; but galactosemia, fructosemia, and various other kidney disorders need to be considered.

Noteworthy is the fact that the kidney tubules reabsorb glucose, but do not reabsorb fructose. This has clinical significance. When we give sugar to a hospitalized patient by intravenous drip, we use dextrose (glucose) because it is not as readily lost through the kidneys as fructose would be. In addition, glucose is much cheaper to use and less irritating to the veins.

Urine tests for sugar usually become consistently positive only when the disease diabetes mellitus with all its symptoms appears. However, in children and adults, a preclinical form of diabetes may be present without sugar in the urine. For this reason, urine testing for sugar is not sufficient to exclude diabetes. This has become an especially important point, because it now appears that individuals with preclinical diabetes may be significantly improved by nutritional means.

The patient with established diabetes follows the progress of his blood sugar by testing his urine for sugar each day. This becomes a way of life for most diabetics who wish to provide the best possible control of their condition.

Commercially available urine sugar tests are available in the form of dipsticks or tablets. Clinitest Tablets[36] are used by the majority of diabetics. It is important to know that these valuable testing tablets contain sodium hydroxide and hence are toxic (caustic) if ingested. They constitute a hazard to small children who could eat them if they are not properly stored out of their reach. Life-threatening burns and strictures of the esophagus may occur as the result of Clinitest Tablet ingestion.[37] All diabetics should be aware of this and take appropriate precautions so that Clinitest Tablets and small children do not find themselves together.

Glucose-specific Tes-Tape[38] may also be used for urine-sugar testing if the patient's doctor so indicates.

[36]Clinitest Tablets are manufactured by Ames Company, Elkhart, Ind. 46514.

[37]"Clinitest Tablet's Peril to Children," a report in *Medical Tribune*, Wednesday, March 12, 1975, by John D. Burrington, M.D.

[38]Tes-Tape is manufactured by Eli Lilly and Company, Indianapolis, Ind. 46206.

STAGES OF DIABETES

Clinical (Overt) Diabetes

Clinical diabetes is full-blown symptomatic diabetes, characterized by increased blood-sugar levels and sugar in the urine. The fasting blood sugar is elevated, and the GTT (glucose tolerance test) is unmistakeably diabetic.

If not treated, clinical diabetes often progresses to ketosis with acetone in the urine and sweet, fruity, breath odor. Weakness and decreased resistance to infection are also seen. When the onset of the disease occurs over a short period of time, the symptoms of acute onset are seen. These are increased thirst, appetite, and urination, frequently with weight loss. Often the unsuspecting parents of a child with the acute onset of diabetes will not appreciate the nature of his sickness until he becomes seriously ill with vomiting, abdominal pain, rapid breathing, and dehydration — all indications of diabetic acidosis and ketosis. Infection may be present somewhere in the body. Such an acute condition may progress to diabetic coma and death unless prompt and intensive medical care is obtained.

Patients with clinical diabetes mellitus who are taking insulin by injection must be ever alert for overdosage of insulin. If a diabetic takes his usual amount of insulin in the morning but fails to eat as much as usual, he may develop low blood sugar. Such an insulin reaction may also be brought on by excessive exercise. Low blood sugar in the diabetic is characterized by mental disturbances; incoordination; "fuzzy" thinking; sweating; pallor; cool, "clammy" skin; tremor; yawning; sleepiness; hunger; convulsions; or coma.

Chemical Diabetes

The individual with chemical diabetes has a preclinical disorder. There are no clear-cut symptoms of diabetes. Tests for sugar in the urine are usually negative. The fasting blood sugar is normal, but the glucose tolerance test is diabetic. Chemical diabetes begins with the first detectable abnormality in the glucose tolerance test or the cortisone glucose tolerance test.[39]

[39]Some authorities recognize a separate category of latent subclinical diabetes in which the GTT is normal, but the cortisone GTT is abnormal. In these individuals, diabetic findings may be noted under stressful conditions such as pregnancy, heart attack, or surgery.

Chemical diabetes is inability of the body to handle a glucose load in a normally expected way. Mild symptoms of hypoglycemia may be present. There is no question that some of these cases proceed to develop into overt diabetes. Whenever possible, treatment of patients with chemical diabetes is important to prevent clinical diabetes.

Some cases of chemical diabetes are definitely reversible. A favorable diet, restoration of body mineral balance to normal, provision of a relatively stress-free lifestyle, daily exercise sufficient to give a conditioning effect, weight control, and provision of meaningful goals for living are likely to be of benefit (see Chapter 12, "Glucose Tolerance Curves").

In my experience, chemical diabetes is not an infrequent finding in children of today's culture.

Prediabetes (Potential Diabetes)

In the prediabetic state, there are no diabetic symptoms, the fasting blood sugar is normal, the glucose tolerance test is normal, but there is a genetic predisposition to diabetes. From the practical standpoint, the prediabetic patient cannot actually be recognized, since there is no reliable way in clinical practice to discover the person who has the gene for diabetes. Degrees of likelihood may be established, however, according to the family history of diabetes.

It is possible that hair-mineral patterns may provide a marker for such genetically predisposed individuals. The patient who has a low, flat type of glucose tolerance test may be prediabetic. When evaluated in a research laboratory, these prediabetic individuals, as a general rule, are found to have a subnormal plasma-insulin response to glucose, amino acid, and tolbutamide administration.

It is also possible that a test for lingual (tongue) sensitivity to sugar may identify the patient with the diabetic gene. It has been shown that diabetics have a low sensitivity of taste for sugar. Thus, they eat more sugar to obtain the same sweetness that someone else would have with a lesser amount. This has been termed, "sweet tooth."

More detailed analysis of glucose tolerance tests and natural tolerance tests (see Chapters 11 and 13) might yield additional information about the prediabetic state. Insulin, growth hormone, and glucagon levels may provide information upon which earlier diagnosis of diabetes can be based. Refinement of hair-mineral analysis or tissue biopsy could possibly speed identification of this stage. Such information must be patiently sought and assembled by researchers and clinicians.

A Spectrum of Stages

As the reader has become familiar with each of the stages of diabetes, he may have noted that a spectrum of involvement has been described. These stages are convenient groupings used to identify and categorize individuals with similar degrees of involvement. As further information is developed about diabetes, more helpful categories may be constructed.

There are apparently some individuals who, no matter how they abuse their bodies or how much sugar they consume, seem never to develop diabetes. Perhaps these individuals would require a severe direct insult to their pancreas or surgical removal of the organ to become diabetic.

Then there are those individuals who show possible early warning signs of diabetes — the tendency to obesity and fluid retention, mothers who give birth to large-sized babies, those individuals with high blood sugar in pregnancy or after surgery, those who show sugar in the urine after excitement or fear, those with functional hypoglycemia, etc.

Then come those who have manifested diabetic glucose tolerance tests without any symptoms of diabetes.

Next in the spectrum would be the maturity-onset obese diabetic who may actually reverse his disorder by virtue of weight loss and diet.

Finally, one recognizes the clinical diabetic who is likely to have his disorder as a fixed problem for the remainder of his life. The juvenile diabetic has an irreversible disorder as far as science knows at this time.

The reader may recognize himself, his relatives, or his friends somewhere in this spectrum. It is hoped that the reader may profit by an awareness of the matters of carbohydrate metabolism. In Chapter 15 of this book, suggestions are given that may be helpful in preventing diabetes and related problems.

DRUG-INDUCED DIABETES

There are certain circumstances in medical treatment which can bring about a diabetic state or a state of reversibly elevated blood sugar. The administration of the cortisone derivatives (glucocorticoids) is one. The use of ACTH is another. Diabetes occurs as a side effect of treatment with these agents. Such a steroid diabetes has been noted in 14% of patients treated with cortisone-like drugs for more than three days. Most of the time, this kind of diabetes goes away when the drug is discontinued. This diabetogenic action of cortisone is used in the cortisone glucose tolerance test, a stress test designed to uncover patients with chemical diabetes (see Chapter 11).

In patients with kidney transplants who receive glucocorticoids to inhibit graft rejection, diabetes has been noted to develop in 3 - 8% of the cases.

Thiazide diabetes is the condition of elevated blood sugar encountered in some patients who are being treated with thiazide diuretics.[40] Many commercially available diuretics contain thiazides or their derivatives.

Diazoxide is the chemical name of a drug which elevates the blood sugar by inhibition of insulin production in the beta cells of the pancreatic isles. Diazoxide is marketed as *Hyperstat* by Schering Pharmaceutical Company, and it is used to treat high blood pressure as well as low blood sugar. Interestingly enough, it is a thiazide drug, but one which does not have a diuretic action.

L-asparaginase is an enzyme which is used in the treatment of childhood leukemia. It brings about an increased blood sugar in experimental animals. In leukemia patients treated with the drug, diabetes mellitus has been reported as a toxic side effect of treatment. Asparaginase may even give rise to diabetes when it is used in low doses.

The mechanism of this sugar elevation is not known, but it may be due to the action of the enzyme on the amino acid asparagine, which is a component of the insulin molecule. Inflammation of the pancreas with damage to the isles of Langerhans could also be a factor.

---

[40]Diuretic: a drug or agent that causes an increase in the flow of urine.

Large doses of diphenylhydantoin (Dilantin: Parke, Davis, & Company), a potent anticonvulsive drug, have been associated with elevated blood-sugar levels. These elevated sugar levels occur, however, only when toxic levels of Dilantin have been present. There is evidence that Dilantin inhibits insulin secretion.

The oral contraceptive drugs have been responsible for impaired utilization of glucose. This means that blood-sugar levels may be elevated in women on the "pill." The fasting sugar level is unchanged, but the levels in a glucose tolerance test are apt to be elevated, especially the 1-hour value. This effect may be due to vitamin deficiencies which are known to occur in women on the "pill." It may also be associated with increased levels of glucocorticoid hormones or growth hormone. There is also an increased resistance to insulin in the peripheral tissues of the pill users. The utilization of glucose may improve when vitamin B_6 is given to women taking the oral contraceptives.

The fact that vitamin therapy may lessen such a diabetic state may indicate that vitamin therapy should be tried in other patients with carbohydrate disorders such as diabetes and hypoglycemia.

An analog of glucose that contains sulfur is 5-thio-dextroglucose. It is being investigated as a male contraceptive agent because of its effect in inhibiting the transport of glucose into the cells of the testicle. The drug raises blood sugar and inhibits appetite when used in doses higher than those used to suppress the production of sperm.

In the majority of patients with mental illness, diabetes or diabeticlike GTT's occur. In view of this, it is interesting to note that virtually all psychoactive drugs affect diabetes. Most tranquilizers aggravate the high-blood-sugar levels of diabetes while most antidepressants tend to lower them toward normal.

Chlorpromazine (Thorazine: Smith, Kline, and French, Inc.) is a major tranquilizer. It has been implicated as a cause of elevated blood sugar and glucose intolerance.

Niacin and niacinamide are major vitamins used in the megavitamin therapy of mental illness. They may also increase blood sugar.

EFFECT OF PREGNANCY ON BLOOD SUGAR

In a way, pregnancy is a diabetogenic state. Pregnancy can be considered to be a stress event that leads to glucose intolerance during the pregnancy.[41] Reference has already been made to the increased need for vitamin B_6 during pregnancy and the relationship of B_6 deficiency to the diabetic state. Some believe that hyperglycemia in pregnancy may be due to estrogen-induced increase in cortisol levels.

However, according to Dr. David Hunter (M.D.), glucose tolerance actually improves during the first 3 months of pregnancy. This may be due to estrogen-induced increase in insulin production.

As pregnancy continues beyond the first 3 months, the placenta manufactures increasing amounts of human placental lactogen (HPL), a hormone resembling growth hormone. HPL antagonizes insulin. Glucose intolerance results, and a diabetic-like state is found when blood-sugar levels are monitored.

As soon as pregnancy is terminated, the glucose intolerance (gestational diabetes) of pregnancy reverts to normal unless the woman also has diabetes mellitus when she is not pregnant.

For many years it has been recognized that the woman who bears oversized newborns is at high risk for the development of diabetes. The excessive infant size appears to be an excellent marker to indicate maternal hyperinsulinism. Excess insulin secretion during pregnancy is stimulated by high levels of blood sugar. The young woman who embarks upon a pregnancy should be aware and alert to the dangers of consuming a diet that contains sugar and that is nutritionally deficient.

INFANT OF THE DIABETIC MOTHER

The pregnant mother with diabetes should be extremely careful to keep her diabetic condition under good control. When the pregnant mother's diabetic condition is out of control, homeostasis is disrupted, frequent bouts of ketosis occur, and blood-sugar levels fluctuate widely. The result is an increased incidence of fetal death; an increased incidence of congenital defects in the

[41]"Normoglycemia: essential in pregnant women" by David J. S. Hunter, M.D. Published in *Consultant,* November 1977, pages 162 - 172.

infant; oversized babies; enlargement of organs; puffy, "tomato-face" babies; other chemical disturbances; respiratory distress with hyaline membrane formation in the lungs; and an increased risk of dying due to low blood sugar. The fetus of the diabetic mother is usually delivered at 36 or 37 weeks of pregnancy by induced labor in order to prevent stillbirth. Three chemical measures are available to assist the physician in assessing the maturity and well-being of the infant; these are the estriol level, the ratio of lecithin to sphingomyelin in the amniotic fluid, and the insulin-glucose ratio in the amniotic fluid. Death in utero and neonatal hypoglycemia occur when there is a high level of insulin and a low level of glucose (hyperinsulinism) in the amniotic fluid.[42]

Experienced pediatricians know full well that the baby of the diabetic mother is a "high risk" baby. The infant bears close watching. He often behaves as though he is an immature infant even though his size is overly big.

Abnormalities of the blood electrolytes[43] in the baby of the diabetic mother are not infrequent. For example, these newborns frequently develop low blood levels of calcium and high blood levels of phosphorous in the first few days of life. This has been shown to be due to inadequate function of the parathyroid glands in the infant of the diabetic mother.

When the mother's diabetes is well controlled, her infant has a greatly improved chance of living and experiencing good health in the newborn nursery.

It has long been observed that the nondiabetic mother who gives birth to an infant of large size (generally more than 9 pounds) is a potential diabetic. Physicians caring for such mothers often order a glucose tolerance test to check for diabetes.

The oversized infant has become that way because of hyperinsulinism. Excess maternal blood sugar has triggered an excess insulin response from the baby's pancreas. Excess insulin stimulates fat synthesis and excessive growth. Beta-cell overgrowth that is responsible for the hyperinsulinism may continue

---

[42]"The Glucose-Insulin Ratio in Amniotic Fluid" by Robert L. Newman, M.D., and Gino Tutera, M.D. Published in *Obstetrics and Gynecology*, Vol. 47, 1976, pages 599 - 601.

[43]Electrolytes: minerals circulating in the blood which have positive or negative electrical charges.

to be a problem after birth. When the supply of glucose from the mother is shut off at birth, the infant may become hypoglycemic.

Hyaline membrane disease killed the prematurely born son of President John F. Kennedy. Hyaline membrane disease, also known as RDS (respiratory distress syndrome) is very frequent in this country. It is known to commonly affect newborns in the developed nations of the world, but it is only rarely encountered in underdeveloped countries whose populations are nourished differently.

Hyaline membrane disease is associated with a lack of pulmonary surfactant. This chemical substance assists in keeping the tiny breathing spaces of the lung (the alveolar air sacs) open for the transfer of oxygen and carbon dioxide gases. Premature infants, stressed babies, and newborns delivered from diabetic mothers are deficient in pulmonary surfactant, and they have an especially high incidence of hyaline membrane disease, with collapsed air sacs.

Insulin depresses the development of pulmonary surfactant in the lungs, while adrenocortical hormone (hydrocortisone) accelerates its development. Babies of diabetic mothers secrete large amounts of insulin, and these infants have deficient surfactant with the high rate of hyaline membrane respiratory distress.

Certain pregnant women have a higher than average risk of delivering a baby with hyaline membrane disease. Treatment of these individuals with corticosteroid drugs is proving effective in the prevention of this condition. This may be so because corticosteroids given in late pregnancy enhance the maturation of the fetal lung.

The balance of insulin and adrenal-corticoid hormones is crucial for optimal function in the body (see Chapter 8). At the end of pregnancy, this balance acquires special meaning, as the life of a newborn is threatened by asphyxia.

Hyperinsulinism is better managed by encouraging, whenever possible, lifestyles that foster optimal balance of insulin and adrenocorticoids. In the meantime, however, the medically supervised use of adrenocorticoids in late pregnancy to redress imbalance appears to be of significant help. Further research is needed, of course, to elucidate the hows and whys of this effect.

TREATMENT OF DIABETES

Modern management of diabetes produces good results. Insulin has been the backbone of treatment for many years. Without insulin, diabetics only lived 2 or 3 years. Antibiotics have assisted greatly in combatting the infections to which diabetics are especially prone. Modern surgery, anesthesia, blood transfusions, and X-ray have all played a role in improving the health of the diabetic. Ordinarily, diabetic individuals live healthy lives with a minimum of complications unless they fail to obey certain restrictions which their disease thrusts upon them.

The goals in treatment of the diabetic child are

1. freedom from ketosis (no acetone in the urine),
2. adequate growth,
3. freedom from low blood sugar,
4. absence of infection, and
5. minimal deviation from social activities considered normal for the peer group.

In addition, it is wise to attempt to keep the urine free of sugar, but not at the expense of these goals that have been listed.

Most physicians believe that these goals are highly desirable. Usually they can be met without strict dietary supervision and without strict parental supervision. All efforts should be made to have the child learn about his disease and its management as soon as possible. If emotional problems exist, immediate attention to these should be obtained. There is no more unhappy situation than a family in which there is a diabetic child and unresolved psychological problems.

Insulin given by injection, usually once a day, is almost invariably needed in juvenile diabetes. Sometimes, giving insulin twice a day is necessary to achieve the best result.

Juvenile diabetes is known to be associated with insulin deficiency. Elevation of growth hormone and glucagon has also been shown to be characteristic of juvenile diabetes. The recent finding that the hypothalamic hormone somatostatin can inhibit growth hormone and glucagon suggests that it may play a role in future treatment.

In the adult or maturity-onset diabetic, the main thrust of treatment is directed toward improving the diet and eliminating excess weight. Often, no further treatment is needed. When it is, the use of the oral drugs to lower blood sugar is frequently effective. The use of these drugs may give a more normal pattern of blood-sugar variation than the administration of insulin.

Certainly the availability of effective medication to control blood-sugar levels should never substitute for sound nutritional measures, proper exercise, and the attainment of positive mental health. The use of drugs should be considered as treatment to be used only if dietary, nutritional supplement, exercise, and stress change is not effective.

Dietary improvement for the diabetic consists of a program which ensures adequate quantities of protein, fat, carbohydrate, and fiber and which eliminates "junk" foods and empty calories. Of great importance is the elimination of any foods, inhalents, or chemicals that may be responsible for causing deviation in blood-sugar levels. Mineral balancing is of utmost importance.

A new test that measures long-term blood-sugar control promises to serve as an index of the adequacy of treatment in diabetes.[44] Measurement of glycosylated hemoglobin levels indicates the diabetic's average blood-glucose levels for the preceding 2 months. The test is based on the fact that glucose molecules join hemoglobin during red cell formation. Once a hemoglobin molecule is glycosylated, it remains in that form. In the non-diabetic, 5 - 7% of hemoglobin is glycosylated. In the diabetic, there is a threefold elevation that reflects the mean of increased blood sugar during the life cycle of the red blood cells.

The Oral Hypoglycemic Drugs

The oral hypoglycemic drugs are of two chemical varieties: sulfonylureas and phenformin. The first group, the sulfonylurea compounds, include at least four different drugs. They are

1. tolbutamide (Orinase: The Upjohn Company),
2. acetohexamide (Dymelor: Eli Lilly and Company),
3. chlorpropamide (Diabinese: Pfizer Laboratories Division, Pfizer, Inc.), and
4. tolazamide (Tolinase: The Upjohn Company).

[44]"New Test Gauges Long-term Diabetes Control," a report appearing in the *Medical Tribune,* February 16, 1977, page 3. This report details the work of Dr. Kenneth H. Gabbay (M.D.), Professor of Pediatrics, Harvard University.

In general, one can consider this listing to indicate the approximate strength of these drugs. That is, Orinase might be considered the weakest; Dymelor, of intermediate strength; and **Diabinase and Tolinase, the strongest. In terms of duration of** action, Orinase is short, Dymelor and Tolinase are intermediate, **and Diabinase has the longest duration of action.**

A new oral hypoglycemic drug, glyburide (Micronase: The Upjohn Company), is much more potent than presently available sulfonylurea preparations. The dose is 2.5 mg. to 5.0 mg. or so, instead of 250 mg. to 500 mg. of the other sulfonylureas. At this time, the drug is only available outside the United States.

The sulfonylureas stimulate the release of native insulin from the islet beta cells. We can readily see that an intact pancreas is necessary for these drugs to act. In fact, the sulfonylureas may be used to measure the functional capacity of the beta cells. Newer information indicates that they increase insulin action and have a salutary effect on insulin receptors and on the beta cell basement membrane.

The sulfonylureas are sulfa-drug derivatives, but they have no antibacterial action. They can lower the blood sugar in normal patients as well as in diabetics. Some persons have abused their health by ingesting these drugs to produce factitiously low blood sugar.

Chlorpropamide (Diabinase: Pfizer Laboratories Division, Pfizer, Inc.) has as one of its side effects the capacity to induce water retention with a lowering of the sodium in the blood (water intoxication). Water retention occurs in about 4% of patients with diabetes mellitus treated with chlorpropamide, and it may be responsible for episodic heart pain (angina pectoris) in these patients at night.[45]

Tolbutamide may also be responsible for water retention, but it is not a common occurrence.

Tolazamide, acetohexamide, and glyburide actually increase the excretion of body water. They are therefore diuretic sulfonylurea agents. They should be appropriately considered for use in the treatment of diabetics who already have water retention.

---

[45]"Actions of Sulfonylurea Agents on Water Metabolism Vary Widely" by Arnold M. Moses, M.D. Published in *Metabolic Therapy*, Vol. 4, No. 1, Spring 1975.

The second variety of oral hypoglycemic drugs is represented by a single substance, the biguanide compound known as *phenformin* (DBI: Geigy Pharmaceuticals; Meltrol: USV Pharmaceuticals). Physicians have used the sulfonylureas and phenformin in combination.

The mechanism of action of phenformin is not completely known, but it apparently makes insulin work more effectively at the cell level. It increases glucose utilization in cells by interfering with the oxidative metabolism of glucose. It also blockades carbohydrate absorption across the wall of the gut and is thought to inhibit glucose release from the liver. Phenformin does not stimulate insulin release, and it does not lower blood sugar unless the person taking it has diabetes. Phenformin probably decreases insulin oversecretion by virtue of its blood-sugar-lowering effect. This in turn reduces the formation of fat and facilitates fat breakdown in the body.

Phenformin has been particularly helpful in the obese individual with elevated blood sugar who has difficulty in weight control. At times, the effectiveness of the substance has been surprising, even in small doses.

In July 1977, The Secretary of the Department of Health, Education, and Welfare ordered an end to the general marketing of phenformin. This action was taken because of the unacceptably high risk of lactic acidosis associated with the use of this drug. At the time this book was written, the drug will still be available for certain selected individuals who meet rather rigid criteria for its use.

The patient most likely to benefit from the oral hypoglycemic drugs is the maturity-onset diabetic whose condition is not satisfactorily managed by dietary regulation and weight loss. Chemical diabetes in children is being treated with the oral agents, but only as an experimental measure to date.

The drugs are also useful in diabetics who are allergic to insulin. They are, of course, preferred to an injection by most persons. One drawback to the use of the oral hypoglycemic drugs is their relative expense. They are 2 - 3 times as expensive as an equivalent amount of insulin. It is also possible that they might become less effective after several years of use.

Use of the oral drugs tolbutamide and phenformin has been linked with a heart disease death rate twice as high as for patients who took insulin or no drug. Whether this statistic is correct or not bears further investigation. At the present time, the evidence is moderately strong that premature death from heart disease is associated with the use of these oral agents.

The antidiabetic drugs insulin and sulfonylureas can lower blood sugar too far when they are taken with guanethidine (a drug to lower blood pressure), oral anticoagulants, chloramphenicol antibiotic, MAO-inhibiting antidepressent drugs, or propranolol (a drug used to correct irregular heartbeat). In the presence of alcohol, some antidiabetic drugs can bring on abnormally low levels of blood sugar, but the moral here is to avoid alcohol! Taken with the thiazide diuretics, the antidiabetic drugs may raise blood sugar instead of lowering it.

As already noted, the anticonvulsant drug Dilantin has a substantial effect in reducing the secretion of insulin in the body. The Dilantin-treated seizure patient may thus have blood-sugar elevations on this basis.

OUTLOOK FOR THE FUTURE

The best treatment for diabetes is the prevention of the disease. Unfortunately, no sure way is yet known to do this, although it is likely that lifestyle change including dietary improvement, physical conditioning, immune engineering, mineral balancing in the body, allergic management, vitamin therapy, and stress-reduction procedures will be effective in many cases.

Until reliable preventive measures are perfected, researchers continue work toward perfecting an "automatic pancreas." This small device would be implanted in the body to monitor blood-sugar levels and release insulin when necessary. Such a device is not yet available for human diabetics, but a working model has been developed and progress is rapidly being made. The artificial pancreas is more appropriately termed the *artificial beta cell* and consists of 5 parts: a sensor for glucose, a computer, a micropump, a power supply, and a refillable reservoir for insulin.[46]

[46]"Status of the Artificial Beta Cell" by Samuel P. Bessman. Published in *Guidelines to Metabolic Therapy*, Vol. 3, No. 4, Winter 1974. Published by the Upjohn Company.

Wah Jun Tze at the Children's Hospital in Vancouver, British Columbia, has developed an artificial pancreas which has successfully controlled blood-sugar levels in diabetic rats, eliminating the need for insulin injections.

There is promise that pancreatic beta cell transplants may be forthcoming for diabetics. Dr. Arnold Lazarow of the University of Minnesota, has created diabetes in rats and transplanted beta cells successfully to control the diabetes. Pancreas from inbred strains of rats or mice is grown in organ culture in such a way that the digestive (acinar) cells of the pancreas disappear while the beta cells grow.

At Washington University School of Medicine in St. Louis, Missouri, pancreatic islet cells have been successfully transplanted from healthy animals to diabetic animals. Monkeys as well as rats have been used in these experiments.

Test animals are first made diabetic by administration of streptozotocin, surgical removal of the pancreas, or both. Then islet cells from a healthy animal are given to the diabetic animal. The administration of these cells into the liver by way of the portal vein has given better results than placing the cells under the skin or in the abdominal cavity.

The diabetic animals have returned to normal or near normal in weight, urine volume, urine glucose, fasting blood sugar, eating and drinking habits, and ketosis.

Islet cells have been successfully transplanted in a primate mammal, the monkey; and in 1976, a University of Minnesota team performed the first human islet-cell transplantation safely and without complications in 7 diabetics. No patient was cured of diabetes, but improvement did occur. Larger quantities of islet tissue are needed. The problem of host rejection of foreign tissue is being combatted, but has not yet been solved.

Rejection of graft tissue by the host is the principal problem in the use of organ transplants. At the present time, immunosuppressant drugs must be used to inhibit the rejection process, and those immunosuppressant drugs have dangers in themselves. Science is closing in on the problem, but there are problems which must yet be solved before new insulin-producing cells are made available to diabetics.

The use of fetal cells, cells grown in tissue culture, or cells treated chemically to alter their antigenicity are ways that tissue rejection may be solved.

The entire matter of transplants and implants for the treatment of diabetes was recently reviewed in an article entitled, "Surgery for Diabetes" (*Medical World News*, January 13, 1975).

There is yet another line of scientific investigation that holds promise for diabetes. Recombinant DNA research[47] has already enabled scientists to transform laboratory-bred bacteria into insulin-producing factories. Genetic material from rats has been inserted into bacteria with the result that the germs then manufacture rat insulin. It is believed that such gene manipulation will lead to a plentiful supply of human insulin and perhaps to a cure for insulin-deficient diabetes.

In a landmark article, Dr. Roger W. Turkington (M.D.) and Dr. Howard K. Weinding (M.D.) have redefined diabetes.[48] They found that the abnormal GTT failed to predict the late complications of diabetes, especially the eye changes (diabetic retinopathy). They found insulin reserve rather than glucose intolerance to be the true criterion for the diagnosis of diabetes. Eye, kidney, and nerve complications occurred only in diabetics whose peak blood-insulin values were less than 60 microunits per milliliter. In the view of Drs. Turkington and Weinding, a hyperglycemic GTT may be evidence for diabetes but is not diagnostic thereof. Diabetes is diagnosed by finding insulin deficiency. A hyperglycemic GTT in the presence of adequate or excessive insulin would indicate insulin resistance, not diabetes.

Diabetes is a large problem and a growing one in American society. The juvenile diabetic is particularly unfortunate because of the severity of his disease and the need for him to take daily insulin injections for the rest of his life. The elimination or marked reduction in sugar intake in our society could be a major stroke in diabetes prevention. Persons who choose to do this will usu-

---

[47]Recombinant DNA research: manipulation of hereditary material (genes) to change original characteristics of an organism to something different.

[48]"Insulin Secretion in the Diagnosis of Adult-Onset Diabetes Mellitus" by Roger W. Turkington, M.D., and Howard K. Weindling, M.D. Published in the *Journal of the American Medical Association*, Vol. 240, No. 9, September 1, 1978, pages 833 - 836.

ally become deeply interested in additional companion measures that enhance nutritional health.

Although it is too early to make definitive statements, it appears that the chemical state of diabetes can be greatly improved (and sometimes eliminated) by dietary improvement, exercise, vitamin and mineral therapy, and reduction of adverse stress. The use of chromium supplementation and other mineral balancing is just beginning to be explored. Antiallergy therapy may be important.

Carefully supervised medical management of established diabetes appears to reduce the frequency of complications in the disease.

It does appear that early diagnosis of diabetes and prediabetes is the key to successful management. Much investigation needs to be done in order to understand the early phases of metabolism which precede frank (clinically recognized) diabetes. In order to detect the future maturity-onset diabetic, physicians should frequently obtain blood-sugar testing and insulin testing in high-risk children. Physicians, too, must communicate to their patients appropriate information regarding diet, exercise, nutritional supplements, and related health measures.

It is hoped that the reader will better understand the problems of the diabetic by having read this chapter.

Further information about diabetes may be obtained by writing the American Diabetes Association in your local area. Your local medical society should be able to help with the address or phone number. Also, you may contact the Juvenile Diabetes Foundation, 3701 Consohocken Avenue, Philadelphia, Pennsylvania 19131. An informative booklet entitled *A Guidebook for the Diabetic* has been published (1972) by the Ames Company, Division of Miles Laboratories, Inc., Elkhart, Indiana 46514.

CHAPTER 11

THE GLUCOSE TOLERANCE TEST (GTT)

The glucose tolerance test (GTT) is an important tool of medicine that is used to detect disorders of carbohydrate metabolism. The GTT is most often carried out for the purpose of diagnosing diabetes. It may also be used to detect low blood sugar. A test of 3- or 4-hours duration is most commonly requested when diabetes is suspected. A 6-hour test is required if one wishes to exclude low blood sugar or the combination of diabetes and low blood sugar.

The reasoning behind the GTT is that a heavy load of sugar given to the patient will test the limits of that person to absorb and metabolize ("burn") sugar. Such a challenge (stress loading) with sugar may permit a diagnosis of diabetes before the patient's clinical condition has deteriorated to the point of obvious symptoms. Thus, the GTT tests the patient's ability to tolerate (manage or use) a large load of glucose.

An oral GTT is one in which the glucose is given by mouth, and blood-sugar values are followed at periodic intervals thereafter to see how well that person absorbs and metabolizes the glucose load. Commonly, the blood sugars are determined just prior to the sugar load, then again at one-half hour after the load, and at each hour thereafter until the test is terminated. The blood-sugar value obtained just before the glucose load is given is known as the fasting blood sugar.

An intravenous GTT is done by giving the glucose in the vein rather than by mouth. The intravenous test gives higher values of blood glucose, because insulin is stimulated less by this technique than it is when glucose is given by mouth. In this book, the term *GTT* always refers to the standard oral GTT.

Administration of a glucose load is the commonest tolerance test performed in clinical medicine. Other tolerance tests may be performed with lactose, xylose, galactose, iron, vitamins, etc.

Some persons have pointed out that the GTT is a rather severe stimulus. According to some, the traditional GTT (using a glucose drink) can be construed as how much damage can be done to the pancreas. Indeed, I have had patients who became temporarily quite ill after undergoing the test. One man developed severe chills and a cold feeling all over his body within five minutes after he drank the sugar load. These symptoms persisted for a full day as did a severe headache, which came on twenty minutes after the start of the test. Reactions such as this are not uncommon. In rare cases, patients have developed convulsions during testing. Fatalities have been reported following GTT's in patients with kidney disorders. These deaths have presumably occurred due to rapid lowering of serum potassium concentration as a result of glycogen deposition. As is the case with any diagnostic procedure, the value of the information gathered from the test must be weighed against the expense, inconvenience, psychic stress, and physiological stress on the part of the patient. Much of the time, however, the procedure is relatively innocuous, considering the valuable information derived from it.

Those persons who have been eating considerable amounts of sugar in their diets appear to manage the glucose load with fewer reactions. It is as though their body mechanisms are geared to cope with a large glucose load. Nevertheless, these persons may be the very ones who develop other problems associated with the chronic long-term intake of sugar.

In an attempt to find a more physiologic or more natural test of a person's carbohydrate metabolism, some physicians are now using a variation of the GTT. In the procedure which I have used, the patient is instructed to eat his usual breakfast, and this meal is used as the load instead of a glucose drink. Blood sugars are then obtained over a 6-hour period. I term this a *natural tolerance test* (NTT) (see Chapter 13). In other procedures, the patient is directed to eat a specific meal instead of what he would customarily eat (see Chapter 13) or the patient's blood sugars may be obtained every 2 hours throughout the day (eight-hour no-load test). Dr. Emanual Cheraskin (M.D.) at the University of Alabama has pioneered in this kind of testing.

Now let us return to the standard GTT. What is actually given to the patient as the sugar load? Glucose itself may be used. When it is, it is usually dissolved in water with lemon juice added to improve palatability. Various commercial preparations of sugar substances are available in prepared quantities suitable for use in the GTT. Table 3 lists many of the available test substances and their ingredients. Since some individuals react adversely to artificial flavorings or other chemical additives, the use of certain commercial preparations could be undesirable. How frequently untoward results occur is unknown. The advantages of these preparations — ready availability, ease of administration, and uniform make-up and quantity — may offset the possible disadvantages, although I prefer the natural tolerance test.

Dr. Carlton Fredericks (Ph.D.) has wisely suggested that a mixture of glucose and fructose be used in testing.[1] Since sucrose (refined sugar) consists of glucose and fructose, this would approximate the sugar load that the patient would experience if he is eating a typical American diet.

How much glucose is given in a GTT? Table 3 shows that 100 grams of glucose is the usual amount contained in these commercially prepared solutions. It has been found that the administration of more than 100 grams of glucose does not significantly alter the test results. Children are given an amount of glucose based on their weight.[2] The loading dose given to persons who deviate considerably from ideal body weight should be based on ideal body weight. Sixty grams of glucose per square meter of body surface area is another figure which has been recommended. The glucose-loading solution should be drunk within a 5-minute time period.

[1]"Hotline to Health." Published in *Prevention* magazine, March 1976, page 48.

[2]Infants between 18 months and 3 years are given 2 grams per kilogram of body weight. Older children receive 1.75 grams per kilogram or 1 gram per pound of body weight to a maximum of 100 grams. Some arbitrarily give children less than 7 years of age 50 grams of glucose. Some sources suggest 1.25 grams per kilogram for individuals more than 12 years, or an arbitrary dose of 100 grams.

TABLE 3

SOME COMMERCIAL PREPARATIONS USED FOR GLUCOSE TOLERANCE TESTING

Product	Made or Distributed By	Contains	Other Ingredients	Unit Size
Flavose	Reliable Chemical Co., Inc.	100 gm. glucose	natural fruit flavor	200 ml.
Orangedex	Custom Labs., Inc.	100 gm. glucose	citric acid, orange juice, oil of orange, color, carbonated water, sodium benzoate	10 oz.
Dextol^{TM}	Scientific Products	100 gm. glucose	citric acid, natural oils of lemons and limes, propylparasept (bacteriostatic and fungistatic) — carbonated or noncarbonated	240 ml.
Dex-Ade	Aloe	100 gm. glucose	lemon-lime flavor, propylparasept	240 ml.
Dexicola®	Curtin Science Co.	100 gm. glucose	carbonated, propylparaben (preservative)	10 oz.
Trutol	Sherwood	100 gm. glucose	orange, lemon-lime, grape, or cola — all carbonated except the grape	10 oz.
Glucola	Ames	rapidly hydrolyzable saccharides in aqueous solution equivalent to 75 gm. glucose	water, kola extract, sodium benzoate, artificial flavors	7 oz.
Harlecola	Harleco	100 gm. glucose	cola flavor	12 oz. pop-top can

At what time of day is the GTT performed? Ordinarily it is carried out in the morning after an overnight fast. Dr. H. J. Roberts (M.D.) of Palm Beach, Florida, as well as some other physicians, however, intentionally request afternoon tests, believing that abnormalities are more likely to be found. An appreciable diurnal variation (day-night difference) in blood-sugar levels is said to be normal. Afternoon blood-sugar levels are known to be higher than morning values. If a person becomes fatter, this morning-afternoon variation decreases, and it is said to be lost in patients whose glucose homeostasis is impaired. In individuals who have low blood sugar, afternoon testing may also show exaggeratedly low values. It appears that afternoon testing is likely to bring out the extremes of glucose imbalance, at least in the early phases of the problem.

In addition to afternoon testing, additional stress maneuvers may be used to intensify or bring out an abnormal response in the GTT. Dr. William Philpott (M.D.), an orthomolecular psychiatrist, has the patient exercise at one time and hyperventilate (breathe rapidly) at another time during the tolerance test. The physician's supervision is mandatory in such stress testing to intervene in case of a serious drop in blood sugar.

In order to better detect drops in blood sugar, drawing blood specimens at $2\frac{1}{2}$, $3\frac{1}{2}$, and $4\frac{1}{2}$ hours may be necessary in addition to the standard hourly values. Some physicians even check blood-glucose levels every 15 minutes. Careful education of laboratory personnel can help them recognize signs of blood-sugar change so they can obtain extra blood specimens when the patient has symptoms.

It has been standard procedure to instruct the patient to eat a high carbohydrate diet[3] for at least three days before the test. The reason for this is that the body "forgets" how to handle (metabolize) carbohydrate when it is not eaten and a "falsely" diabetic test may result ("starvation diabetes"). It may also be quite possible to obtain a depressed GTT curve when carbohydrate has not been included in the diet. Thus a diagnosis of diabetes could be missed.

A growing number of physicians question the wisdom of this traditional carbohydrate priming. I believe that more useful information is obtained about the status of the patient "as is" by

[3]For the adult, this would be 250 - 300 grams of carbohydrate per day.

performing the GTT or NTT without prior carbohydrate loading.

During the GTT, a urine specimen is checked for sugar and acetone at the time of each blood drawing. No medications known to affect blood sugar should be taken before the test. Oral contraceptives, cortisone derivatives, thiazide diuretics, and high doses of niacin may increase blood glucose. Aspirin and other salicylates can alter blood-gluocose values. Chewing gum or smoking may affect the test as may some antibiotics, barbiturates, tranquilizers, and antidepressants. Caffeine in coffee, tea, or diet drinks may alter blood-sugar levels. The antihistamine Pyribenzamine may decrease blood sugar. Other antihistamines may also possibly have an effect. LSD and marijuana tend to decrease blood sugar.

If the patient is a female, the time of the menstrual cycle should be taken into account. In the immediate premenstrual days, the blood sugar is apt to be lower than at other times. On days 13 through 18 of the cycle, the blood sugars are apt to be their highest.

The physician who orders a 5- or 6-hour test will gain much additional information from the test if the patient's symptoms and signs are recorded during the test. This can be done by having the parent, laboratory technician, or physician in attendance record for each hour of the test the presence of yawning, sleepiness, sweating, nausea, pallor, abdominal pain, vomiting, dizziness, overactivity, and the like. A checklist may be used. (See Appendix A for a descriptive method of evaluating the GTT curve as well as for a representative symptom checklist.) When the GTT is reviewed, these symptoms and signs can then be correlated with the blood-sugar levels. I have been repeatedly impressed by the presence of symptoms and signs at times when a *change* in blood-sugar level occurs. The majority of time, the symptoms and signs occur when the blood sugar *descends*, but not infrequently they occur at a time when the blood sugar *rises* or is at peak levels.

It is quite helpful to instruct the patient to write down a sentence or two about how he feels at the time each blood is drawn. He should also be instructed to sign his name to each statement. This accomplishes several things: It provides samples of the patient's thought patterns as well as his handwriting as the test progresses. The patient's writing, thoughts, and feelings can then be correlated at a later date with blood-glucose levels.

Some physicians have gained much information by having a patient draw a picture at the time of each blood drawing. I have found this approach quite useful. In a future publication I will report these findings.

Now what is done with the values that are obtained in the GTT? The blood-glucose levels may be charted on a graph to permit easier visualization of the glucose curve. Such a graph is shown in Figure 51.

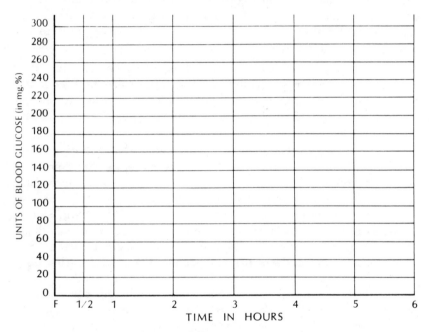

**Figure 51. Graph for plotting the results of the glucose
tolerance test (GTT)**

See the text paragraph immediately below for an explanation of this graph.

The vertical column is marked off in units (mg.%) of blood sugar, and the horizontal part is marked off in hours. The blood-sugar values are entered at the appropriate places on the graph, and a line is drawn joining each point on the curve.

An individual whose blood values are these: fasting, 85; ½ hour, 115; 1 hour, 145; 2 hours, 90; 3 hours, 85; 4 hours, 75; 5

hours, 85; and 6 hours, 90 would have a curve like that shown in Figure 52. This glucose tolerance curve would generally be considered normal.

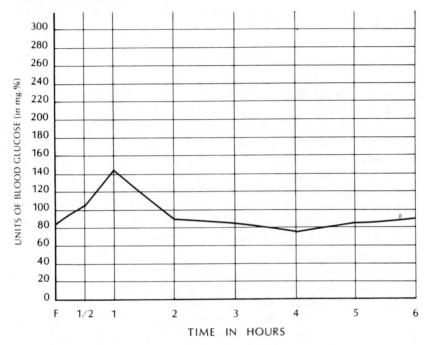

Figure 52. Graph of GTT
This curve would generally be considered to represent a normal GTT.

By glancing at the curve, the physician obtains a visual image of this aspect of the patient's carbohydate metabolism. He, of course, gets considerable information from the absolute values for the blood sugar, but the relationships of one sugar level to another may be more evident when displayed in this graphic fashion.

Blood glucose can be determined by varying methods. Unless otherwise specified, the blood-sugar values given in this book refer to whole blood, "true" blood-sugar methods. Glucose concentration in whole blood is lower than in plasma.

Automated equipment using plasma or serum gives values which are usually 15% higher than whole-blood methods. Blood drawn from veins has a glucose level as much as 30 mg.% lower than that obtained from capillaries or arteries. Blood taken from the finger, toe, or earlobe is capillary blood and would give higher values than blood drawn from the veins. One can see that it is important to obtain blood specimens from the same source throughout the GTT and to run the glucose determination by the same procedure. In practice, blood obtained from arm veins is used the majority of the time.

A detailed discussion of laboratory tests for glucose in blood and urine and drug interference with these tests has been provided in an article by Dr. M. Michael Lubran (M.D., Ph.D.).[4]

A reduction in blood sugar may take place in blood that sits in the test tube after it is drawn from the patient. Pseudohypoglycemia due to artifactual[5] lowering of blood-glucose values (in vitro autoglycolysis) must be considered whenever values are low.[6] It is known that whole blood allowed to stand will show a progressive drop in glucose concentration. The glucose decline is correlated with the duration and temperature of incubation and with the white blood cell count of the blood.

Autoglycolysis occurs much more rapidly in blood from patients with leukemia than in normal individuals. Not until the white blood cell count exceeds 60,000 per cubic millimeter do leukemic bloods consume more glucose than normal controls. Lowered blood-sugar values in cerebrospinal fluid have long been known to be correlated with glucose utilization by bacteria as well as by white blood cells in the fluid.

The addition of an antiglycolytic chemical (sodium fluoride, for example) to whole blood can prevent artifactual lowering of glucose values. Separating the plasma or serum from blood as soon as it is drawn also circumvents the problem. A delay in transporting or processing blood by the phlebotomist, transportation personnel, or the laboratory technician could account for factitial lowering of blood-glucose values. Most of the time, however, well-run medical laboratories have established routines that avoid this problem.

[4]"Drug Interference with Laboratory Tests: Glucose" by M. Michael Lubran, M.D., Ph.D. Published in *Drug Therapy*, December 1974.

[5]Artifactual: caused by an action outside the body.

[6]"Leukocytosis and Artifactual Hypoglycemia" by Thomas J. Goodenow, M.D., and William B. Malarkey, M.D. Published in *Journal of the American Medical Association*, Vol. 237, No. 18, May 2, 1977, pages 1961 - 1962.

The glucose level of the blood taken before the ingestion of the glucose load (on an empty stomach) is known as the fasting blood sugar (FBS). This FBS by itself has much significance. The FBS is clearly elevated in diabetes and often is quite low when there is an organic cause of low blood sugar, such as an insulin-secreting tumor of the pancreas. A normal FBS does not exclude diabetes. The ideal FBS is thought to be about 85 mg.%. From a statistical standpoint, the higher the FBS, the higher are the values following the glucose load.

By following the FBS levels alone, much information can be obtained about a patient's well-being. Values which fall well below 85 mg.% (for example, in the 70's) may indicate a need for dietary improvement or other treatment. Values which fall well above 85 mg.% (for example, 100 to 110 mg.%) usually call for reduction of sugar and/or other carbohydrates in the diet, but other treatments may also be required.

Many physicians now obtain an FBS and a blood-sugar test two hours after a meal. This is done to see whether food has caused a significant elevation in the second blood-sugar value. This kind of testing is often used as a screening procedure to substitute for a GTT. Normally, the two-hour blood-sugar value should approximate the value for the fasting sample. If only one blood specimen is to be obtained, a sample two hours after a meal may be preferable to the fasting specimen.

In the early part of the glucose tolerance curve, the blood sugar rises due to glucose absorption from the gut. During this time, insulin secretion is stimulated, the use of glucose in the tissues increases, glycogen breakdown in the liver is inhibited, and glycogen formation in the liver is increased. Ordinarily, there is a rise to a peak which averages about 50% above the fasting level at 30 - 60 minutes. In my experience, healthier children and adults appear to have a glucose peak at 60 minutes rather than at 15 or 30 minutes.

At the time when the blood sugar reaches its peak, glucose is being used up in the body as rapidly as glucose is entering the blood from the intestine. The maximum rise should not exceed 160 - 170 mg.% regardless of the initial level. Some authorities recognize 200 mg.% as the upper limit of normal, but I believe that 160 - 170 mg.% is the peak value which should not be exceeded.

When the blood sugar starts to fall, the use of glucose in the body (enhanced by insulin) exceeds the glucose entering into the body. Gut absorption at this time is minimal or completed. By the end of 2 hours, the blood-sugar level should have returned to fasting levels or near fasting levels. This return within 2 hours to approximately the fasting level is of more importance in excluding diabetes than the finding of a nonelevated peak value.

At the 4-hour time, or thereabouts, a dip of the blood sugar some 10 or 15 mg.% below the fasting level may occur. This is due to a lag in the production of the counterregulatory factors[7] which maintain a steady sugar level. Decreased glucose output from the liver and increased glucose usage are not reversed as rapidly as the blood sugar falls. There is a tendency for a greater dip to occur, the higher the initial rise.

The recovery of the blood sugar after the dip is due to increased liver output of glucose and decreased usage of sugar in the body. A rise in blood sugar at 5 hours may be due to the activity of the counterregulatory factors.

All urine specimens analyzed during the GTT should normally be free of sugar and acetone. When the fully developed disease diabetes occurs, the urinalysis shows the presence of sugar usually in large amounts. Acetone may also be present. As explained in the previous chapter, however, it is quite possible to have chemical diabetes without sugar in the urine. Some persons show sugar in the urine under stress (anxiety, etc.), and these individuals may be those with a hereditary predisposition to diabetes. Usually sugar appears in the urine when the kidney threshold for sugar excretion is exceeded. This is variable but most often is around 170 mg.% blood sugar. It is probable that this level of 170 mg.%, the usual kidney threshold, represents the upper limit of normal for blood-sugar elevation after ingestion of a sugar load.

The diagnosis of diabetes is usually made when these blood-sugar conditions prevail (whole-blood, true glucose values):

fasting:	110 mg.% or more (100 - 109 is suspect)
1 hour:	160 - 180 mg.% or more (170 is commonly used)
2 hours:	120 mg.% or more
3 hours:	110 mg.% or more

[7]Glucagon, epinephrine, growth hormone, glucocorticoids.

An alternate method for diagnosis is the summation technique.[8] The fasting, 1-, 2-, and 3-hour whole-blood-sugar values are added up. If the total is greater than 500 mg.%, the diagnosis of diabetes is made. In another modification (Danowski), the fasting, ½-, 1-, and 2-hour blood-sugar levels are totaled and interpreted as follows:

500 or less:	not diabetic
501 to 650:	limited "chemical" diabetes
651 to 800:	more severe chemical diabetes
800 or over:	overt diabetes

As I indicate in Chapter 12 and Appendix A of this book, a description of the glucose tolerance curve is often more helpful than labels such as normal or abnormal.

For a discussion of low blood sugar, the reader is referred to Chapter 14 of this book. At this time it can be said that blood-sugar values below 80 are probably less than optimal, below 70 indicates probable hypoglycemia, below 50 is diagnostic of hypoglycemia, and below 40 makes one suspicious of an insulin-secreting pancreatic tumor or other organic lesion.

In pregnancy, the GTT tends to be diabetic largely because the blood-sugar levels do not return to normal as rapidly as in the nonpregnant state.

The patient who has a fever should not have a GTT, because fever impairs the utilization of glucose in the body. The presence of infection itself may do the same.

The individual who has been bedridden also fails to properly utilize glucose. Physical inactivity brings about a state in which the body cells are resistant to insulin. Physical activity, on the other hand, improves the utilization of glucose. Diabetics who take insulin require less insulin when they are physically active than when they are sedentary.

In our present culture, man's sedentary lifestyles are rapidly qualifying him for the designation: Homo sedentarius. One witnesses children in front of the TV set on the average of 45 - 54 hours per week! In lectures, Dr. Howard Coleman (O.D.) of Rum-

[8]"Diabetes Mellitus: When It Does and Doesn't Exist" by E. A. Haunz, M.D. Published in *Consultant,* January 1975. This material is used with the permission of Dr. Haunz as well as *Consultant.*

ford, Rhode Island, has used the term "gasoline fanny" to describe our modern citizenry, scurrying here and there in a wide variety of mechanized vehicles. Others have wondered whether man's legs will become vestigial because of lack of use.

Modern man with his technological triumphs continues on in his dizzy pace, consuming food with less and less nutritive value. The combination of inactivity and malnutrition appears to underly the increasing presence of diabetes, a disorder characteristically found in "civilized" populations.

As a person grows older, it appears that his blood-glucose levels become somewhat higher. Whether this is related to obesity, physical inactivity, altered diet with age, etc., or whether it is strictly an effect of age, is not clear. I suspect it is not due to aging alone.

Three types of blood-fat disorder (hyperlipidemia III, IV, and V) are often associated with an impaired ability to utilize glucose. These disorders are also associated with obesity. In Types III and IV, elevated triglyceride levels are due to carbohydrate eaten in the diet.

Overactivity of the adrenal cortex (Cushing's syndrome) or overactivity of the thyroid gland may also give rise to elevated blood-sugar values in the GTT.

When the absorption of sugar from the gut is suspected to be faulty, an intravenous GTT can be done. In this test, the load of sugar is given in the vein, and the rate of disappearance of sugar is followed for several hours thereafter. The expected normal is the elimination of glucose at a rate of 2% per minute. Diabetics reduce the sugar level at a slower rate. Sixty-eight minutes after the glucose load, the blood glucose should be no more than 28 mg.% above the fasting value; and at the end of 2 hours, the value should be within 5 mg.% of the fasting value.

A variation of the GTT is the 1-hour, 2-dose test (Exton-Rose test). This test is less influenced by prior dietary habits than the standard GTT. In this test, a fasting blood specimen is drawn after which 50 grams of glucose is given by mouth. A half hour later, the blood is obtained again, and an additional 50 grams of glucose is given. After a half hour, the final sample is drawn.

In the 1-hour, 2-dose test the fasting level should be between 60 - 100 mg.% (whole blood, true glucose) and the half-hour specimen should not rise more than 75 mg.% above that level. The final sugar should not exceed 5 mg.% above the second. All urines should be sugar-free.

The normal values for this 1-hour, 2-dose test show us that each successive dose of sugar produces a smaller rise in blood sugar than the preceding. This is good evidence that the *handling of carbohydrate in the body is strongly related to past dietary experience.* "Starvation diabetes" has already been mentioned and explained as a sluggishness of the body in handling carbohydrate when it has not been recently exposed to it. This is something like the athlete who needs a warm-up period to reach peak effectiveness when he has not been exercising.

What about the opposite situation: the individual who continually exposes his body to a barrage of carbohydrates throughout the day, day in and day out? In many such individuals, it appears that the body becomes "superprimed" to metabolize carbohydrate. Such an individual often has a flat GTT, a curve that represents relative hypoglycemia, a curve that falls instead of rising after glucose ingestion, or a curve with erratic characteristics.

Dietary factors, then, in the days and weeks before a GTT may definitely affect the outcome of the test. Diet plays a strong role in carbohydrate metabolism. A high intake of carbohydrates may tend to depress the blood-sugar levels (at least in early stages). A high intake of fats may tend to elevate the sugar values. In "starvation diabetes," the stored body fats are utilized for energy, and the effect is similar to a high-fat diet.

From these observations, we can conclude that carbohydrate stimulates the production of insulin, while fat tends to depress it. Fifty-six percent of protein is converted to carbohydrate in the body. Because of this, we would expect protein to stimulate insulin release and, indeed, this is so. Protein, however, appears to stimulate insulin release much more gently and far less than carbohydrates themselves. Individual variation, however, is great.

In individuals with normal GTT's, blood-insulin levels make a prompt appearance within the first 30 minutes. They then rise rapidly, thereafter, reaching a peak at 30 - 60 minutes, then falling slowly to normal levels. In general, the blood-insulin curve parallels the glucose curve rather well in the normal individual. In maturity-onset diabetics, there is an absence of the early insulin release, as well as an overall delay in the secretion of insulin. At the end of 2 hours, however, the insulin level is distinctly higher than normal. This sluggish rise, delayed peak, and excessive secretion, along with a slow decline, are characteristic of maturity-onset diabetics.

In patients with juvenile diabetes, the blood levels of insulin are very low at all times.

In view of the dietary effect upon the GTT, we must be cautious about the use of blood-sugar norms which have been obtained on individuals whose diets were different than the individuals being tested. It may be wiser to think in terms of *usual* values for a particular population with a particular set of dietary habits. As the diet of that population changes, so may the values for the GTT's.

Since children's diets vary so greatly one from another, GTT variation may be even more profound in children than in adults.

It is difficult to know what normal blood-insulin values are in our present American culture with its high dietary load of sugar and other refined carbohydrates. Studies on individuals who avoid these dietary elements need to be carried out.

Terminology that *describes* a GTT, rather than assigning a designation of normal or abnormal, may be most helpful in evaluating an individual, his diet, and his state of carbohydrate metabolism.

If a glucose tolerance test is judged to be normal, it does not necessarily mean that it is ideal or optimal for that individual. Remember, we derive our norms from supposedly healthy people, but these persons undoubtedly have a host of minor or nonevident health problems which are not screened for when the subjects are selected. We need to develop a set of norms from absolutely healthy individuals who have a minimum of dental cavities, periodontal disease, hernias, hemorrhoids, varicose veins, dandruff and acne, postnasal drip, constipation, fatigue, etc. Not to do so is to encourage mediocrity in health matters.

The finding of a normal GTT at a certain point in time does not tell us that the GTT was normal in prior years or that it will be normal in ensuing years. A "normal" GTT could represent one phase of carbohydrate metabolism going, for example, from a low, flat curve to a diabetic or dysinsulinism type of curve.

The following factors may elevate the blood sugar:

Drugs:
  Oral contraceptives
  Salicylates (aspirin, etc.)
  Niacin
  Diuretics (thiazides and ethacrynic acid)
  Phenylephrine (Neo-Synephrine)
  Cortisone derivatives
  Adrenalin and allied agents
Severe Infections
Central nervous system disorders
Surgical Procedures
Physical trauma
Burns
Overactivity of adrenal cortex
Overactivity of thyroid
Overactivity of anterior pituitary
Neuroblastoma tumor
Pheochromocytoma tumor

Notice that the cortisone derivatives may give rise to elevated blood sugar. This effect has been taken advantage of in the cortisone GTT.[9] In this test, the patient is given a cortisonelike drug and subsequently is given a standard GTT. The cortisone GTT has value in discovering latent diabetes or predicting the later development of the disease.

Another test which may be used for the detection of diabetes is the tolbutamide test. The reader will remember that tolbutamide is an oral hypoglycemic drug which stimulates the production and release of insulin from the pancreas. In the normal person, 20 minutes after the intravenous injection of 1 gram of tolbutamide, the blood sugar drops 25% or more. In the diabetic,

[9]According to the method of Stephan S. Fajans, M.D., and Jerome W. Conn, M.D., cortisone acetate is given orally in 2 doses, 8½ hours and 2 hours before the GTT. If the patient weighs less than 160 pounds, 50 mgm. per dose is given. For patients above 160 pounds, 62.5 mgm. per dose is given. The upper limits of normal at 1, 1½, and 2 hours are blood-sugar values of 180, 160, and 150 mg.%. The more abnormal values that a patient has, the closer he is considered to be to the diabetic state. (See "An Approach to the Prediction of Diabetes Mellitus by Modification of Glucose Tolerance Test with Cortisone" by Stephan S. Fajans, M.D., and Jerome W. Conn, M.D. Published in *Diabetes*, Vol. 3, No. 4, July - August 1954, pages 296 - 304. Also see "Comments on the Cortisone Glucose Tolerance Test" by Stephan S. Fajans, M.D., and Jerome W. Conn, M.D. Published in *Diabetes*, Vol. 10, No. 1, January - February 1961, pages 63 - 67.)

the drop is 15% or less at 20 minutes. At 30 minutes, a drop of 23% or less indicates diabetes.

The GTT is a helpful tool to the physician. It has served as the standard diagnostic instrument for establishing or excluding the clinical problem of diabetes. The test gives much more information when it is carried out for a duration of 6 hours. It may give considerable information about low blood sugar and a person's individual pattern of carbohydrate metabolism. In the future, readily available clinical determinations of growth hormone, adrenal hormones, insulin, and other factors, along with blood sugar, will enhance the information that the clinician derives from the test.

Low blood sugar is discussed in Chapter 14.

In the next chapter I will discuss the glucose tolerance curve (the graphic representation of the GTT) and give examples of many patterns.

CHAPTER 12

GLUCOSE
TOLERANCE CURVES

In this chapter I wish to further acquaint the reader with GTT curves. The wide variety of responses to an oral glucose load appears to be as individualistic as fingerprints. However, unlike fingerprints, the GTT curve may vary from time to time in response to nutrition, hormonal status, stress, and other factors.

I wish to show in this chapter that a GTT can be individually described and interpreted rather than merely labeled as normal or abnormal. Reference to Figure 53 will acquaint the reader with the wide variety of GTT curves that may be encountered. Physician readers will find a helpful method of descriptive evaluation of the GTT curve in Appendix A of this book.

In connection with GTT curves, let us ask: "What is a desirable curve and, indeed, what is the optimal state of carbohydrate tolerance for each individual?" Although the complete answer to these questions is not available, at least asking the questions may permit us to make progress in obtaining a more and more accurate answer. Certainly, nutritional factors are major items which shape an individual's GTT curve. Attention to optimal nutrition for an individual may lead us to an understanding of desirable and optimal carbohydrate metabolism. Practical nutritional suggestions are outlined in this book in Chapter 15.

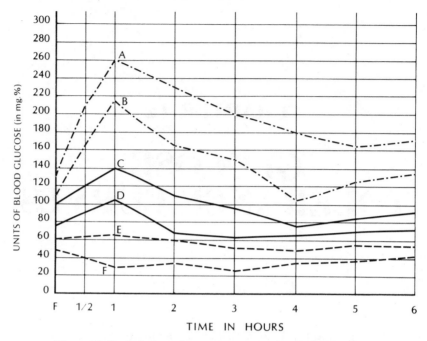

Figure 53. Oral glucose tolerance test curves to show a continuum ranging from diabetes through normal to low blood sugar

Curves A and B are diabetic. Curves C and D could be categorized as normal. Curves E and F fall into the pattern commonly described as hypoglycemic, although some would term Curve E as a low, flat curve included in the broad group of normals. Notice that Curve F descends instead of rising after a glucose load. Many other variations occur which are not depicted here. For example, a curve may be diabetic in the early phase and hypoglycemic in the later phase. These curves were constructed by the author for purposes of illustration.

Hormonal and nervous factors are, likewise, important determiners of individual glucose tolerance. It is probable that an individual's nutritional input directly influences his hormonal and neurological function.

Dr. J. W. Tintera (M.D.)[1] established what he believed to be the desirable range for GTT curves (see Figure 54). This representation of the desirable range for GTT curves is an attempt to delineate what is "ideal."

[1]"The Hypoadrenocortical State and Its Management" by J. W. Tintera, M.D. Published in *New York State Journal of Medicine*, Vol. 55, No. 33, July 1, 1955.

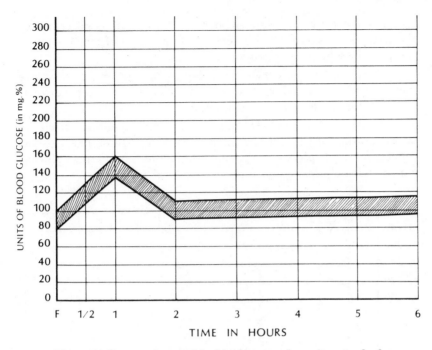

Figure 54. Ranges of normal for blood-sugar values after standard glucose tolerance testing, according to Tintera

See text, pages 234 - 235.

Since Dr. Tintera's range of desirable blood-sugar levels has distinct and somewhat narrow limits, it should be used and interpreted with caution until further evidence is available to substantiate or refute it. A curve almost within this range or partially within this range could, for example, be desirable or optimal for a particular individual. If the deviation from his ranges is large, it is likely that that curve is less than optimal for that individual. GTT curves C and D in Figure 53 are thought to be desirable.

Dr. Hugh Powers (M.D.) has characterized GTT curves in three varieties[2] (see Figures 55, 56, and 57). The curve that has a high peak and a precipitous drop (Figure 55) is termed the psychosomatic curve by Dr. Powers.

[2]Address to Houston Hypoglycemia Society, April 19, 1971, by Hugh W. S. Powers Jr., M.D. Also, "Dietary Measures to Improve Behavior and Achievement" by Hugh W. S. Powers Jr., M.D. Published in *Academic Therapy*, Vol. IX, No. 3, Winter 1973 - 1974.

The flat curve at a low sugar level (Figure 56) is described as the listless curve by Dr. Powers. It is the type of curve seen in the patient who is fatigued and zestless.

The third type of curve, according to Dr. Powers, is the curve frequently associated with irritability. This stair-step curve is too high for too long a time (Figure 57).

Dr. Juan Wilson (M.D.) has written an excellent article on interpretation of the glucose tolerance curve.[3] He emphasizes the functional and physiologic viewpoint and correlates glucose curves with clinical conditions.

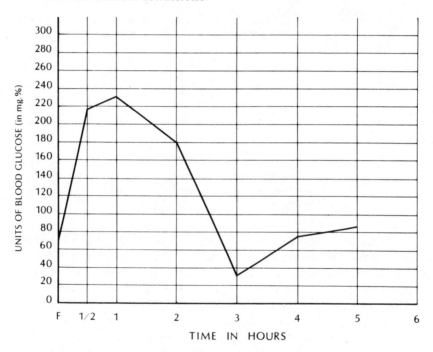

Figure 55. Oral glucose tolerance test curve described as psychosomatic by Dr. Hugh Powers (M.D.)

Notable in this type of curve is the high peak and rapid decline. This great magnitude of rapid change from high to low levels may be very upsetting to body homeostasis.

[3]"Physiological and Psychological Implications of the Glucose Curve" by Juan Wilson, M.D. Published in *Journal of the International Academy of Metabology, Inc.*, Vol. III, No. 1, 1972.

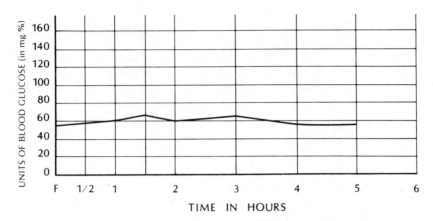

**Figure 56. Oral glucose tolerance test curve described as listless
by Dr. Hugh Powers**

Notable in this type of curve is the lack of adequate rise in the blood-sugar level. This is basically a flat curve. It could be due to inadequate or slow absorption.

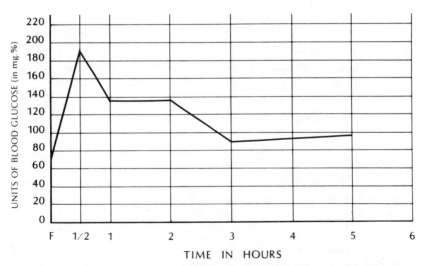

**Figure 57. Oral glucose tolerance test curve described as irritability by
Dr. Hugh Powers**

Notable is the excessive peak and the prolonged decline which occurs in a stair-step manner. This can be termed a prediabetic curve.

I will now present some GTT curves that were obtained on patients in my practice. The blood-sugar values have all been determined by the whole-blood, venous, true-blood-sugar technique. This selected sampling of curves with brief case studies will acquaint the reader with a wide variety of GTT's gathered from actual patients.

Hopefully, the presentation of these curves will stimulate others to seek and better define optimal values for carbohydrate tolerance.

GTT curves which fall into the category of low blood sugar will be found in Chapter 9 and 13 and in Appendix B.

In the case studies that follow, the deviations in the blood-sugar values may not have a one-to-one relationship with the in-

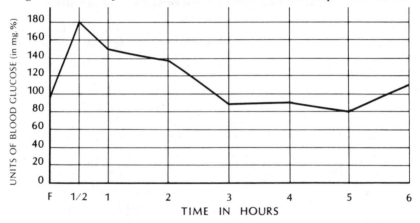

Figure 58. A glucose tolerance test curve

GTT of a 9-year-old boy with low attention span, hyperactivity, irritability, and allergies.

This GTT could be considered a fairly normal curve but it could also be interpreted as prediabetic.

The value of the use of a symptom checklist is demonstrated by this case. At 3, 4, and 5 hours, the boy was "miserable." He felt very tired, nauseated, and hungry. He "revived" during the last hour. Note that his adverse symptoms correlated exactly with the low points in the sugar curve, and his "revival" correlated with the rise in blood sugar at 6 hours.

Some would say this boy has "low blood sugar." He certainly does not have low blood sugar in absolute terms, because the blood-sugar levels do not ever drop below 80 mg.%. The boy does show us symptoms which appeared when his blood sugar was lowest, and thus he could be said to have relative or symptomatic low blood sugar. It is possible to have a prediabetic, diabetic, or "normal" GTT curve with prominent symptoms at the time of blood-sugar drop.

dividual's clinical problems. It is likely, though, that the instabilities of carbohydrate metabolism and the clinical disorders are related to the same core factors such as diet, emotions, physical activity, mineral and vitamin inadequacy, metabolic disturbances, etc.

The reader who wishes to consider a wide variety of GTT curves should consult Appendix B.

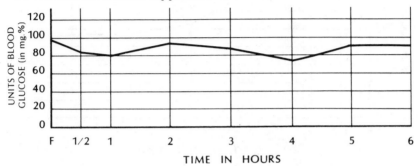

Figure 59. A glucose tolerance test curve

This 13-year-old undergrown girl has a GTT which shows a decline after glucose loading. The curve is generally flat. This curve could be due to lack of absorption of glucose, excessive insulin release, or enhanced susceptibility to normal amounts of insulin.

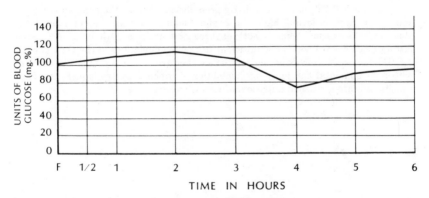

Figure 60. A glucose tolerance test curve

GTT of an 11-year-old girl with moderate overweight and underachievement in school.

The curve is generally flat with a dip at 4 hours. The flat GTT curve has been considered to fall within the broad range of normal by many academic physicians. This curve could also be interpreted as slow and minimal glucose absorption from the gut, excess insulin secretion, or enhanced reaction to insulin.

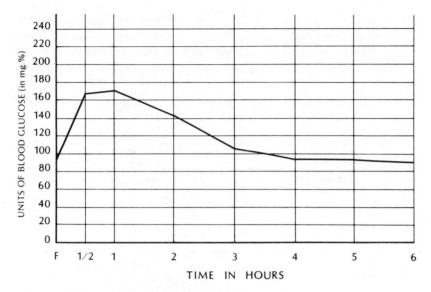

Figure 61. A glucose tolerance test curve

A 5½-year-old girl with a history of "spilling" sugar in the urine on several occasions. There was a strong family history of diabetes.

The child was also a chronic bedwetter and manifested delay in milestones of development. Her lagging development is evidenced by her drawing of a person (see Figure 62) which is quite immature for her age.

This girl's GTT curve is distinctly prediabetic, because the peak values at ½ and 1 hour are borderline high, the 2- and 3-hour values are higher than desirable, and the fasting level is not reapproached until the 4th hour. A normal fasting-blood-sugar level is not sufficient to exclude deviations in the GTT curve.

Deviant GTT curves of one sort or another are commonly found in children with developmental and learning disabilities — when they are looked for. They usually indicate disturbed nutrition, often with allergy.

Figure 62. Drawing of a person done by the 5½-year-old girl in Figure 61 with a prediabetic GTT

Lagging development is indicated by the lack of body, hands, nose, and hair, as well as by the misplacement of the arms.

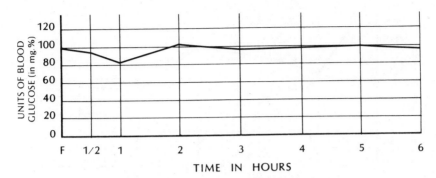

Figure 63. A glucose tolerance test curve

A 12-year-old boy with chronic school difficulties in reading, writing, and math.

This boy also had recurrent allergic nasal and sinus congestion with cough. He craved sweets throughout his life. Fifteen percent of his circulating white blood cells were eosinophils ("allergy cells").

The GTT shows a curve which *descends* after the administration of the glucose load. When not descending, the curve is a flat one. This could be considered to be a form of low blood sugar, probably due to hyperinsulinism. The fasting and 6-hour levels of 100 and 96 respectively, although not abnormally high, may be somewhat higher than ideal.

This boy when first seen was described as a "sweet-a-holic" by his mother. After elimination of sugar in his diet, the school work improved, and his head congestion disappeared. Whenever he did happen to eat sugar, thereafter, it was quite noticeable that the boy became sleepy.

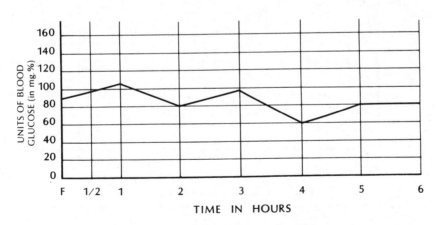

Figure 64. A glucose tolerance test curve

A 7-year-old boy with hyperactivity, incoordination, learning disorder, and severe allergic rhinitis.

The GTT curve is flat, sawtooth, and has a dip to 60 mg.% at 4 hours.

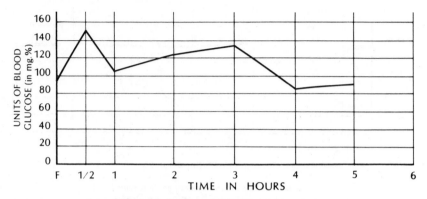

Figure 65. A glucose tolerance test curve

An 11-year-old boy with difficulty in reading who repeated 3rd grade and attended several remedial private schools.

The GTT curve is a saw-tooth curve at a fairly normal level. An irregular saw-tooth curve indicates erratic absorption, erratic insulin release, or erratic blood-sugar-regulating compensatory factors.

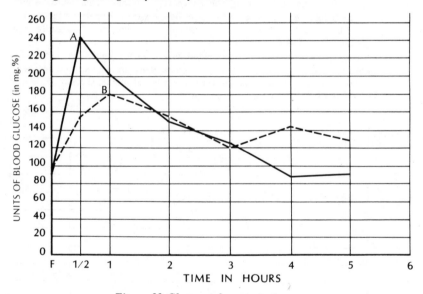

Figure 66. Glucose tolerance test curves

A 12-year-old boy with severe headaches and allergic congestion of the nose and sinuses.

Curve A was obtained in April. It is a diabeticlike GTT with a rapid drop. In September, after restriction of sugar in the diet and with supplemental vitamin treatment, Curve B was obtained. Although the improvement in Curve B is gratifying, still more improvement in the GTT curve is expected as time passes.

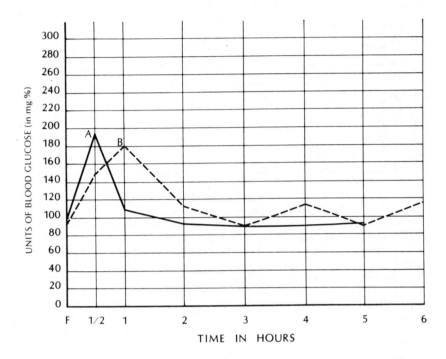

Figure 67. Glucose tolerance test curves

A 10-year-old asthmatic boy with recurrent vomiting and frequent episodes of fatigue and hunger that occurred about 2 hours after a meal.

The GTT on 12/72 (Curve A) shows an early peak at ½ hour, an excessive peak (195 mg.%) and excessively rapid drop, the lowest point of the curve at 3 hours, the late part of the curve below the fasting level, and an absence of a rise after the 3-hour low level. The boy felt sleepy and sickish at 3 and 4 hours, and he felt "bad" at 5 hours.

As soon as the 12/72 test was obtained, this boy was treated with restriction of sugar in the diet, frequent feedings, and vitamin-mineral supplements.

The GTT on 11/73 (Curve B) shows a later peak at 1 hour, a slightly lower peak, a smaller drop between 1 and 2 hours, and a compensatory rise after the 3-hour low value. During this test the boy felt well.

During the year of treatment, he had almost no asthma and no vomiting spells, and almost no episodes of fatigue and hunger.

If the peak glucose level is reached at 15 or 30 minutes, it is my experience that the child has a diet which contains large amounts of sugar and other refined carbohydrates.

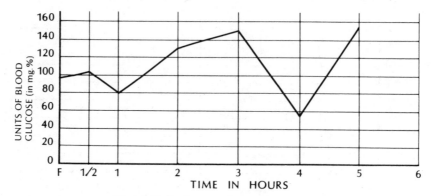

Figure 68. A glucose tolerance test curve

A 6-year-old boy with marked mood swings, bedwetting, hyperactivity, temper tantrums, and aggressive, violent, destructive behavior. There was also allergic rhinitis and conductive hearing loss, requiring drainage tubes in the middle ears.

This curve is markedly erratic (saw-tooth). The glucose load gives essentially no rise in blood sugar, but sets off a series of reactions swinging from a drop in blood sugar, to a high rise, to hypoglycemia, to a high rise at the termination of the test.

Clinical behavior and glucose tolerance curves may exhibit similar wide swings.

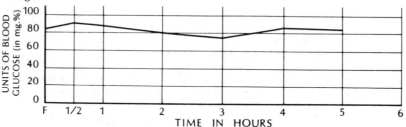

Figure 69. A glucose tolerance test curve

A 10-year-old boy with bedwetting, tantrums, sleepwalking, nightmares, fatigue, and hyperactivity. Hay fever was prominent with 9% eosinophils ("allergy cells") in the blood white-cell count.

This GTT curve is flat.

Erratic clinical behavior may be associated with an erratic GTT curve (see previous case, Figure 68), but it may also be seen in association with a very stable GTT.

A flat curve such as this is described by some medical authorities as normal. The question remains: What are optimal physiological conditions for this patient? This type of curve usually changes to one which shows a rise and fall of blood sugar within normal limits, when sugar and other refined carbohydrate are severely limited in the diet and the diet otherwise improved. The eosinophil count is also commonly noted to decline. Often, specific food sensitivities must be discovered and treated.

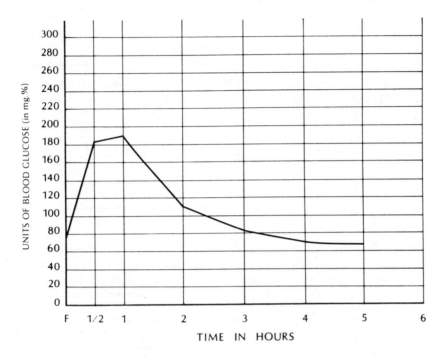

Figure 70. A glucose tolerance test curve

An 11-year-old boy with mental retardation. Ten percent eosinophils ("allergy cells") in the peripheral blood white-cell count.

The GTT is distinctly prediabetic. The drop from a peak of 190 to a low of 68 suggests relative hypoglycemia. Notice also the large drop between 1 and 2 hours. The 4- and 5-hour values are below the fasting level. This has been pointed out as a feature of hypoglycemia.

A combined diabetic and low blood-sugar curve has been termed dysinsulinism. It probably represents faulty timing of insulin release.

Elevation of the eosinophil count in the blood may be used as an index of the presence of an allergic state. It also appears to reflect the diet which is high in sugary foods as well as artificial colors, flavors, and preservatives.

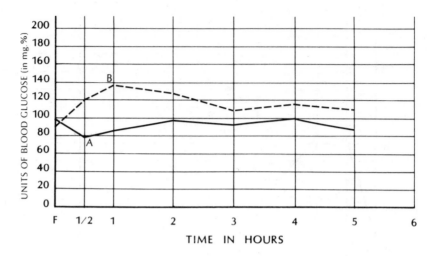

Figure 71. Glucose tolerance test curves

This 10-year-old boy was hyperactive and unable to pay attention in class. Because he was disruptive, he had spent two years at home receiving homebound instruction.

He was found to be a bright boy with above average I.Q. who was greatly in need of socialization with children of his age. He was growing steadily further behind in academics despite the use of Ritalin medication.* He was becoming progressively more overweight.

A severe allergic nose and sinus condition had been present since early infancy.

A severe visual problem consisting of very weak focusing power and unstable visual fusion was found.

The boy craved sweets and was constantly seeking a source of sugary food. He urinated every half hour or more often.

A GTT (Curve A) had been obtained one year previously and had been reported as normal. The curve shows that the boy's blood sugar *drops* after the ingestion of the sugar load rather than rising. This is probably due to a hair-trigger hypersensitive pancreas mechanism (hyperinsulinism). Note also that the fasting glucose of 98 mg.% is somewhat high in a relative sense.

Treatment with allergy desensitizing injections, elimination of sugar from his diet, the use of learning lenses, and provision of special education have resulted in improved academics and health. He has lost weight and no longer urinates frequently. His attention span has markedly increased, he is not now described as hyperactive, and he no longer takes Ritalin medication.

After 6 months of treatment, a GTT was obtained (Curve B). One can readily see that at this time the blood sugar *rises* instead of falling in response to the glucose load. Further improvement is expected.

*Ritalin (methylphenidate) — Ciba Pharmaceutical Company, Division of Ciba-Geigy Corporation, Summit, N.J. 07901 — is a stimulant drug often used for children who have attentional inconstancy and hyperactivity.

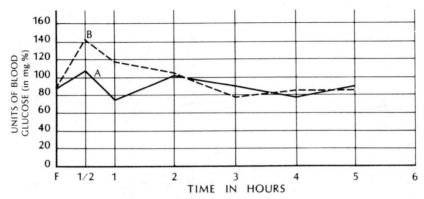

Figure 72. Glucose tolerance test curves

An 8-year-old hyperactive boy described as a "loner." Intermittent asthma and chronic nasal allergy were present.

The GTT of 12/72 (Curve A) has an early dip (1 hour) and is somewhat flat before and after the dip. Following vitamin-mineral treatment and dietary improvement, GTT Curve B was obtained. There has been a pronounced change toward normal with a slight dip at 3 hours.

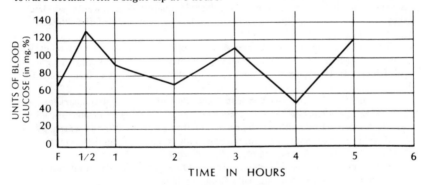

Figure 73. A glucose tolerance test curve

A 9-year-old boy with grand mal and petit mal seizures, as well as bedwetting. Nine percent eosinophils in the blood.

This curve is saw-tooth. There is a dip at 2 hours, and a more profound dip to 50 mg.% occurs at 4 hours.

Hypoglycemia as a pathologic condition is said to occur when blood-sugar levels reach 45 mg.% or below. It is quite probable, however, that this widely fluctuant curve does not represent an optimal state of physiology for this person. Certainly, homeostasis has "flown the coop" in carbohydrate metabolism such as this. The level of sugar in the brain may be lower for a more prolonged time than it is in the blood.

The boy's seizures were markedly reduced in frequency when dietary improvements were made. Patient's with convulsive seizures frequently are found to be hypoglycemic.

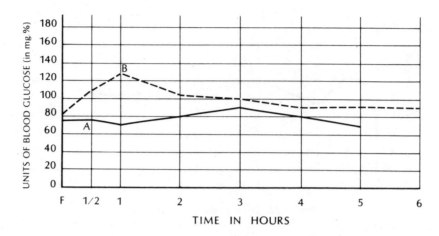

Figure 74. Glucose tolerance test curves

GTT curves of a 4-year-old girl who had come in for a well-child check-up and was found to be very obese and to have an irregularity of the heartbeat. The child's weight had been normal from birth until 3 years of age, when the mother said the child began to consume considerable quantities of sweetened drinks, candy, cookies, etc. The girl's blood count showed that 11% of the white blood cells were eosinophils ("allergy cells").

The GTT curve marked Curve A was obtained before any treatment was given. The curve is flat and declines slightly in response to the glucose load. Dietary instructions were then given to eliminate all sugar from the child's diet.

One month later the girl had lost 7 pounds and the irregular heartbeat was gone. The child looked and acted "like a different child." Her mother stated she was much more active and that she no longer wanted to sit still and watch TV.

Within another month or two, more pounds were lost. The differential white blood count at that time showed only 1% eosinophils.

After return to a normal weight, the child had no difficulty in maintaining a normal weight by exclusion of sweets from the diet.

The GTT done 6 months after the first one (Curve B) shows that the curve has changed its configuration and is no longer a flat or depressed curve.

Cases such as this one make it appear that the flat or declining curve is one manifestation of a less than optimal state of health.

This case is a dramatic example of the effect of a high sugar diet on the health of a child. Cases such as this are not at all rare in my experience, although they may be less extreme than this one.

I have also observed that the consumpton of high quantities of sugar may produce hematuria (blood in the urine). It is also probable that frequent and/or high-level sugar intake promotes urinary infection. I have had considerable success in eradicating recurrent urinary infections by improvement of the diet, particularly by eliminating sugar (sucrose) and the associated artificial colors, flavors, and preservatives that are usually consumed along with sugar in prepared food products.

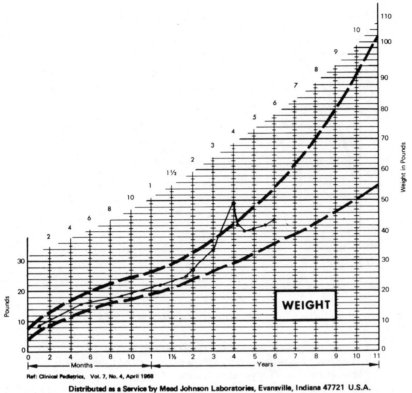

Ref: Clinical Pediatrics, Vol. 7, No. 4, April 1968

Distributed as a Service by Mead Johnson Laboratories, Evansville, Indiana 47721 U.S.A.

Lit. 257R, 1/71

Copyright 1968, 1971 by Alexander and Rudolph A. Jaworski

**Figure 75. Graphic demonstration of the weight of the child
whose GTT curves were presented in Figure 74**

The interrupted lines represent percentile distribution of weight in the American population. The upper interrupted line is the 97th percentile and the lower interrupted line is the 3rd percentile. The continuous line represents the patient's weight from birth to six years.

At the age of 4 years, the child was markedly overweight, had an irregular heartbeat, and had a history of high dietary intake of sugar. Eleven percent eosinophils were found in the blood. Within a 6-month period after restriction of sugar and other refined carbohydrate, the child's weight was normal, the heartbeat was regular, the eosinophil count in the blood had dropped to 1%, and the child was much more active physically.

It is likely that sugar, other refined carbohydrates, or the foods and chemicals commonly associated with them when consumed in high quantity induced an allergic reaction with increase of eosinophils in the blood.

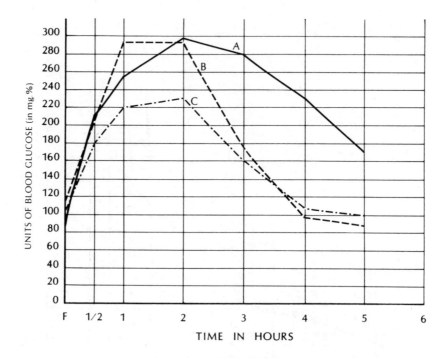

Figure 76. Glucose tolerance test curves

An 11-year-old boy with severe reading problems and hyperactivity.

Initial curve in May (Curve A) is diabetic. The follow-up curve in August of the same year (Curve B) is also diabetic, but the curve returns to normal levels at 4 and 5 hours. Between May and August, his dietary sugar was restricted and brewer's yeast was given. Allergy treatment commenced in September 1973. Further improvement is demonstrated in the GTT curve of November of the same year (Curve C).

Reduction of adverse stress, whether it be nutritional, allergic, visual, or environmental may be important in trying to improve carbohydrate metabolism. Brewer's yeast contains a glucose-tolerance factor which may assist in normalizing states of carbohydrate imbalance.

It is expected that further improvement in the GTT curve will be forthcoming.

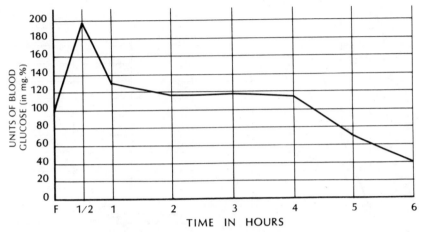

Figure 77. A glucose tolerance test curve

This 12-year-old boy with asthma and hay fever eats no breakfast but has a heavy intake of sweets throughout the day. He is underexercised. Eosinophils were 16% in the blood.

The GTT curve is prediabetic due to excessive rise and tendency to sustained high levels. The final dip to relatively low blood-sugar levels might suggest a failure of the liver to provide glucose to the bloodstream. The fasting blood sugar of 100 mg.% is probably higher than desirable.

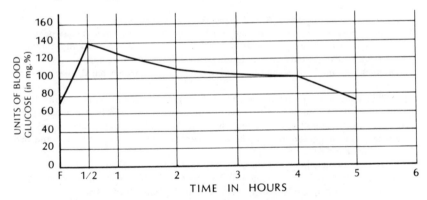

Figure 78. A glucose tolerance test curve

This 7-year-old girl is hyperkinetic and has minimal brain dysfunction. She also has nasal allergy. This child has a fairly "normal" glucose tolerance test. Nevertheless, she develops extreme hyperactivity, aggressive behavior, and inappropriate laughter after eating any sugar-containing food. Such symptoms could be due to cerebral "allergy" to sugar, a NeuroAllergic Syndrome.

Note the drop in sugar level at the 5th hour. (No 6th hour specimen was obtained.) Blood-sugar decline late in the GTT curve may indicate inadequate stores of glycogen in the liver or failure to release glucose from glycogen stored in the liver.

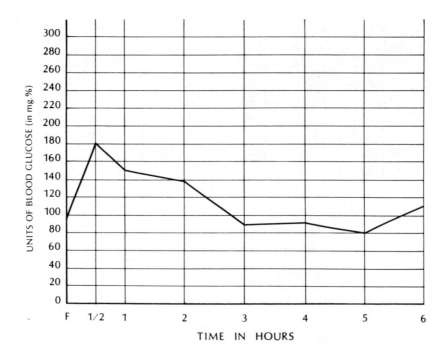

Figure 79. A glucose tolerance test curve

An 11-year-old hyperactive boy with short attention span, behavior problems, and a chronically poor appetite. Psychological counseling at a child guidance clinic for a 1-year period was not followed by improvement. There was a history of being a "hellion" since birth. Findings typical of chronic nasal allergy were present.

This GTT appears fairly normal at first glance. However, the peak of 180 mg.% is borderline high, and this peak is reached within the first one-half hour. The decline is somewhat prolonged, the peak low being reached at 5 hours and the total drop being 100 mg.%. The increment of drop between the 2nd and 3rd hour is considerable.

During the last hour of this GTT, the boy was "miserable" with nausea and sleepiness. He "revived" during the last half hour. These symptoms correlate well with his sugar levels, which were lowest at 5 hours and which rose during the last half hour.

This curve is a good example of how symptoms may be present even though the absolute sugar level is not pathologically low.

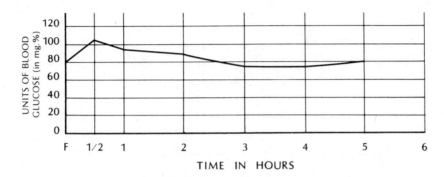

Figure 80. A glucose tolerance test curve

A 9-year-old girl with headaches and nasal allergy.

The GTT shows a somewhat low rise and an essentially flat curve thereafter. Is this GTT normal for this individual?

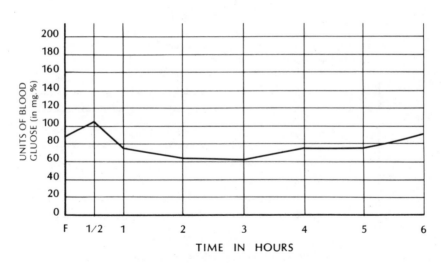

Figure 81. A glucose tolerance test curve

A 10-year-old boy was hyperactive and impulsive, and he had recurrent respiratory infections. He had a great thirst and a craving for sweets. He was made hostile by Ritalin medication. His intelligence was high.

His GTT curve is a generally flat curve with very little rise and has an early and excessively prolonged dip. Does this child have low blood sugar? Read Chapter 14 in this book for a discussion of low blood sugar.

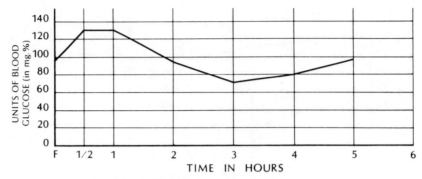

Figure 82. A glucose tolerance test curve

An 11-year-old boy was barely getting by in school. He also had recurrent headaches, obesity, and incoordination.

This GTT would be termed normal or near-normal by most medical authorities. The symptom checklist, however, showed nervousness, nausea, pallor, yawning, stomachache, headache, malaise, urinary frequency, and difficulty in swallowing at the 3rd and 4th hours, corresponding with the lowest points of the curve.

If this boy has psychological symptoms, why does he choose to have them just at these times?

One must wonder about the metabolism of this child and its relationshp to his symptoms and learning difficulties. We are just beginning to open the door to an understanding of the chemical basis of behavior. Often, it is related to nutrition, allergy, and maladaptive reactions.

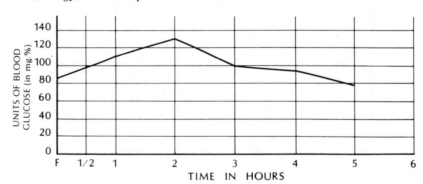

Figure 83. A glucose tolerance test curve

An 8-year-old hyperactive boy in the 3rd grade.

This GTT curve tends to be flat with a delayed peak value. The boy had a headache at the beginning of the test. He displayed highly imaginative and creative art work at 2 hours, but this diminished to sleepiness and passive reading at 4 and 5 hours, when he was also very hungry.

This kind of observation of different behavior with different levels of blood sugar is common if looked for.

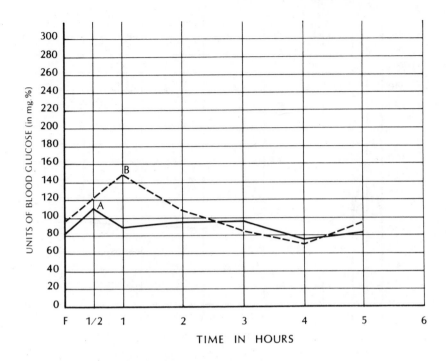

TIME IN HOURS

Figure 84. Glucose tolerance test curves

A 9-year-old boy with asthma and socially withdrawn behavior.

The GTT curve of 1/15/73 (Curve A) shows a tendency to be flat, with blood sugars in the low normal range. The curve of 7/26/73 (Curve B) shows an essentially normal pattern. Between these two dates, the boy was treated with allergic desensitization injections, psychological guidance, and nutritional improvement measures. There may be more than one way to bring about alterations in carbohydrate metabolism.

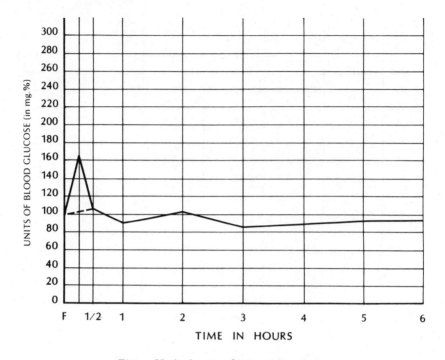

Figure 85. A glucose tolerance test curve

This is the GTT curve of an 11-year-old boy with sleepwalking, nervousness, a school problem, and many fears. He craved sweets and had headaches, hand tremors, and poor memory.

The GTT shows a flat curve throughout except for the large rise in blood sugar at 15 minutes. The increase from a blood sugar of 100 to 164 within 15 minutes and the prompt fall to 106 by ½ hour constitute a severe change in homeostasis.

The usual GTT does not measure a 15-minute specimen, hence this early rise in blood sugar is often not detected. The fact that the early rise occurs indicates that the absorption of sugar is intact. In fact, the absorption of sugar may be exceptionally rapid. My clinical impression indicates that the optimal situation is for peak glucose to be reached at 1 hour. Routine measurement of blood-sugar values every 15 minutes would give us a more complete picture of glucose metabolism.

The broken line between the fasting and the ½ hour points depicts the GTT curve that would have been recorded if the 15-minute value had not been taken.

CHAPTER 13

THE NATURAL TOLERANCE TEST (NTT)[1]

Homeostasis is the steady state. Man's ability to survive, centers around his ability to maintain homeostasis. Blood pressure, body temperature, fluid balance, and many other functions must be maintained and controlled at a relatively steady state if the human body is to remain in top working worder. Blood-sugar levels are no exception.

The state of health is characterized by a particularly steady state. The state of disease is characterized by wide fluctuations in clinical and physiological measures. In between are many states of partial health or partial disease which have varying degrees of fluctuation in clinical and physiological measures.

Dr. Emanuel Cheraskin (M.D.) and his co-workers at the University of Alabama have investigated the departures from body homeostasis and their relationship to disease. Some of their investigations and philosophies are to be found in these books: *Diet and Disease; Predictive Medicine, a Study in Strategy;* and *Psychodietetics: Food as the Key to Emotional Health* (see "References," pages 489 and 490).

The purpose of a diagnostic GTT is to temporarily change the steady state in an individual in regard to blood-glucose levels and

[1] A natural tolerance test (NTT) is a form of glucose tolerance test (GTT) in which food is used instead of a glucose drink.

to measure the ability of the individual to restore homeostatic equilibrium after the glucose is absorbed. In the GTT, a stressful stimulus is used to magnify the physiological responses of the individual.

One can also obtain valuable information about a patient when variations in blood glucose are measured in the patient who is not stressed by the administration of a large glucose load. When close attention is paid to small deviations in blood-sugar values, a tolerance test can be performed by using the patient's usual meal as the stimulus or load. Since the patient eats his usual meal, and since no artificial loading dose is given, I have termed this kind of test a *natural tolerance test* (NTT). Some physicians request the individual to include some carbohydrate food in the test meal. Others request a specific quantity of carbohydrate food.

A test meal used by Dr. Arthur Kaslow (M.D.)[2] consists of 3 oranges, 6 dates, and 1 ripe banana. This provides 103 grams of nonprocessed carbohydrate and thus is equivalent to the standard GTT using 100 grams of glucose.

At this time, the NTT has not been widely used in clinical medicine. When additional study of the NTT is carried out, it may be possible to reliably diagnose diabetes, hypoglycemia, and other carbohydrate disorders by this test. At this time, the NTT does give a great deal of information about the stability of sugar homeostasis in an individual.

When the NTT is carried out, blood sugars are obtained at the same intervals as in the 6-hour GTT. I have also often obtained a specimen 15 minutes after the meal. The contents of the patient's meal (usually breakfast) are recorded, and the patient is observed by the parent, physician, nurse, or laboratory technician for signs and symptoms that may occur during the test. The blood-sugar values are plotted on the same graph used for the GTT. If an afternoon test is performed, the patient's usual lunch serves as the stimulus or load.

Experience with the NTT indicates to me that what we eat profoundly affects the hormonal and chemical balance in our bodies, as well as our outward behavior. One individual may develop an elevated blood sugar when he eats a certain food, but another person may show a lowered blood sugar when he eats the

[2]2428 Castillo Street, Santa Barbara, Calif. 93105.

same food. Much research is needed to define normal, usual, and optimal patterns for the NTT.

It appears quite likely that avoidance of dietary sugar results in improved stability in NTT curves. It may also be that avoidance of other refined carbohydrates, pasteurized and homogenized milk, contaminated animal products, excess quantities of fruit, and most cooked food could improve carbohydrate metabolism in many individuals.

Correlation of specific foods and quantities of foods with blood-sugar patterns needs to be carried out. It is hoped that this book with its preliminary observations will stimulate the development of this research. In the meantime, one must be aware of the profound individual variation that characterizes a person's reaction to food.

When one carefully considers the implications of the natural tolerance tests presented in this book, he may obtain a new perspective about food and its effects on the human body.

Most of us take for granted the "simple" act of eating and presume that everything is proceeding along satisfactorily just because we eat. It appears that most profound changes in our body chemistry are occurring "under our very noses" as we proceed from soup to nuts. Is our blood sugar rising or falling as we eat? How long is stability maintained? What foods disrupt it most? What other changes are occurring as a result of blood-sugar change? These mysteries of eating, gastrointestinal signals, absorption, hormone release, enzymatic digestion, etc., are only now beginning to come to our intimate awareness. There is so much yet to be learned about ourselves!

In Chapter 9, some NTT's were shown to illustrate hyperinsulinism. I am now going to present some other examples of NTT's (see Figures 86 through 109).

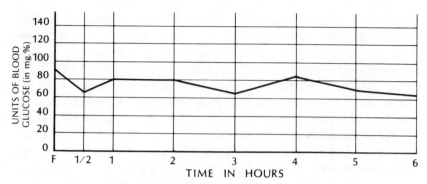

Figure 86. A natural tolerance test curve

A 4½-year-old boy with nightmares, hyperactivity, and markedly negative behavior. Tests for adrenal-cortex function were low for age.

Breakfast consisted of pancakes with butter and syrup, bananas, milk, and 3 ounces of orange juice.

This natural tolerance test (NTT) shows decline of blood sugar, dropping from a fasting level of 90 to 66 at ½ hour. There is another decline at 3 hours. This pattern suggests hyperinsulinism and erratic glucose homeostasis.

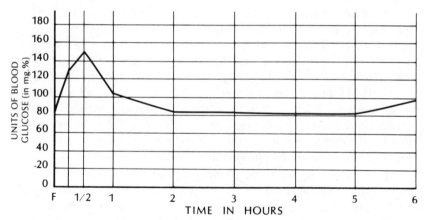

Figure 87. A natural tolerance test curve

A 9-year-old boy with hyperactivity, temper tantrums, daydreaming, and destructive behavior. Breakfast consisted of Sugar Smacks cereal, toast and butter, and chocolate milk.

This natural tolerance test shows a curve that probably is of a diabetic type, considering that no standard glucose load was given. Since the boy does not have sugar in the urine, excessive thirst, excessive hunger, weight loss, and ketosis, he does not have clinical diabetes. Such chemical diabetes often improves toward normal when treated with dietary improvement and vitamin-mineral supplements. Usually, a program of regular daily exercise is also needed, and weight control is of great importance.

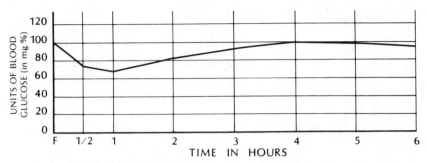

Figure 88. A natural tolerance test curve

An 18-year-old boy with a ravenous appetite but little weight gain.

Breakfast consisted of orange juice, egg, cereal with sugar, milk, toast with jelly, and Danish pastry.

The natural tolerance test shows a rather profound drop from a higher than desirable fasting level of 100 to 74 at ½ hour and 68 at 1 hour. Hyperinsulinism is suggested. This type of curve could result from excessive insulin response or from inadequate glucose absorption from the gut in the presence of "normal" insulin release. Often a hypersensitivity to normal amounts of insulin seems to account for this type of curve. The usual cause of such supersensitivity appears to be the repetitive ingestion of sugar in large amounts and deficiency of nutrients.

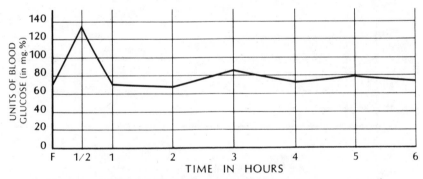

Figure 89. A natural tolerance test curve

An 8-year-old girl with learning disability and low frustration tolerance. Breakfast consisted of 2 pieces of toast, butter with honey, and 4 ounces of chocolate milk.

The NTT shows a rise from 70 to 134 mg.% following this breakfast. The level of 70 is probably lower than desirable for the fasting blood sugar and the rise to 134 may be higher than desirable following a meal. This type of response is characteristic of the child who would probably have a diabetic curve on a standard GTT.

This girl's breakfast as well as the pattern of the NTT, indicate that her diet probably contains an excessive amount of refined carbohydrates on a chronic basis.

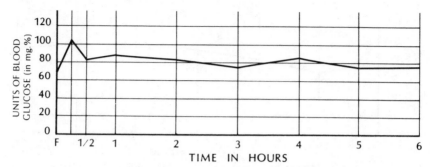

Figure 90. A natural tolerance test curve

A 9-year-old boy with chronic hyperactivity and decreased attention span. A smear of the boy's nasal mucus was "loaded" with eosinophils ("allergy cells"), indicating a likelihood of an allergic basis for his hyperactivity.

Breakfast consisted of blueberry pancakes with syrup and butter, and hot chocolate with whipped cream.

In this NTT, the fasting blood sugar of 68 is probably lower than desirable, and the rise from 68 to 103 within 15 minutes may be excessive.

The majority of patients that I see with learning or adjustment problems have fasting blood sugars which are above 90 or below 80 mg.%. It is common to find the lower values. Persons who adapt well to family, school, and society are thought to be those whose blood-sugar levels are most frequently between 80 to 90 mg.%.

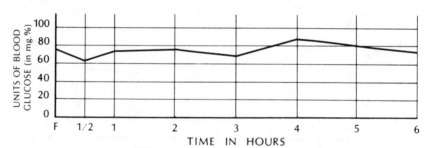

Figure 91. A natural tolerance test curve

This 14-year-old boy was seen because of hay fever, learning disorder, headaches, stomachaches, vomiting, and fatigue. A lower than normal level of adrenocortical secretions was shown by laboratory tests. Breakfast consisted of corn flakes with milk, 2 slices of white toast with margarine, fried potatoes, strawberry jelly, and salt and pepper.

The fasting blood sugar of 74 and the 6-hour sugar of 72 are probably less than optimal values. The blood sugar goes down after eating and does not fully recover until 4 hours after the meal. It is possible that omission of strawberry jelly would have improved these values. It is possible that a breakfast of eggs, meat, or fish, replacing the cornflakes and white bread, may have resulted in an improved curve.

(See Chapter 8, pages 132 - 134, for a detailed case study of this patient.)

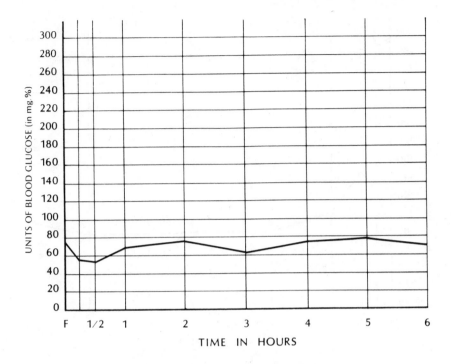

Figure 92. A natural tolerance test curve

NTT of a hyperkinetic 5-year-old boy whose physical appearance suggested poor nutrition. Bad dreams, recurrent colds, intermittent hearing loss, and constipation were additional symptoms.

Breakfast consisted of 6 oz. of orange juice, 1 egg, grits, bacon, 1 piece of white toast, 4 ounces milk, and 2 bites of pancakes with syrup.

Notice the drop of 20 points from 74 to 54 mg.% within 15 minutes after eating! This drop continued to 52 mg.% at ½ hour. Note also the 6-hour value of 70 mg.%. Often the fasting and the 6-hour values will be quite similar, indicating whether the values are tending to be optimal (80 - 90 mg.%), low, or high. In this case, the levels are thought to be somewhat lower than desirable.

If this boy's diet were altered, it is likely that this distinct blood-sugar depression following food could be eliminated. In some patients, however, treatment with hormones, vitamins and minerals, allergy injections, visual therapy, or psychological counseling is required to bring about desired change. All such treatments, however, should only be carried out on specific indication under the direction of a supervising physician. Physicians interested in nutrition and preventive medicine are often members of the International Academy of Preventive Medicine (871 Frostwood Drive, Houston, Texas 77024) or The International College of Applied Nutrition (P.O. Box 386, LaHabra, California 90631).

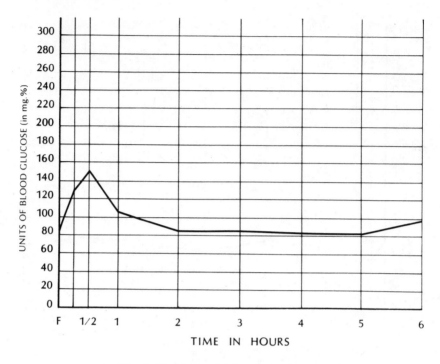

Figure 93. A natural tolerance test curve

A 9-year-old severely hyperactive boy. Thirteen percent of this boy's circulating white blood cells were found to be eosinophils ("allergy cells"). Two years prior to this exam, there were 9% eosinophils.

Breakfast of Sugar Smacks cereal, white toast and butter, and chocolate milk resulted in this NTT. The curve has a diabetic appearance. The reader should remember that *no* glucose load was given to this boy other than the "food" contained in his own breakfast. Some would argue that sugar, processed cereals, white bread, and chocolate milk hardly qualify as food.

This boy had a headache at 1 hour. Symptoms such as this may be often seen an hour or two after eating in those individuals whose blood-sugar values deviate significantly up or down from 80 to 90 mg.%.

An allergy to sugar or chemicals often exists in cases like this. Removal of sugar, artificial colors, and artificial flavors from the diet often greatly improves the hyperactivity and is followed by return of the blood eosinophils to normal levels. In other cases, specific food-allergy testing and treatment is needed.

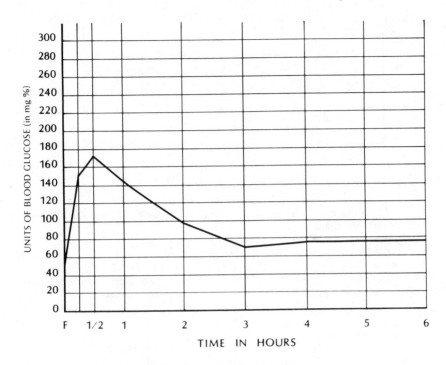

Figure 94. A natural tolerance test curve

An 8-year-old girl with learning disorder.

Lunch was eaten for the NTT and consisted of fried haddock (breaded), green beans, jello, and water. This curve falls into the broad category of the diabetic state (chemical diabetes). The intensity of the blood-sugar rise is shown in the enormous and rapid change from a fasting level of 56 mg.% to a sugar level of 150 mg.% at 15 minutes after lunch!

It would be desirable to retest this patient, withholding the jello in her lunch, to see how the glucose curve would change. As a result of the experience I have had with the NTT, I am growing increasingly respectful of the role of food in affecting body chemistry. We are often completely unaware of the magnitude of the biochemical changes taking place within us after eating.

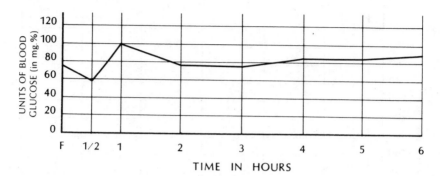

Figure 95. A natural tolerance test curve

A 7-year-old boy with reading, writing, and coordination difficulties; also asthma. Father alcoholic.

Lunch consisted of milk, 2 slices white bread, cheese, beef patty, pea soup, crackers, jello, coffee, lettuce, tomato, and margarine.

This NTT shows an early, sharp decline in blood sugar after lunch, with a subsequent erratic course at 1 and 2 hours. Notice that the glucose curve is more stable after the effect of food has worn off — that is, after 2 hours.

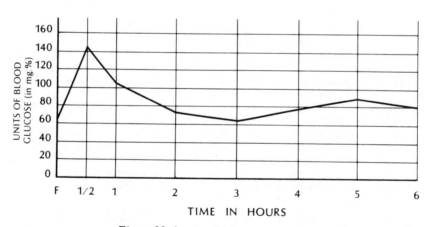

Figure 96. A natural tolerance test curve

A 15-year-old 9th grader with chronic behavior problem, rebellious attitude, and brushes with the law.

Breakfast consisted of 2 glazed doughnuts, 1 glass of milk, 1 slice of ham, and 2 pieces of bacon.

This NTT curve shows a large elevation of the blood sugar at ½ hour after breakfast. Such a rise in blood glucose following this meal probably indicates a state of chemical diabetes. The child was sleepy at 3 hours, coinciding with the low point of the curve.

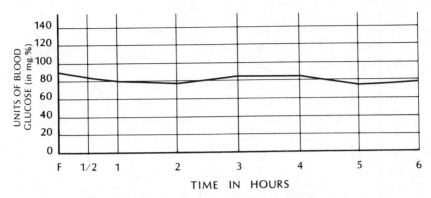

Figure 97. A natural tolerance test curve

An 18-year-old boy with bad temper, chronic behavior problem, severe impatience, and "short fuse."

Breakfast consisted of 2 eggs, toast, butter, jelly, and milk.

The relatively mild descent of blood-sugar values in the first 2 hours following a meal suggests this boy's symptoms may not be primarily related to chemical imbalance in glucose homeostasis. Nevertheless, the fact that the blood sugar does not rise after a meal may be suggestive of a maladaptive response. A meal with a greater amount of sugar may have produced a greater lowering of blood sugar. At times, qualitative symptomatic effects occur which are not reflected in the glucose curve.

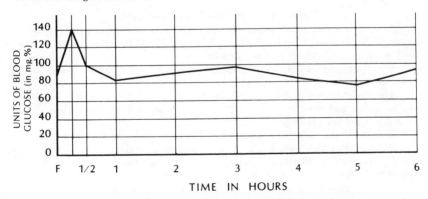

Figure 98. A natural tolerance test curve

An 8-year-old 2nd grade boy with behavior problems, learning disability, hyperactivity, and chronic fatigue.

Breakfast consisted of egg, sausage, toast with jelly, and chocolate milk.

Note the quick rise in blood sugar at 15 minutes following breakfast. This indicates very rapid absorption of sugar. A repeat test without jelly and chocolate in the meal could be done to evaluate the effect of diet on blood sugar. If the 15-minute specimen had not been obtained, the curve would then be semiflat and lack of absorption might be suspected.

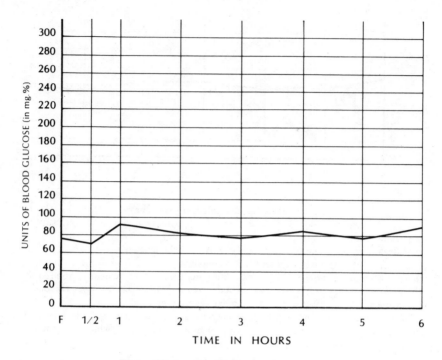

Figure 99. A natural tolerance test curve

A 6-year-old girl with inattentive behavior, who was doing poorly in first grade. This girl lived in a family which allowed the child to eat very little "junk food" such as candy, sweetened soft drinks, etc.

Breakfast consisted of egg, fried potatoes, white toast with butter and apple jelly, and ½ cup hot chocolate.

The early decline in the sugar values is definite but not especially large. If her breakfast had been without hot chocolate and apple jelly, the decline at ½ hour may have been eliminated.

This NTT has less deviation than many of the other curves seen in maladapting children. Nevertheless, a seemingly small deviation may have major significance for the individual. Notice that the fasting blood sugar is probably somewhat lower than desirable.

Analysis of the hair minerals in this girl revealed a high level of mercury and a depletion of copper. Treatment of these mineral imbalances was followed by improvement in her school work. A repeat NTT was not done.

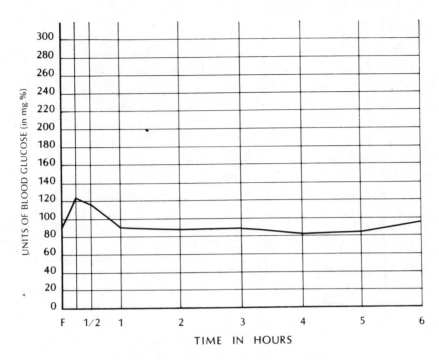

Figure 100. A natural tolerance test curve

An 11-year-old girl with a mild-moderate degree of mental slowness. Severe nasal allergy. Abnormal EEG which reverted to normal after allergy desensitizing injections.

Lunch consisted of peanut butter and jelly sandwich, milk, apple, and cookie.

The relatively mild elevation of glucose levels in this lunch suggests that this girl has moderately good glucose tolerance (utilization) with little disturbance of glucose homeostasis caused by her lunch. The elevation of blood sugar at 15 minutes and ½ hour may well have been less if her lunch had contained less sugar-containing foods. In some cases, best homeostasis is achieved by also reducing the sugar in natural foods such as fruits and juices.

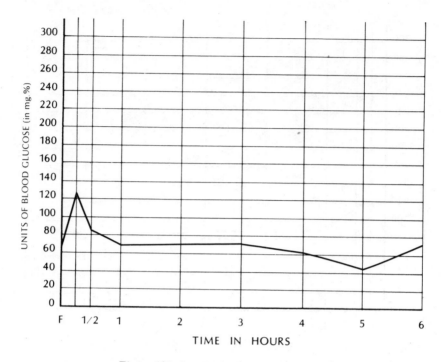

Figure 101. A natural tolerance test curve

A 5-year-old girl with motor incoordination and hyperactivity.

Breakfast consisted of cranberry juice, milk, and a small bowl of dry cereal (Total).

Notice the jump from a fasting level of 66 mg.% to a 15-minute blood sugar of 124 mg.%! This considerable rise, indicative of a relatively unstable carbohydrate mechanism, is followed by a distinct hypoglycemic dip at 4 and 5 hours. The rapidity of the initial rise, as well as the absolute increase of 58 mg.% within a 15-minute period, are clues that hypoglycemia might be expected at some time in the testing period.

Sugar in juice, lactose in milk, preservative chemicals in dry cereal, starch in cereal, and the effect of protein all need to be studied to determine the effect on clinical behavior and glucose levels.

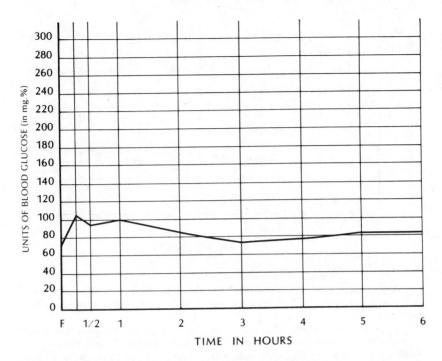

Figure 102. A natural tolerance test curve

An 8-year-old obese girl, "sick all the time" with abdominal pains, nausea, leg aches, irritability, pallor, runny nose, short attention span, and nosebleeds. A blood count showed a great excess of "allergy cells" (eosinophils) in the blood, and a smear of the nasal mucous likewise showed large numbers of "allergy cells."

The girl's blood-sugar response to a breakfast consisting of scrambled eggs, toast with butter, orange juice, and milk is shown.

Headache, irritability, and runny nose were prominent at 3 hours, when the blood sugar was at its lowest point.

Note the fasting sugar at 72 mg.%, a level considered to be probably less than optimal. The blood sugar rose from 72 to 105 within 15 minutes and remained above the fasting level for more than 2 hours. This is a diabetic trait and suggests that the child may have a sluggish insulin response or insulin resistance.

Whenever allergic children experience a lowering of blood sugar, their allergic symptoms may appear or worsen. Frequent feedings of food which is free of sugar and other refined carbohydrate tends to diminish the allergic symptoms. The relationship of food allergy to disturbances in glucose homeostasis needs to be clarified.

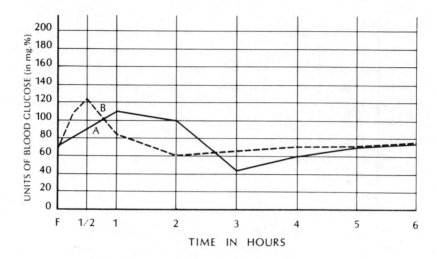

Figure 103. Tolerance test curves

A 9-year-old girl with educational and developmental retardation, auditory and visual perceptual disorders, fatigue, irritability, short attention span, and poor memory.

Glucose tolerance test at age 7 years (Curve A) displays a probably lower than desirable fasting blood sugar at 70 mg.%, and a profound dip in blood sugar to 43 mg.% at 3 hours. The GTT ends up at levels of 70 and 74 mg.%, which are also probably less than desirable.

This girl was treated by megavitamin therapy and by dietary restriction of sugar and other refined carbohydrates, with improvement in her learning, perceptual difficulties, and behavior. Nevertheless, she sought additional help at 9 years of age because of weakness, fatigue, and overactivity.

The results of a natural tolerance test at 9 years of age are shown by Curve B.

For this NTT she ate breakfast consisting of scrambled eggs, ham, bacon, toast, grits, butter, and orange juice. The blood sugar rose from 70 mg.% to 122 mg.% at ½ hour, with the sharpest rise in the first 15 minutes.

Notice that the fasting blood sugar is still probably lower than desirable at 70 mg.%. A slight hypoglycemic dip occurs at 2 and 3 hours. The girl experienced weakness at ½ hour, when the blood sugar was at its highest peak. When the sugar level was at its lowest peak at 2 hours, she was very sleepy. From 3 to 6 hours she became markedly overactive.

This case shows pathologic hypoglycemia on the initial GTT as well as symptomatic functional hypoglycemia 2 years later in response to the ingestion of food. The rapid rise of the blood sugar and the height reached after eating breakfast may indicate a prediabetic tendency at 9 years of age.

The presenting complaints of this child were reproduced during the natural tolerance test and coincided with periods of high and low blood sugar. It is valuable to use a symptom checklist when a GTT or NTT is performed.

Many children like this girl require treatment of an existing allergic condition in order to become symptom-free.

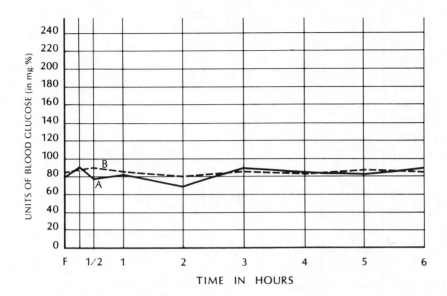

Figure 104. Natural tolerance test curves

Natural tolerance test of a 10-year-old girl with irritability, hyperactivity, and a behavior problem. Curve A is the initial one.

Breakfast consisted of 8 ounces of orange juice, 6 ounces Raisin Bran cereal, and 6 ounces of skimmed milk.

The blood-glucose values fluctuate over a range of 20 mg.% (low of 72 and high of 92 mg.%) in the 3 hours after this breakfast. Once the effect of food has dissipated, the blood sugars remain remarkably level.

This NTT is probably a more desirable curve than many others which rise higher or drop lower. Nevertheless, even fewer fluctuations and less range of blood-sugar change may be desirable.

Curve B is the NTT obtained 3 months after the first one. The child had been treated by restriction of sweets and other refined carbohydrates during this 3-month period and given vitamin and mineral supplements appropriate for her.

Her behavior problem had improved greatly, she was less hyperactive, and her irritability had disappeared.

Her breakfast for the second test consisted of hamburger, toast with butter, and 6 ounces of tomato juice.

The NTT curve now shows the stability that was at this time characteristic of her overall behavior. It appears that, at least in this girl, what you eat determines to a large degree what you do and how you do it.

At times, laboratory tests fail to indicate the presence of a process which is suspected from the clinical condition of the patient. It is important for physicians to be guided by laboratory tests; but in the long run, to treat the patient rather than the laboratory test. Appreciation must be given to small changes in laboratory tests which may be significant but which may fall within the wide ranges of what is termed normal.

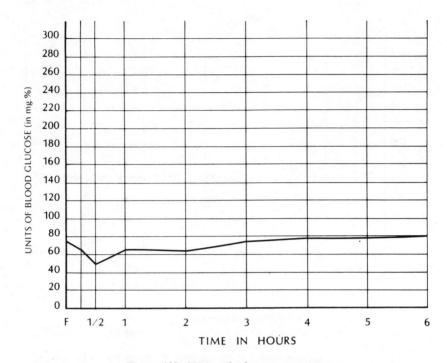

UNITS OF BLOOD GLUCOSE (in mg.%)

TIME IN HOURS

Figure 105. A natural tolerance test curve

A 7-year-old boy with headaches, hyperactivity, sinus trouble, fatigue, short attention span, excessive talking, and dizziness.

Breakfast consisted of 1 slice of white toast with butter and jelly, ½ cup of boxed cereal with marshmallows, and ½ pint of milk.

Notice the hypoglycemic pattern with declining blood sugar at 15 minutes, peak low of 50 mg.% at ½ hour, and relatively low values until the 3rd hour, when the effect of food had disappeared.

The hair mineral pattern for this child is shown in Figure 106.

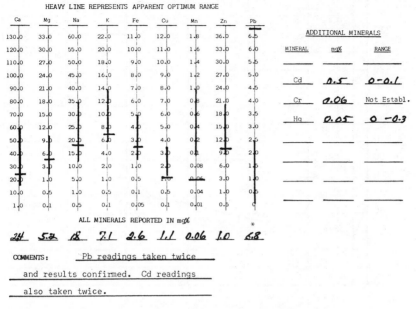

Figure 106. Hair mineral analysis of boy in Figure 105

Ca	= calcium	K	= potassium	Mn	= manganese	Cd	= cadmium
Mg	= magnesium	Fe	= iron	Zn	= zinc	Cr	= chromium
Na	= sodium	Cu	= copper	Pb	= lead	Hg	= mercury

This boy has a significant elevation of lead and cadmium in the trace mineral analysis of hair. Zinc, manganese, chromium, and copper are lower than desirable, and additional calcium may be needed.

When nutrient minerals are low, toxic elements tend to accumulate. Lead and cadmium interfere with enzyme systems in the body. The situation is intensified by nutrient deficiency. Altered body physiology comes about because of toxic excess and/or nutrient depletion.

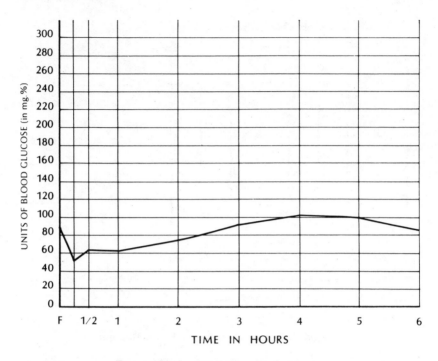

Figure 107. A natural tolerance test curve

A 9-year-old severely hyperkinetic child whose behavior is markedly improved when he takes Ritalin medication.

This NTT was obtained when the boy was taking Ritalin. His breakfast consisted of 2 slices of toast, 3 pieces of bacon, fried potatoes, 1 glass of milk, and a small glass of orange juice.

The rather severe depression of blood sugar that occurs 15 minutes after breakfast is followed by a longer duration blood-sugar depression which does not return to the fasting level until the 3rd hour.

It is highly likely that this boy's hyperactivity will be improved by dietary alterations. Commonly it is possible to do away with the use of Ritalin or to decrease the dose, as a result of dietary improvement. The effect of Ritalin on glucose tolerance and natural tolerance needs to be studied. In my experience, the patients who take Ritalin have rather pronounced hypoglycemia.

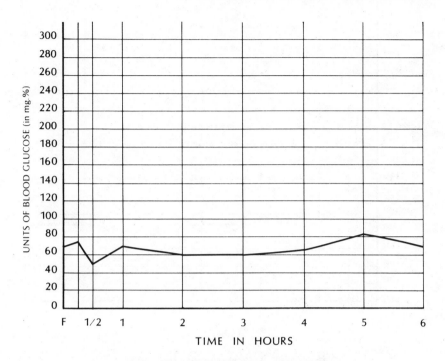

Figure 108. A natural tolerance test curve

A 13-year-old boy with hyperactivity and learning disability. His diet was heavy in sugary foods.

Breakfast consisted of 3 pancakes with butter and maple syrup and ½ cup of coffee.

This NTT shows a fasting blood sugar of 68, well below the probably desirable optimal level of about 85 mg.%. After an initial brief rise at 15 minutes, the sugar drops to 50 at ½ hour! The level then remains fairly low at 60 mg.% at 2 and 3 hours.

This boy's NTT shows he has a distinct hypoglycemic problem — at least in response to pancakes and maple syrup. Perhaps he is starting to drink coffee because of its tendency to temporarily combat low blood sugar. Coffee contains caffeine, which tends to temporarily elevate the blood sugar and stimulate the nervous system. The temporary effect of feeling better is often followed by a letdown in energy and in blood-sugar level. In my experience, those individuals who consume a large amount of coffee often have rather pronounced hypoglycemia.

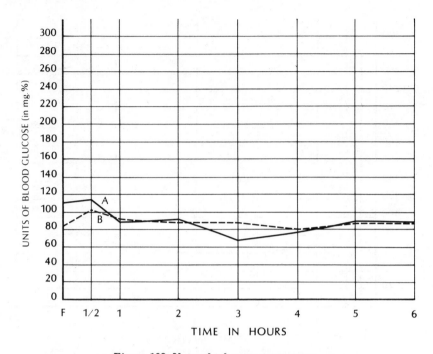

Figure 109. Natural tolerance test curves

Natural tolerance tests done 1 year apart on a 16-year-old poorly nourished boy with learning disability. The first curve, Curve A (solid line), was obtained before treatment. The second curve, Curve B (dotted line), was obtained after one year of dietary change and individualized vitamin-mineral supplements.

Breakfast for the first curve consisted of Frosted Flakes cereal, milk, bacon, and toast. For the second curve, he ate French toast with syrup and butter, bacon, and milk.

The first curve is higher than optimal at the beginning, declines after breakfast, and progressively falls to a low point at the third hour. A change from 110 to 68 occurs within 3 hours, and the third hour low-point value is 42 mg.% lower than the fasting value. The second curve demonstrates a lower fasting value, a rise after breakfast, and then a very slow, gradual descent to a low at the fourth hour. The fourth-hour low is only 2 mg.% below the fasting value.

The quantitative changes noted in NTT's are small compared with those seen in GTT's. A qualitative change, however, such as seen in these examples, is frequently noted. Although the changes of improvement are quantitatively small, these changes are usually associated with reduction of clinical symptoms and improved health.

CHAPTER 14

LOW BLOOD SUGAR (HYPOGLYCEMIA)

WHAT IS IT?

First of all, the reader must realize that low blood sugar means different things to different persons. The lay person, the physician in clinical practice, and the academic research investigator may each have his own view of what low blood sugar is. These views, at times, can be widely varying. They often engender much emotion, as an individual with one belief may castigate or deride another for holding a different belief.

Low blood sugar is an insufficient supply of glucose in the blood and tissues to meet the energy needs of an individual at any particular moment. It is best thought of as one aspect of a disturbed metabolism. Low blood sugar has multiple origins.

Nearly everyone would agree that low blood sugar means a level of sugar in the blood that is below some accepted value. Just what that normal value is, is a matter of some controversy. Discussion regarding this matter contributes to the emotional heat that characterizes the issue of hypoglycemia amongst the medical scientist who functions in an "ivory tower," the clinical physician, and the man on the street.

If one goes by the accepted norms provided by clinical laboratories, low blood sugar would be present when blood sugar is below 60 or 70 mg.% (whole-blood, true-glucose method). Of course, one must remember that these norms have been derived from everyday "men on the street" and, as such, may not represent optimum or desirable values.

To most academically oriented physicians, low blood sugar is said to be present when the value for the sugar (glucose) in the blood is below 40 or 45 mg.%. Dr. Rachmiel Levine (M.D.) wrote in the *Journal of the American Medical Association* that blood-sugar levels of about 40 mg.% or below begin to cause symptoms.[1] (Low blood sugar has a different definition in the newborn [see pages 304 and 337].) Hardly anyone would question the existence of low blood sugar when the value is below 40 - 45 mg.%. Dr. Levine points out, however, that gradual lowering of the blood sugar may produce no symptoms even with a blood sugar as low as 5 - 10 mg.%. I believe this *can* be true, but a person is not apt to be efficient and productive at such severely low blood-sugar levels. In fact, one would be more like a cold slab of stone than a warm, vibrant person. The point is well made, however, that the rapidity of blood-sugar fall is a most important variable as to whether a person has obvious symptoms associated with low blood sugar.

For descriptive purposes, I will term blood-sugar values below 45 mg.% as *absolute low blood sugar*. It is set apart from the much more common state, *functional low blood sugar*, in which blood-sugar levels are apt to be between 45 and 78 mg.%.

Many medical experts in the field of carbohydrate metabolism strongly endorse and promote the concept of absolute low blood sugar. They believe that most other cases of so-called low blood sugar are actually psychological problems, nervous disorders, misdiagnoses, or figments of the imagination. Many low-blood-sugar experts in academic institutions believe that patients whose blood-sugar levels are in the 60's or 70's probably represent part of a normal distribution curve.

Some physicians who only recognize the existence of absolute low blood sugar will, however, hedge a bit and admit that low blood sugar may exist at levels of 50 mg.% or so.

Lay individuals are encountered at times who believe that

---

[1]"Hypoglycemia" by Rachmiel Levine, M.D. Published in *Journal of the American Medical Association*, Vol. 230, No. 3, October 21, 1974.

low blood sugar refers to a set of symptoms such as weakness, fatigue, lack of drive, dizziness, etc. Although each of these symptoms may be characteristic of low blood sugar, the diagnosis of low blood sugar should not be applied to them unless low-blood-sugar values have been demonstrated in association with the symptoms. In addition, symptoms of low blood sugar should be relieved by sugar if the diagnosis of hypoglycemia is to be established. The sugar may be given to a patient by mouth or in the vein as a therapeutic trial.

Some symptoms and signs of low blood sugar are:

weakness	tremors
dizziness	anxiety
yawning	personality change
headache	hunger
sleepiness	abdominal discomfort
mental confusion	pallor
sweating	lethargy
cool skin	coma
palpitations	convulsions

A particular person has his own individual set of symptoms that characterize his response to low blood sugar. One individual may, for example, yawn and grow sleepy; another may become hypercitable; another, hungry and pale; yet another may have a convulsion. In general, the lower the blood sugar, the lower is the blood pressure and the lower the body temperature.

Night tremors and sweats, bad dreams, restless sleep, abdominal pain, and behavior problems may be associated with low blood sugar. A dull morning headache, low in the back of the head, is especially characteristic of hypoglycemia.

A not infrequent cause of confusion in diagnosis occurs when a patient has symptoms of hypoglycemia relieved by eating, yet his blood-sugar tests fail to reveal lowered sugar levels. It is entirely possible for some individuals to have low levels in between the times when blood tests are drawn. Such persons would have a very unstable, erratic, saw-tooth GTT curve if additional blood-sugar values were determined.

Another explanation is as follows: Under some circumstances, the level of glucose in the brain may be lower than it is in the blood.[2] An individual may thus have neuroglycopenia (inadequate glucose for brain function) even though he may not have low blood sugar by the criteria of some people. This phenomenon may be operative far more often than has been realized. Undue reliance on blood values at the expense of clinical treatment may not assist some persons to better function and optimal health.

Failure to monitor the rate of rise and fall of blood sugar and failure to correlate symptoms and performance with blood-sugar levels will also result in failure to consider blood sugar an important factor in a patient's clinical problem.

The truth is that there are no hard and fast criteria between normalcy and functional hypoglycemia. This is what sometimes maddens those who must live and function with neat, tidy, diagnostic categories. There is a continuum between clearly abnormal and clearly optimal blood-sugar levels. As I have and will repeatedly indicate, the blood-sugar level at any moment is important not only for how low or high it is but also for where it has been and where it is going and what else is going on at the same time.

Hypoglycemia may not be like pregnancy — where you either have it or you don't. It is often more like weight. Everybody has some, but it becomes a problem when there's too much of it.

The entire population of the developed nations has been sensitized to sugar over the past 200 years. Individuals manifest this sugar-sensitive state by varying degrees of carbohydrate dysmetabolism, the commonest form of which is functional hypoglycemia.

A normal blood-sugar level or normal GTT does not necessarily exclude functional hypoglycemia. Normal levels may indeed, however, fail to substantiate the diagnosis. On another day with a different diet, or under different circumstances, lower blood-sugar levels might appear. Also, a "normal" blood-sugar level may not be an optimal level for the individual in question.

The responsible physician will test and possibly retest before excluding low blood sugar. He most assuredly should obtain

[2]"Blood Glucose: How Reliable an Indicator of Brain Glucose?" by Jean Holowach Thurston. Published in *Hospital Practice*, September 1976, pages 123 - 130.

a record of symptoms correlated with the changes in blood-sugar levels.

The responsible physician will insist, however, on some laboratory evidence before he concludes that low blood sugar is present. As noted, the level of blood sugar at which the diagnosis is made is somewhat controversial, but the presence of symptoms alone is usually not sufficient to establish the diagnosis.[3] It is true that primary psychological difficulties, allergies, mineral imbalance, and other disorders may present a clinical picture identical to that commonly encountered in the patient with functional hypoglycemia.

A physician who rigidly follows "the academic line" recognizes low blood sugar only when values are below 45 mg.% (absolute low blood sugar).

Physicians who are somewhat less rigid in their interpretation of blood-sugar values recognize the existence of low blood sugar when blood-sugar levels are below 60 or 70 mg.%.

Other physicians with a more functional viewpoint are usually those physicians in clinical practice. They believe that significant low blood-sugar states may exist when blood-sugar values fall in the 60's or 70's. This kind of physician is apt to also have a keen appreciation for the relative viewpoint. He understands that low blood sugar may exist at almost any blood-sugar level — *depending on the preceding sugar level and the rate of fall.* For example, if a patient's usual fasting blood sugar is 115 mg.%, a level of 85 mg.% could represent low blood sugar for that person at a particular moment.

The most functional view promotes the belief that relative hypoglycemia is present if a person's blood sugar drops more than 25 mg.% in 1 hour and is productive of symptoms. Dr. Harvey Ross (M.D.) indicates that a fall of more than 50 mg.% in 1 hour accompanied by symptoms of low blood sugar or a blood-sugar value more than 20 mg.% below the fasting level indicates hypoglycemia.[4] Dr. Ross states that one person may get

[3]Hypoglycemia is only one of several sugar-related disorders. Allergy or maladaptive reaction to sugar (or other foods) may give rise to numerous symptoms without necessarily altering blood-sugar levels. Dysinsulinism and chemical diabetes are more common than realized. Diabetes is rapidly increasing in the American population.

[4]"Hypoglycemia" by Harvey M. Ross, M.D. Published in *Journal of Orthomolecular Psychiatry,* Vol. 3, No. 4, Fourth Quarter of 1974.

along very well with a blood sugar of 70 mg.%, whereas another person may need 90 mg.% to have a normally functioning brain.

The functional or relativistic physician recognizes that some individuals have symptoms of low blood sugar at levels which do not produce symptoms in other individuals. The presence of symptoms may depend on personality factors, environmental setting, etc. As is the case with all good medical care, the physician should look at the patient and his environment as well as the laboratory test.

Everything I know about human individuality tells me that there are persons who just do not fit into established diagnostic categories. A glimpse at medical history shows us that poor health conditions are described first in their extreme form and only later are milder variants discovered. The medical student studies the textbook picture of a health problem, but rarely encounters it in practice. So it is, I believe, with low blood sugar. Mild variants of this state probably occur which may not exactly fit presently established diagnostic criteria.

The physician who has a functional viewpoint and thinks in relative terms believes that subnormal blood-sugar values are usually the result of a rebound effect from eating sugar. He usually recognizes that the low blood sugar is only one part of a nutritional-endocrine-nervous complex. He is apt to recognize the part played by the endocrine hormones in regulating glucose homeostasis. He commonly believes that hypoglycemia which follows glucose intake is related to ineffective or insufficient adrenal-corticoid-hormone opposition to insulin. The functionally oriented physician, more than anyone else, is aware that low blood sugar is a lot more than a laboratory test. He and his patients, however, should be reminded that low blood sugar is a sign, a valuable sign, of some other condition. Most commonly, functional low blood sugar is a sign that the individual's total lifestyle needs urgent attention in regard to diet, exercise, mental viewpoints, and perhaps nutritional supplementation.

It is indeed true that emotional disorders are associated with symptoms of weakness, fatigue, dizziness, irritability, depression, disinterest in sex, etc. Often overlooked, however, is the fact that emotional disorders, themselves, may be due to nutritional depletion, nutritional imbalance, cerebral allergy, or yes, at times, functional hypoglycemia. Deficiencies of magnesium, iron, zinc, vitamin B$_6$, pantothenic acid, niacin, and other nutrients may be responsible for mental syndromes which appear to be the primary disorder. Hypoglycemia is often associated with nutritional deficiencies, and allergic stress may be a contributing factor in the production of symptoms.

It has become apparent that protein deficit may produce the identical symptoms that characterize functional states of low blood sugar. An individual with hypoglycemia may have, as his primary problem, insufficient protein intake or faulty protein digestion.

A valuable procedure for diagnosing hypoglycemia is the measurement of adrenocortical hormone (hydrocortisone) during the GTT. A blood glucose drop that elicits an increase in blood hydrocortisone indicates hypoglycemia.

In thinking about blood sugar, we must entertain these questions: Is 60 mg.% adequate and optimal for the learning-disabled child trying to cope with a subject he doesn't understand? Is a level of 60 mg.% optimal for the allergic child encountering a load of offending pollen? What is the *best* blood sugar for the overworked, tired executive? The patient with an unresolved visual problem? The anemic housewife? The mother of too many children, too soon? The weary airline pilot landing a superjet?

Finally, we must always keep this question in mind: Of what larger metabolic problem may hypoglycemia be a part?

The answers to these questions may not be clearcut. Nevertheless, an organized rational plan of investigation (differential diagnosis) should be carried out. What we do know suggests that functional low blood sugar is associated with nutritional and nervous disorders and that avoidance of sugar is usually a helpful measure in treatment.

The importance of an adequate blood-sugar level is underscored by the treatment of intestinal parasites in people by use of a drug known as mebendazole (Vermox: Ortho Pharmaceuticals). Mebendazole irreversibly blocks glucose uptake in nematode worms. This leads to a depletion of glycogen in the parasite, resulting in hypoglycemia and to insufficient energy for worm survival.

My clinical experience leads me to agree with Dr. Alan Nittler (M.D.), who has presented the following five criteria, any of which indicates the presence of hypoglycemia.[5] These criteria relate to the standard 5- or 6-hour GTT and are:

1. The blood glucose must rise to the half-hour and on up to the one-hour level. In other words, there must be at least one hour of increased energy because of the glucose intake.
2. The difference between the fasting and the lowest sugar levels must not exceed 20 units.
3. There must be no levels lower than the normal low level established for the test used. If the test used states 70 - 110 as the normal range, then there should be no levels below 70.
4. The drop from the high point to the low point should be about 50 units.
5. The one-hour level must be at least 50% greater than the fasting level.

These criteria constitute an attempt to separate those patients with functional hypoglycemia from those with more desirable blood-sugar patterns on the GTT.

HISTORY

The introduction of insulin for the treatment of diabetes in the 1920's first brought to the physician's attention the existence of low blood sugar. Dr. Seale Harris (M.D.), in a classic paper in the *Journal of the American Medical Association*, described low blood sugar in 1924.[6] Thus, low blood sugar as a recognized disorder is only 57 years old in 1981.

In 1927, the first recognized case of an insulin-secreting tumor of the pancreas was reported.[7] As the years have gone on,

[5]"Hypoglycemia and the New Breed of Patient" by Alan Nittler, M.D. Published in the *Journal of the International Academy of Metabology*, Vol. V, No. 1, March 1976, pages 40 - 45.

[6]"Hyperinsulinism and Dysinsulinism" by Seale Harris, M.D. Published in the *Journal of the American Medical Association*, Vol. 83, September 6, 1924, page 729.

[7]"Carcinoma of the Islands of Pancreas: Hyperinsulinism and Hypoglycemia" by R. M. Wilder, Ph.D., M.D., F. N. Allan, B.Sc., M.D., M. H. Power, and H. E. Robertson, Ph.D. Published in the *Journal of the American Medical Association*, Vol. 89, No. 5, July 30, 1927, pages 348 - 355.

it has become evident that the commonest form of low blood sugar is that described as benign, functional, reactive, nervous, or sugar-induced. In 1936, Dr. J. W. Conn (M.D.) showed that a high protein, low carbohydrate diet was an effective treatment for this form of low blood sugar.[8] This diet has remained the cornerstone of treatment in low blood sugar to the present. Currently, however, the disadvantages of chronic, high-protein diets are being examined, and the benefits of low-protein intake are being evaluated.[9]

In recent years, a veritable deluge of books and articles has appeared dealing with low blood sugar (hypoglycemia). The functional or relative concept of low blood sugar has been elaborated in books by Drs. E. M. Abrahamson (M.D.), Clement G. Martin (M.D.), Peter J. Steincrohn (M.D.), J. Frank Hurdle (M.D.), and a number of others. The writings and lectures of Dr. Carlton Fredericks (Ph.D.) on this subject are extensive and were among the earliest to appear (see "References," pages 489 and 490).

These individuals have pointed out that symptoms of weakness, confusion, sleepiness, lack of energy, etc., may be associated with blood-sugar values which do not necessarily fall into the absolute range of low blood sugar. Many authors emphasize the rapidity of the drop as an important factor in producing symptoms.

Much of the academic medical establishment believes that the diagnosis of hypoglycemia has been erroneously extended to include persons with symptoms which were termed psychosomatic symptoms 10 to 20 years ago.

Any discussion of functional hypoglycemia is not complete without mention of Dr. Stephen Gyland (M.D.). Because his story is so instructive, it will be told here. This information as well as complete access to Dr. Gyland's writings have been supplied by his widow, Mrs. Ruth Gyland, and the material is used with her consent.

[8]"Advantage of High Protein Diet in Treatment of Spontaneous Hypoglycemia, Preliminary Reports" by J. W. Conn, M.D. Published in the *Journal of Clinical Investigation*, Vol. 15, 1936, pages 673 - 678.

[9]*Live Longer Now: The First One Hundred Years of Your Life* by Jon N. Leonard, J. L. Hofer, and N. Pritikin. Published in 1974 by Grossett & Dunlap, New York, N.Y. Also, *Hypoglycemia: A Better Approach* by Paavo Airola, Ph.D. Published by Health Plus Publishers, P. O. Box 22001, Phoenix, Ariz. 85028.

Dr. Stephen Gyland was a general practitioner of medicine in Tampa, Florida (see Figure 110). Although only some 20 miles away across Tampa Bay, I, growing up in St. Petersburg, knew nothing about him or his struggle at the time. In recent years, I have been privileged to meet and know his widow, daughter, son-in-law, and their family.

Figure 110. Stephen Gyland, M.D.
Even in a still picture, the warmth and caring of this man come through. It would not be difficult for such a man to help a fellow human being, and the use of a nutritional mode of therapy would intensify the therapeutic effect.

Dr. Gyland suffered for many years with functional hypoglycemia. In the early 1950's, his illness worsened; and he became a tottering, confused, alcohol-consuming shell of his former self. He was sustained during this time by the love and care of family and friends, who had known him as the generous, compassionate physician he had always been.

In his worst days, Dr. Gyland had the following problems: staggering, mental confusion, lapses of memory, trembling, severe rheumatoid arthritis, depression, racing heart, blurred and double vision, blackout spells (relieved by food), abnormal brain-wave tracing, and binges of alcohol consumption. For some 20 years before this, he had had severe tension headaches and recurrent abdominal distress.

At the age of 59 in 1953, he looked 90 years old, according to his wife. His blood cholesterol was 476; and he developed a foot drop, requiring a brace.

Dr. Gyland and his wife did a great deal of "doctor shopping," trying to find a medical explanation for the health problems that threatened his existence as well as his medical practice and family stability. He was told at various times that he had a brain tumor, cerebral arteriosclerosis, and diabetes.

Three-hour glucose tolerance tests had shown that his blood sugars were very high. The diagnosis of diabetes was repeatedly made, and he was treated with insulin for a number of years with no improvement. It became evident that food, not insulin, was the factor that brought him around when he was unconscious or otherwise especially sick.

Dr. Gyland's son, in medical school, hinted at the possibility of low blood sugar a number of times; but always, diabetes was considered to be his problem.

On one of his visits to a medical center, a 6-hour GTT was performed. It showed elevated blood-sugar values early in the course of the test with lowered values later on in the test. Dr. Gyland, himself, now plunged into the medical literature to find out all that he could about low blood sugar.

Insulin was discontinued, and he followed the hypoglycemia diet recommended by Dr. Sidney Portis (M.D.). This diet removed sweets and alcohol but permitted coffee. Accordingly, Dr. Gyland continued to drink 6 to 8 cups of coffee daily. Another 6-hour GTT revealed the same blood-sugar pattern. This time, the lowest blood-sugar value was recorded at 29 mgm.%.

It was at this point that Mrs. Gyland began her search of the medical literature for help. She ran across some paragraphs in a

1950 Psychiatry volume.[10] A University of Pennsylvania psychiatrist had written a tribute to Dr. Seale Harris (M.D.) and also quoted from Dr. E. M. Abrahamson (M.D.). In this material, Mrs. Gyland learned that Dr. Harris had found that caffeine caused the release of stored sugar from the liver. In a patient with an overactive pancreas (hyperinsulinism), this was the equivalent of eating sugar by mouth.

Within one month of stopping coffee, Dr. Gyland was well. Continuing on the basic low-blood-sugar diet and without coffee, his headaches, arthritis, mental confusion, blackouts, racing heart, and foot drop were all things of the past. Within 3 months he had resumed his practice. Six months later, while on a high-fat diet, Dr. Gyland's cholesterol had dropped from 476 to 250.

This man of good cheer, doing for others, reclaimed a place among the living. In addition, Dr. Gyland's daughter was greatly helped by a similar dietary approach.

In 1953, Dr. Gyland and his wife met Dr. Abrahamson in New York; and a few years later, Dr. Gyland paid a visit to the aging Dr. Seale Harris in Birmingham. Dr. Gyland also wrote the psychiatrist who had been quoted in the encyclopedia, but the letter came back marked "deceased." This man's writing had been the signpost recognized by Mrs. Gyland as the road to health. The printed page bearing his words had been in Dr. Gyland's office for many years. It was not, however, until a 6-hour GTT was made and correctly interpreted that the diagnosis of low blood sugar was made. When this diagnosis became available, then the correct treatment was able to be found in the scanty literature that dealt with this problem.

In 1953, Dr. Gyland wrote in the "Queries and Minor Notes" section of the *Journal of the American Medical Association* (Vol. 152, No. 12, July 18, 1953). The writing was entitled, "Possible Neurogenic Hypoglycemia." It was written about hyperinsulinism and the importance of a diet test (high protein, high fat, low carbohydrate, no sweets, no caffeine, and frequent feedings). Dr. Gyland referred to the original articles of Dr. Seale Harris (*Jour-*

[10]"Psychosomatic Interrelationships: Treatment in Functional Disorders" by L. H. Twyeffort in *The Cyclopedia of Medical Surgery & Specialties*, page 407: "Since caffeine stimulates glycogenolysis, coffee and such drinks were prohibited. Sudden rise in blood sugar has the same stimulating effect on the insulin apparatus as would ingested sugar via the stomach." The book was published in 1950 by F. A. Davis Co., Philadelphia, Pa.

nal of the American Medical Association, Vol. 83, No. 10, September 6, 1924, page 729, and *Annals of Internal Medicine,* Vol. 10, No. 14, October 1936, page 514).

In 1957, Dr. Gyland addressed general practitioners at a meeting of the American Medical Association in New York. His paper was published in 1958 in the *Brazilian Medical Journal,* Resenha. In 1958 at the meeting of the Southern Medical Society in New Orleans, he presented a paper entitled, "Functional Hypoglycemia as a Cause of Neurodermatitis." Before-and-after slides accompanied the paper. This paper and the slides have since been given to the medical school library at the University of South Florida, Tampa, Florida.

After becoming aware of functional hypoglycemia, Dr. Gyland became a clinical expert in the disorder. He managed to gather together his clinical experiences in a manuscript that has never been published. Through the courtesy of Mrs. Gyland, I have been privileged to read the entire manuscript. The first section written by Dr. Gyland is appropriately titled, "I've had it!" In the paragraphs that follow, I will summarize and discuss some of Dr. Gyland's findings.

Dr. Gyland studied 1,307 cases of functional hypoglycemia. Clinical studies of this magnitude are not easily come by. The sheer bulk of clinical data assembled and organized by Dr. Gyland is an excellent example of clinical research that can be carried out in an office setting.

There were 839 females and 468 males in the group that was studied, and they ranged from 1 to 84 years of age. The most prominent symptoms were as follows: nervousness; weak, faint, or dizzy spells; tremors; sweats; irritability, often leading to marital problems; vertigo; lack of energy; undue fatigue and exhaustion unrelieved by sleep; choking spells; headaches; abdominal distress (gas, heartburn, belching, cramps, etc.); drowsiness; blackout spells; muscular pains and cramps; twitching and jerking of muscles; insomnia; inability to concentrate; constant worries; anxiety; depression; crying spells; mental confusion; visual abnormalities such as blurred vision, spots before the eyes, seeing double, and a feeling of sand in the eyes; forgetfulness; indecision; incoordination; numbness; palpitation and rapid heart beat; excessive sighing and yawning; allergies; itching and crawling sensations; antisocial or unsocial behavior; cold feet and hands; and accident-prone behavior.

Dr. Gyland's manuscript contains these words: "In functional hypoglycemia, it is not only the patient who suffers. It is often worse for those who love and care for those who are ill. All of these patients have a change of personality and usually do not realize how desperately ill they are." Dr. Gyland goes on to emphasize: "Overfatigue and a loss of sleep will usually bring the symptoms back. Often women with hypoglycemia improve under treatment, then try to do in one day the housework which they had been unable to do for months — result: tomorrow a bad day."

Among Dr. Gyland's 1,307 cases, there were 47 cases of post-delivery shock or neurosis. These were chronically nervous females who sustained mental breakdowns after the birth of a child. They all responded well to treatment.

Five hundred seventy-five of the 1,307 had some clinically significant allergic condition. In treatment of the hypoglycemia, *all allergies to foods disappeared,* while allergies to pollens and inhalents became less severe. Most of the asthmatics got entirely well. Two hundred thirty cases of neurodermatitis or atopic eczema all cleared up in 2 to 4 months, regardless of the severity or duration of the illness.

Two hundred three of the hypoglycemic cases were found to have early symptoms of rheumatoid arthritis. All arthritis pains disappeared after treatment for their hypoglycemia. Among the 66 whose chief complaint was rheumatoid arthritis of long standing, 41 were free of pain in less than 1 year. The deformities, however, persisted.

Nine hundred eighty had chronic headaches. All, including those who bore the diagnosis of migraine headaches, got well when their hypoglycemia was properly treated.

Twenty-six cases had been diagnosed as diabetics but were found to have hypoglycemia or dysinsulinism. Most were taking insulin; some were on oral hypoglycemic drugs. All were getting progressively worse in regard to symptoms (daily weak spells with tremors and sweats). Dr. Gyland pointed out that the use of blood-sugar-lowering drugs is dangerous in these patients, and the treatment of choice is the dietary treatment of hypoglycemia. This advice comes from a man who, himself, grew worse under insulin treatment and who became healthy when proper dietary treatment was invoked.

Dr. Gyland noted that every alcoholic he observed had a severe hypoglycemia, and drank to relieve his illness, his inferiority complex, his nervousness and emotional instability, and his temper and hostility. He points out that the Alcoholics Anonymous organization does a good job but has recommended the use of coffee and candy in management. This only makes the sick alcoholic more sick, and he drinks again.

Eighteen cases were epileptic, many uncontrolled by anticonvulsant medicine. Their seizures usually came on when a meal was late, when they were tired, or when they had become very emotionally upset. The vast majority got well with Dr. Gyland's treatment (dietary change).

In these 1,307 hypoglycemic patients, the severity of the clinical state bore no constant relationship to the blood-sugar levels. Dr. Gyland shared the belief that I do that more frequent blood-sugar evaluations during the glucose tolerance test would disclose larger numbers of hypoglycemics. Ideally, blood-sugar values should be determined every 15 minutes.

The low, flat, glucose tolerance curve is said to be a normal variant by many academically oriented physicians. In Dr. Gyland's cases, those patients with low, flat curves were all quite ill, and all became well on the standard hypoglycemia dietary treatment which used high protein, high fat, low carbohydrate, no sweets, no caffeine, frequent feedings. The low, flat curve may be associated with inadequate absorption of sugar, pituitary or adrenal insufficiency, or overactivity of the parasympathetic nervous system. Dr. Sidney Portis (M.D.) altered the low, flat curve by treatment with atropine, a drug which blocks the parasympathetic nervous system.[11]

In Dr. Gyland's functional hypoglycemics, most of the fasting-blood-sugar values were between 70 and 110 mg.%. Only 7 were below 60 mg.%. Thus one can see that the fasting blood sugar was usually not abnormal according to the established normal values. When Dr. Gyland's findings are looked at from the standpoint of optimal fasting-blood-sugar level, however, one finds that 70% were above or below the presumed optimal range

[11]Reported on pages 115, 117, and 121 of *Body, Mind, and Sugar* by E. M. Abrahamson, M.D., and A. W. Pezet. Published in 1951 by Avon Books, New York, N.Y.

of 81 to 90 mg.%. (Nineteen percent were below 80 mg.%, 30% were between 81 and 90 mg.%, 27% were between 91 and 100 mg.%, and 22.8% were more than 100 mg.%.)

Of considerable importance in interpreting a GTT is the amount of blood-sugar drop below the fasting level. A 3-hour GTT does not provide this information. Although there are no fixed norms for this, there is a considerable body of opinion that the larger the drop below the fasting value, the greater is the evidence of hypoglycemia. In Dr. Gyland's cases, the drop in blood sugar below the fasting level on GTT was as follows:

below 10 mg.% 2.9%
10 to 20 mg.% 15.6%
21 to 30 mg.% 39.8%
31 to 40 mg.% 25.0%
41 to 50 mg.% 11.4%
51 to 60 mg.% 4.4%
61 to 70 mg.% 76%
71 to 80 mg.% 07%
81 to 90 mg.% 07%

Approximately 82% of the cases had a sugar drop of more than 20 mg.% below the fasting level. This is fairly good evidence that Dr. Gyland was indeed dealing with functional hypoglycemic patients.

The low point of the blood-sugar curve is also an important figure. Very often, the low point depends upon the height to which the blood sugar rises in the preceding minutes or hours. In Dr. Gyland's cases, 40% of the subjects had a low point below 61 mg.% on the curve and 70% were below 71 mg.%.

Dr. Stephen Gyland died in 1960. A *memoriam* dedicated to him is reproduced in Figure 111. He died of complications following surgery on an aneurysm (ballooning out) of the abdominal aorta. Although the origin of this lesion is not known with certainty, it may be that the many years of dysinsulinism with alternately high and low blood sugar may have contributed to its development. Zinc, copper, magnesium, and chromium are metals that play an important role in the bodily economy. Chromium is of particular importance for vascular integrity, and copper has to do with the elasticity of blood vessel walls. Whenever the

In Memoriam

DR. STEPHEN GYLAND

Stephen Gyland was born February 9, 1893, at Wesby, Wisconsin, the son of Nels and Sesli Gyland.

 He finished Viroqua High School, Viroqua, Wisconsin, and took his A.B. degree at St. Olaf College in Northfield, Minnesota. He was a Sergeant in the Army Medical Corps in World War I. He had two years at the University of Wisconsin Medical School, then transferred to the University of Pennsylvania, receiving his M.D. degree in 1924.

In January, 1926, Dr. Gyland located in Tampa. He quickly established a large and successful general practice. He endeared himself to his patients as he was never too tired to visit a sick patient at home and the fee was of secondary consideration.

In 1949 he was taken by an illness that totally incapacitated him by 1951, and lasted until 1953. Having intensive studies done on him at three of our leading clinics and without a definite diagnosis or encouragement, he returned to Tampa, ill and discouraged, but not defeated.

With only time on his hands, he read everything pertaining to his illness. He discovered in Dr. Seale Harris' work on nutrition (1924) symptoms that fit into the pattern of his illness. After numerous blood sugar studies and much reading, he was convinced he had a functional hypoglycemia. He worked out a diet and medication for himself and after three months returned to practice, a well man.

Thereafter, he enthusiastically devoted full time to the study and treatment of functional hypoglycemia. He presented a paper at the American Medical Association meeting in 1957 in New York on "Six Hundred Cases of Functional Hypoglycemia". This paper was published in full in "Resena", a Brazilian Medical Journal. In 1958 he gave a paper at the Southern Medical Association meeting on functional hypoglycemia as a cause of neurodermatitis.

An annual physical examination and X-rays revealed an aneurysm of the abdominal aorta. Dr. Gyland was advised to go to Houston for consultation and was operated on March 28. On the fifth day post-operative he fell from bed in the intensive care unit and fractured a hip. He passed away two weeks later on April 15 and was buried in Tampa on April 19.

He was a member of the American Academy of General Practice, the Hillsborough County Medical Association, the Florida Medical Association and the American Medical Association.

He is survived by his wife, Mrs. Ruth Gyland; a daughter, Mrs. Sally Angemeier; and a son, Dr. Stephen Gyland, Jr., of Jacksonville, Florida.

To his family we extend our deepest sympathy — and to "Steve", a job well done.

BE IT RESOLVED that a copy of this memorial be made a part of the permanent records of the Hillsborough County Medical Association, and a copy furnished to Mrs. Stephen Gyland.

Respectfully submitted,

LEE T. RECTOR, M.D.

Figure 111. Memoriam notice for Dr. Stephen Gyland

This notice appeared in *The Bulletin*, Hillsborough County Medical Association, Volume 9, June 1960, page 24 and is used with the permission of The Hillsborough County Medical Association, Tampa, Florida.

steady state of body chemistry is disrupted, trace-metal imbalance is apt to come about. Patients such as Dr. Gyland may be better helped today through improved techniques of treatment that are available to us.

Although Dr. Gyland's studies are of great interest for students of hypoglycemia, it must be admitted that they are open to criticism. Some people might question his claims of almost universal success. Some will point out that there was no control group of patients who were not treated with the special diet.

Such criticisms are justified, but they do not, however, detract from the fact that the dietary therapy that Dr. Gyland used not only restored Dr. Gyland's health but also helped many other people.

I believe Dr. Gyland's successes were due to dietary treatment as well as to his strong personal belief in that treatment. It is reassuring to know that such favorable responses can occur through an organized system of therapy.

Dr. Stephen Gyland was a pioneer in campaigning for reduction in the use of sugar and caffeine. Today we are still in need of pioneers. We need many, many persons like Dr. Gyland to improve the dietary habits of our nation. The story of Dr. Gyland's illness and his return to health by dietary alteration should be widely known. Perhaps it will assist those members of our society who are casual about their health to seek help in adopting more favorable lifestyles.

WHERE DO WE STAND?

In today's society, there are increasing numbers of persons who are health conscious and who look to physicians for guidance in matters of health. Physicians are experts in disease, but too often they know little about subjects such as functional low blood sugar or, indeed, nutrition in general. When meaningful discussion in regard to nutrition, low blood sugar, allergy, etc., is not forthcoming from physicians, persons often seek other sources of information.

Persons frequently obtain such information from health food stores. Health food stores in present day society have emerged from obscure shops to thriving businesses. The health food "nuts" of 10 - 20 years ago are more and more acceptable to more and more individuals. Along with some misinformation, health food stores do their best to provide an eager public with nutritional information. Meanwhile, their nutritional products are sold to keep people well, while pharmaceutical products (drugs) prescribed by doctors treat them when they are sick. Since the sale of all of these products represents big business, it is no wonder that there are opposing sides in regard to the regulation and licensing of nutritional products and drug products.

Academic medicine does not generally look with favor on health-food products and the education of the public through health food stores and allied organizations. Similarly, the academic physician generally looks with disfavor on the concept of functional hypoglycemia and the many books describing it. Academic medicine very rightly knows that misinformation may be given to persons in such books and by personnel in health food stores. This could be a serious health matter.

I believe most managers of health food stores are trying to do the best they can in an awkward situation. Most of them would welcome communication and interaction with physicians, but the opportunity for interchange of information is not commonly available. Often the camps are divided so rigidly that no middle ground can be attained.

For many persons, the health food store has become a convenient place to go to obtain information about low blood sugar. Most of the books about functional low blood sugar will be found there. Sometimes the clerks can provide the name of a doctor who is interested in low blood sugar; and more importantly, the clerk may know a doctor whose record of success in treatment of this problem is good.

Where then do we stand? Certainly, we stand amidst controversy. We have on the one hand, a rather rigid academic medical camp, and on the other hand, a somewhat looser-thinking group of lay health promoters. The academic medical establishment insists on a firm definition of a disease entity with definite cut-off points which establish the presence of the disorder, or which exclude it. The lay health group recognizes partial and incomplete cases and does not insist on fixed criteria. The medical approach is basically to utilize drugs to treat established disease. The lay health group emphasizes the preventive approach through nutrition, so drugs will not be required. Certainly one cannot quarrel with definition of disease or the goal of prevention. Many in the academic medical group believe, however, that the loose, "blunderbuss," "nutritional," noncontrolled approach to prevention is not only undesirable but actually harmful.

Neither should one quarrel with the objective of treating established disease, for anyone who is sick desperately wants to get well. Obviously, we should combine the best of the preventive approach with the best of the therapeutic approach, used as seldom as needed, for maximum benefit to all. Rigid position-taking, unfortunately, too often impedes badly needed amalgamation of viewpoints.

Persons who frequent health food stores often alter their state of awareness about nutrition by their visits. Some individuals who read the many books available on low blood sugar seem to develop that disorder — at least in their mind's eye. Others will discover a name, at last, for their disorder of long-standing. Still others grow healthy by their involvement with the nutritional information and products that they have received.

When a doctor confirms that a person has hypoglycemia, it can be a sense of relief for the patient. The diagnosis says to him, "I'm feeling lousy, and it's comforting to know there's some reason for it." In addition, the patient knows that hypoglycemia — as bad as it is — can be worked with and improved. It's not necessarily a one-way street to the morgue or the funny farm.

I do believe that functional, relative low blood sugar exists and often goes undetected and untreated. It is true that physicians do not find much information in their medical literature about this condition. As Dr. Carlton Fredericks (Ph.D.) has said, "Physicians slip into the men's room so they won't be seen reading my books." Some physicians, on the other hand, say that Carlton Fredericks is an interesting person and very convincing — but a quack. This interesting man, with razor-sharp mind, in recent times has been President of the International Academy of Preventive Medicine. Much of the information that he has assembled on nutrition through the years is the basis for a healthy diet in America today.

There is no question that certain individuals latch on to the diagnosis of hypoglycemia to explain various symptoms of emotional origin. I also believe that vitamin, mineral, and hormonal imbalances commonly give rise to symptoms which are described and treated as hypoglycemia. It is also true that certain individuals fail to receive the standard medical intervention that they need because of excessive involvement with the health movement. At times, non-M.D. practitioners do patients disservice by not obtaining a properly conducted medical evaluation.

It is probably true that some functionally oriented physicians and some lay persons are overdiagnosing hypoglycemia. It is more likely that many medical authorities are missing cases of significant hypoglycemia that do not fit their diagnostic criteria.

Most important is the type of person(s) encountered by Mr. John Q. Public when he enters a health food store. Desirable is the store manager who knows that natural whole foods are indeed health-giving and that a reasonable level of nutritional supplements are needed by many. Desirable, too, is the manager who knows that natural methods must be supplemented, at times, by medical expertise. Most desirable is the manager who realizes that every person is unique and that what works for one may not for another. Responsible health food store managers and their staffs encourage their customers to seek care and guidance from nutritionally oriented physicians.

In recent years, a number of medical authorities have indicated that hypoglycemia is not rare in the newborn infant. Dr. Harry S. Dweck (M.D.) states: "The highest incidence of hypoglycemia in man is during the neonatal period."[12] Although the accuracy of this statement is in doubt, it is true that hypoglycemia has been sought out and recognized in newborns with increasing frequency in recent years. At the present time, the physician's index of suspicion is high for hypoglycemia in the newborn period.

Nursing personnel in hospital nurseries — especially in newborn intensive-care nurseries — are trained to be alert to the symptoms and signs of hypoglycemia and to test for it "early in the game." Members of the medical team caring for newborns recognize that stressed (high risk) newborns are particularly prone to develop low blood sugar. The harmful effects of untreated hypo-

[12]"Neonatal Hypoglycemia and Hyperglycemia, two unique perinatal metabolic problems" by Harry S. Dweck, M.D. Published in *Postgraduate Medicine*, Vol. 60, No. 1, July 1976, pages 118 - 124.

glycemia (death, brain damage, mental retardation, learning disorder) have been recognized. Early diagnosis and treatment is advocated to prevent such problems.[13]

It is somewhat ironic that the frequency of hypoglycemia is being recognized by the medical establishment in infants who cannot speak directly for themselves. Perhaps we can accept the diagnosis of hypoglycemia in newborn infants because we believe they are too young to be neurotic.

Then, too, we should ask, "Why is a newborn hypoglycemic? What is there about the mother that has to do with this condition? Is she nutritionally deficient? Does she have a problem with carbohydrate balance? Homeostasis? Endocrine function? Nervous system integrity? Is she underexercised? Overweight? Unhappy? Medicated?" Let us remember, the developing fetus is a prisoner in the mother's womb. He is entirely at the mercy of the mother's nutrient supply for 40 weeks.

Adverse stress in a newborn infant is frequently accompanied by metabolic disturbances. Low blood sugar may only be one manifestation of physiologic upheaval. Low blood calcium and electrolyte disorders are other evidences of disturbed homeostasis.

Adverse stress in adults is also frequently accompanied by metabolic disturbances. Low blood sugar (usually of a functional nature) is recognizable and treatable when health-care personnel realize the existence and frequency of the problem. I believe hypoglycemia is more frequent in adults than it is in newborns, because there are so many years of opportunity behind an adult to develop habits of wrong living and wrong thinking.

There is no substitute for careful, thorough, medical evaluation of a patient in poor health. Differential diagnosis, properly done, is essential to proper treatment. Many persons seeking health information, particularly along nutritional lines, have not obtained what they want from their physicians. Because of this, these persons have turned to health food stores and other lay information groups. At times, all medical care is given up and replaced with some blind faith in a nutritional or religious concept.

It is unfortunate that physicians have not had a better exposure in their training to nutritional concepts. It is also unfor-

[13]"Clinical Aspects of Neonatal Hypoglycemia" by Gjoermund Fluge. Published in *Acta Paediatrica Scandinavica*, Vol. 63, 1974, pages 826 - 832.

tunate that some persons completely cast aside the many advantages of modern medical care. Faith and belief in a nutritional or religious concept can work wonders for some individuals in some circumstances, but this is no reason to throw out a system of medicine that is the backbone of our society in the treatment of disease and which grows more knowledgeable every day.

Advances in the biochemical treatment of illness are making rapid strides. The reader may wish to consult Dr. Nathan Kline's book *From Sad to Glad, Kline on Depression*[14] for an excellent summary of the chemical treatment of mental conditions. Dr. George Watson's book *Nutrition and Your Mind*[15] is provocative in its exposition that persons with mental and emotional illness are found to be one of two basic types: fast oxidizers or slow oxidizers. The fast oxidizer is frequently the kind of patient described in this book whose blood sugar *declines* after breakfast or after an oral glucose load. The slow oxidizer is often the patient with a diabetic type of GTT or NTT.

The field of medicine and the science of nutrition have much to say to one another that has not yet been said. The controversy between academic medicine and the lay health movement points up the difficulties we have as a nation in talking to ourselves.

Dr. Carlton Fredericks (Ph.D.) indicates in his lectures that hypoglycemia is an excellent example of cultural lag. He believes that the "milder" form of this disorder, functional hypoglycemia, will eventually be recognized as a common disorder by the medical establishment — somewhere 20 years down the road.

Again and again, rigid position-taking, failure to compromise, and lack of appreciation of another's viewpoint emerge as factors which divide the camps of health and disease. Communication suffers as arrogance, pride, and ego-building thrive.

Joint research on disease prevention needs to be carried out, hopefully by a research team comprising scientists and clinicians of varying beliefs. The difficulties of truly eliminating bias in statistical studies makes such joint projects extremely important.

[14]*From Sad to Glad, Kline on Depression* by Nathan Kline, M.D. Published in 1974 by G. P. Putnam's Sons, New York, N.Y.

[15]*Nutrition and Your Mind, The Psychochemical Response* by George Watson, Ph.D. Published in 1972 by Harper and Row Publishers, Inc., New York, N.Y.

Unfortunately, rigid attitudes, fears, and contentment with the status quo usually prevent such desirable cooperative research.

The reader of this book who is in need of additional information about his health should first consult his local physician. The reader should be sure to inform the physician that information about nutrition and preventive medicine is desired. If satisfaction cannot be obtained, consultation with a physician who is interested in nutrition and preventive medicine could be arranged.

The Adrenal Metabolic Research Society of the Hypoglycemia Foundation[16] is a tax-exempt charitable organization. This group popularizes the concept of hypoglycemia and emphasizes the frequency of the problem, its relationship to adrenocortical insufficiency, and its dietary treatment.

The International Academy of Preventive Medicine[17] is a rapidly growing organization devoted to nutrition and the prevention of disease. The International College of Applied Nutrition[18] has a booklet on nutrition that can be helpful. The International Academy of Metabology[19] deals extensively with hypoglycemia and nutrition. The Academy of Orthomolecular Psychiatry[20] as well as other societies of orthomolecular medicine are developing responsible memberships that provide a forum for basic research and clinical investigation. Chapter 15 of this book is devoted to nutrition and preventive care.

BLOOD-SUGAR LEVELS

The fasting-blood-sugar level (FBS), when elevated, is a good indicator of the presence of diabetes. When the fasting level is low (by rigid criteria), it is most often indicative of an organic cause of low blood sugar. I have come to realize also that an FBS below the optimal value of 80 - 90 mg.% is indicative of a less than optimal

[16]The Adrenal Metabolic Research Society of the Hypoglycemia Foundation, 153 Pawling Avenue, Troy, N.Y. 12180.

[17]The International Academy of Preventive Medicine, 871 Frostwood Drive, Houston, Tex. 77024.

[18]The International College of Applied Nutrition, P. O. Box 386, LaHabra, Calif. 90631.

[19]The International Academy of Metabology, Inc., 1000 E. Walnut Street, Suite 247, Pasadena, Calif. 91106.

[20]The Academy of Orthomolecular Psychiatry, 1691 Northern Boulevard, Manhasset, N.Y. 11030.

diet or of a sometimes subtle endocrine gland dysfunction.

In any consideration of blood-sugar values, it is paramount to consider where the blood sugar has been and where it is going. Absolute blood-sugar levels may be less important than the degree of change in blood-sugar levels and the symptoms that it produces.

When one refers to normal sugar levels, he must keep in mind the population from which the levels were derived. The values listed as normal are those values derived from persons who are free of *evident* disease. These individuals do not have rheumatic fever, tuberculosis, cancer, etc. In this sense they are healthy. If one looks more closely, however, at the "normal" individuals from whom blood sugar normal levels are derived, he would find that these persons have a host of health problems usually considered insignificant. Such matters as dandruff, acne, postnasal drip, constipation, dental cavities, periodontal disease, heartburn, hemorrhoids, varicose veins, insomnia, nervousness, allergy, uterine fibroids, menstrual disturbances, etc., may be present in the "healthy normal" individuals from whom "normal" values for blood sugar have been derived.

When normal values for blood sugar are given, the range of values is very wide. It includes those individuals whose values fall within 2 standard deviations of the mean. The use of such a wide range for blood sugar may not be in the best interest of the individual patient.

Instead of thinking about a normal range of blood sugar, perhaps we should direct our attention to the *optimal* blood-sugar level for an individual. Dr. Emanuel Cheraskin (M.D., D.M.D.) and his co-workers at the University of Alabama have championed this viewpoint.[21] Dr. Cheraskin states: "The body, physiologically and metabolically, is designed to operate within a narrow range of fluctuation. . . . Optimal metabolism is within narrow limits." As previously indicated, the optimal level of blood sugar appears to be about 85 mg.% with a range of perhaps 80 to 90 mg.%. See Figure 112 for a summary of blood-sugar levels and various labels that may be applied to the patient demonstrating these levels.

---

[21]*Predictive Medicine, A Study in Strategy* by E. Cheraskin, M.D., D.M.D., and W. M. Ringsdorf Jr., D.M.D., M.S. Published in 1973 by Pacific Press Publishing Association, Mountain View, Calif.

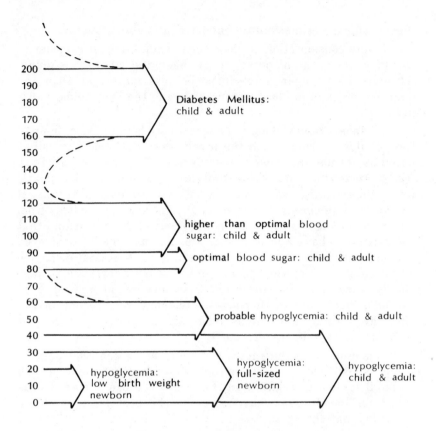

Figure 112. Blood-sugar values (in mg.%) and departures from normal

The diabetic values refer to those obtained on GTT or NTT. All others refer to fasting values or to values on GTT or NTT.

Although individual levels of blood sugar are helpful in evaluating the function of an individual, of more assistance are repeated blood-sugar values taken at frequent intervals, especially when they are correlated with diet.

The optimal level of blood sugar is thought to be about 85 mg.%, but may be somewhere between 80 and 90 mg.%.

A fall in blood sugar of 25 mg.% or more within a 1-hour period may be excessive and is usually associated with symptoms. Criteria for the diagnosis of hypoglycemia may be found in the text on page 286.

In newborns, blood-sugar values below 40 mg.% are suspect, although more rigid criteria for the diagnosis of hypoglycemia are shown. It may be that lower levels are used for diagnosis in newborns, because infants cannot communicate with us as readily to inform us about symptoms they may be experiencing.

In the case of blood-sugar elevations, it has been found that afternoon testing is apt to reveal higher levels of blood sugar than morning testing. A similar situation prevails in the case of low blood sugar. If there is a tendency for hypoglycemia, it may be exaggerated in afternoon testing compared with morning testing.

In a study by L. J. Kryston at Hahneman Hospital in Philadelphia, hypoglycemic symptoms were better correlated with insulin levels than with blood-sugar levels.[22]

The availability of the radioimmunoassay for serum-insulin levels[23] provides the clinician with considerable information about carbohydrate metabolism in clinical conditions. He is able to know whether a patient's insulin response to a glucose load is normal, partial, delayed, or absent.

As with blood-sugar levels, it is vital to be aware of the desirable levels for blood insulin. I am currently investigating blood glucose-insulin relationships. Suffice it to say at this time that the range of normal insulin levels given by the laboratory must not be construed as optimal. Again, the persons from whom these norms have been derived have been everyday Americans, consuming everyday American diets. Unfortunately, these diets are usually high in sugar and other refined carbohydrates and have not proven to be conducive to optimal health.

ORIGIN OF SYMPTOMS

The symptoms and signs of hypoglycemia were enumerated earlier in this chapter.

The patient with hypoglycemia may have many symptoms or he may have relatively few. In some cases, he appears to be without any symptoms although the quality of his everyday performance may be in question.

When symptoms are present, their origin may be due to one of the following:

[22]Paper presented at the Fourth Annual Meeting of the Academy of Orthomolecular Psychiatry, Detroit, Mich., May 4, 1974.

[23]The Phadebas Insulin Test (Pharmacia Labs., Inc.) is a relatively simple, fast, and convenient test for serum insulin, which may be performed in any clinical laboratory.

1. inadequate glucose for brain metabolism,
2. inadequate adrenal corticosteroid hormones for
   cellular metabolism,
3. release of adrenalin from the adrenal gland,
4. an accumulation of lactic acid, or
5. various mixtures of the above.

The brain's need for glucose has already been discussed. Adrenocortical insufficiency was discussed in Chapter 8.

Adrenalin produces symptoms of excitation, with an increase in blood pressure and heart rate, and a release of glucose from glycogen from the liver.

According to Dr. Carlous Mason (D.Sc.),[24] lactic acid accumulation occurs when sugar breakdown is excessive, resulting in an overload of chemical channels at the entrance to the Kreb's Cycle, with resultant excess lactic acid formation.

As pointed out in Chapter 6, the 3-carbon fragment (pyruvate) is formed from the glycolytic breakdown of glucose. When the chemistry is not right for this pyruvate to enter the Kreb's Cycle, lactic acid may be formed and may accumulate in cells.

One of the principal reasons for excess lactic acid formation is inadequate oxygenation of tissues. Decreased oxygen supply may come about because of blood sludging, atherosclerotic narrowing of blood vessels, weak heart action, impaired breathing, and muscular disuse. These are all commonly encountered in the American population.

Lactic acid inside cells, according to Dr. Mason, leads to the retention of sodium, calcium, and water within cells, and decreased cellular function due to these changes.[25]

The improved quality of life that nearly always accompanies a graduated and regulated program of physical conditioning, may have its origin in improved oxygenation and reduction of lactic acid in cells.

[24]Dr. Carlous F. Mason (D.Sc.), biochemist, Mato Laboratory, P.O. Box 7006, Riverside, Calif. 92503.

[25]*Hypoglycemia and the Methyl Approach* by Carlous F. Mason, D.Sc., Biochemist. Pamphlet published by the Mato Laboratory, P.O. Box 7006, Riverside, Calif. 92503. (No date given.)

It has been shown that lactic acid injected into individuals can reproduce the symptoms of anxiety and/or hypoglycemia.[26]

The problem of lactic acidosis has been increasingly recognized of late as a significant clinical problem.[27] It is known to occur in some cases of diabetic acidosis, phenformin treatment, cardiopulmonary bypass, diarrhea with dehydration, and after strenuous physical exercise.

Lactic acid is an end product of anerobic metabolism. The measurement of lactate in the blood is not readily available to the practicing physician. Much further investigation needs to be carried out to establish the role of lactic acid in hypoglycemic conditions.

HYPERINSULINISM

Plainly speaking, *hyperinsulinism* means too much insulin. But the matter is not quite that simple. Too much insulin for what? In relation to what? Based on what dietary standards? As determined in whose lab? By what method? Hyperinsulinism, like hypoglycemia, may be absolute or relative.

Accurate determination of blood-insulin levels has not been clinically available as long as blood-sugar levels. Thus we have less information on blood insulin than we do on blood sugar.

One laboratory indicates that the normal fasting level of blood insulin is 5 to 25 microunits per milliliter. With the GTT, the peak insulin occurs normally at 15 to 30 minutes and does not exceed a value of 150 microunits per milliliter. Anything above 150 is hyperinsulinism — *according to the population on whom the insulin norms were determined!* Undoubtedly, this population was consuming sugar at a high level as is characteristic of most of our population. As far as I know, determination of insulin levels on persons who eat no sugar has not been carried out. Conclusion: At this point, we cannot be too exact about hyperinsulinism with regard to measurements. More information is needed.

It has long been theorized that hypoglycemics are hypoglycemic because of a hyperactive or hyperreactive pancreas. In

[26]"The Biochemistry of Anxiety" by Ferris N. Pitts. Published in *Scientific American*, February 1969, pages 69 - 75.

[27]"Acidosis in the Diabetic: Ketosis or Lactic Acidosis" by Jeffrey A. Passer, M.D. Published in *Current Medical Digest*, Vol. 41, No. 12, December 1974.

other words, these persons develop low blood sugar because their pancreas oversecretes insulin. It has been pointed out that hypoglycemics often have a "hairtrigger" pancreas. These descriptions do, indeed, vividly portray the events that seem to occur in many of the patients with low blood sugar, especially that variety known as *reactive functional hypoglycemia*. (See Chapter 9, pages 157 - 161, for discussion of hyperinsulinism.)

Furthermore, it has been held that frequent eating of high carbohydrate food, especially simple sugar (which is so rapidly absorbed), gives a repetitively high blood sugar which leads to the hyperresponsive pancreas. The effect of caffeine in encouraging the development of blood-sugar elevation with subsequent insulin discharge has already been discussed in Chapter 9. Dr. Stephen Gyland's health problems in relation to caffeine and low blood sugar have been related earlier in this chapter.

There is evidence that maintenance of high blood sugar does stimulate or drive the beta cells of the pancreas to produce insulin. As a matter of fact, continued long-term stimulation of this nature may produce demonstrable damage to these cells (high-output failure). In experimental animals, prolonged hyperglycemia can destroy beta cells.

Permanent diabetes can be produced in animals by administration of growth hormone. This substance produces blood-sugar elevation, which in turn causes hypersecretion from the beta cells with subsequent exhaustion and the destruction of these cells.

Furthermore, as emphasized by Dr. Henry A. Schroeder (M.D.), elevation of blood glucose is accompanied by a rise in blood chromium and an overall net loss of chromium to the body through kidney excretion.[28] Chromium is known to have an intimate relationship with carbohydrate metabolism and atherosclerosis. Diabetics and persons with atherosclerosis are known to be chromium deficient.

One can see that there is good reason to believe that hyperfunction of the beta cells passes into the stage of functional exhaustion and degeneration with the development of diabetes. Thus, the concept of pancreatic overreaction (hyperinsulinism and hypoglycemia) is quite plausible.

[28]*The Trace Elements and Man* by Henry A. Schroeder, M.D. Published in 1973 by the Devin-Adair Company, Old Greenwich, Conn.

Dr. R. M. Ehrlich (M.D.), in discussing the treatment of the child with diabetic acidosis, makes this statement: "Any episode of pallor, perspiration, drowsiness, or significant deterioration in the level of consciousness following the initiation of insulin therapy should be considered hypoglycemia and treated with oral or intravenous glucose."[29] One could consider that nondiabetic persons who do not receive insulin from a syringe are receiving insulin from their own internal "syringe," their pancreas. Just as the physician may "overtreat" as he seeks to find the proper insulin dose, the pancreas of an individual may release too much insulin for the amount of blood glucose. As in the case of the diabetic who undergoes an insulin reaction, the person with hyperinsulinism is in need of care. An induced diabetic insulin reaction requires sugar. The common variety of hyperinsulinism (usually associated with eating sugar) requires sugar avoidance.

In some diabetics (obese, maturity-onset type) there is frequently an elevated level of insulin in the blood according to standard criteria. In these individuals, it may be that an acquired resistance to insulin has developed while the beta cells are hypersecreting.

Classical hyperinsulinism due to an insulin-secreting pancreatic tumor or hyperplasia of the pancreas has been described in Chapter 9. In these rare cases, the administration of a high sugar diet might be indicated until surgery on the pancreas can be accomplished.

Hyperinsulinism may also be used as a term to describe the body's hypersensitive reaction to normally secreted amounts of insulin. In many of these cases, deficiency or ineffectiveness of adrenal-corticoid hormones appears to underly the insulin sensitivity. The concept of balance and imbalance in hormonal relationships has been discussed in Chapter 8.

Functionally oriented physicians are becoming increasingly aware that hypoglycemia is commonly a prelude or forerunner of diabetes. It is likely that a basic disturbance in carbohydrate metabolism is common to both hypoglycemia and diabetes. Such a "carbohydropathy" underlying both disorders may be further understood as techniques of basic research improve. In conversation, Dr. H. J. Roberts (M.D.) has used the term *dysglycemia* as

[29]"Diabetes Mellitus in Childhood" by R. M. Ehrlich, M.D. Published in *Pediatric Clinics of North America,* Vol. 21, No. 4, November 1974.

a general term to describe deviations in carbohydrate metabolism. At the present time, a simplified view of carbohydropathy, or dysglycemia, visualizes multiple states of imbalance between the alpha and beta cells of the pancreas, with other variants formed by gut, liver, brain cortex and hypothalamus, autonomic nervous system, and endocrine factors.

Dr. William Philpott (M.D.)[30] believes that the basic disturbance in metabolic carbohydrate disorders is a hypersensitivity reaction to foods, inhalents, or chemicals in the environment, associated with nutritional deficiencies. My experience leads me to agree that such matters may be responsible for disturbances of homeostasis in carbohydrate metabolism. Whether allergy and maladaptive allergylike reactions are primary disturbances, however, remains in question.

CLINICAL CONDITIONS

I believe that the vast majority of Americans today have varying degrees of illness or departure from optimal health that sooner or later will lead to indentifiable disease. There are many reasons for this, but most prominent are the deterioration of the diet in the last 100 years, the sedentary nature of our populace in the mechancial age of convenience, and our exposure to pollutants in the environment.

Amongst these adults and children of poor health and incubating disease, there are certain individuals who are weaker than their peers. Their weakness has come about because of hereditary traits, biochemical experiences in utero during pregnancy, postnatal traumatic and infectious advents, allergy, psychosocial disadvantage, etc. These weaker individuals are the vulnerable persons who are identified first as unhealthy and maladapting to their culture.

Learning disabled children are youngsters who have failed to successfully adapt to the demands of their school system. My experience indicates that a common characteristic of many LD

[30] "Immunological Deficiency in Schizophrenia: Schizophrenia as a Variant Syndrome in the Nutritional Deficiency Addiction — Diabetes Mellitus — Infection Disease Process," an article by William H. Philpott, M.D., in *A Physician's Handbook on Orthomolecular Medicine*, edited by Roger J. Williams, Ph.D., and Dwight K. Kalita, Ph.D. Published in 1977 by Pergamon Press, New York, N.Y., pages 151 - 155.

(learning disabled) children is the finding that their blood-sugar values deviate significantly from optimal values. Blood-sugar levels are either higher or lower than desirable. GTT's and NTT's depart considerably from the patterns that I judge to be optimal. The majority of those that are deviant, show lowered blood-sugar values, but a significant number have chemical diabetes. These dysglycemias may be attributable to chronic adverse nutritional habits, hereditary traits, vitamin deficiency, mineral imbalance, underactivity, chronic stress overload, and inadequate parenting.

The experience of the New York Institute for Child Development, Inc., is enlightening.[31] This center obtained data on 261 hyperactive children of appreciable severity. The children were of middle class origin and ranged in age from 7 to 9 years. The majority of these children would also be classified as LD because of the nature of their hyperactivity and short attention span with distractibility and impulsive behavior.

Seventy-four percent of the hyperactive children had abnormal GTT's according to standard medical criteria. Sixty-five per cent of the abnormal GTT's could be classified as functional or relative hypoglycemia, as the term is used in this book. Fifty percent of the abnormal GTT's were designated as low, flat curves. Of great interest is the finding that 86% of the hyperactive children had abnormal elevations of the blood eosinophil count (more than 6%). This indicates the strong likelihood that an allergy was present.

If one takes the broad, flexible view of hypoglycemia, he would have to agree with statements which indicate that in our society, hypoglycemia is far more common than diabetes, itself a very common health disorder.

Hypoglycemia has been noted with alarming frequency in alcoholics, schizophrenics, and juvenile delinquents. Some physicians believe that any nuropsychiatric disorder may be associated with hypoglycemia. Dr. Harry M. Salzer (M.D.) has emphasized relative hypoglycemia as a cause of neuropsychiatric illness.[32] Dr. Sidney Portis (M.D.) championed the concept that emotional frus-

[31]"Glucose Tolerance and Hyperkinesis" by L. Langseth and J. Dowd. Published in *Food and Toxicology* in 1978.

[32]"Relative Hypoglycemia as a Cause of Neuropsychiatric Illness" by Harry M. Salzer, M.D. Published in *Journal of the National Medical Association*, Vol. 58, No. 1, January 1966.

tration is associated with increased parasympathetic (vagal nerve) activity, hyperinsulinism, and hypoglycemia.[33]

Dr. Carlton Fredericks (Ph.D.) and others have pointed out that hypoglycemia is involved in many cases of asthma, exactly as it is the cause or the aggravating factor in other types of allergic reactions. Dr. Fredericks found 25% of a randomly selected group of asthmatics to have hypoglycemia.[34] He also points out that diabetics (whose blood sugar is usually high) rarely have asthma and that when asthmatics develop diabetes, the asthma has been known to disappear.[35]

An interesting anthropological study indicates that the Quolla Tribe in the Andes Mountains of South America may owe its vicious, violent behavior to hypoglycemia.[36] This murderous tribe has a high alcohol consumption, chews coca leaves, eats a large amount of candy, and has a poor protein intake. Caffeine and coca stimulate the release of adrenalin, which stimulates the liver to pour out sugar into the bloodstream. Alcohol inhibits new glucose formation from protein in the liver. Sugar itself floods the system with glucose, triggering insulin release with subsequent rebound lowering of blood sugar.

In view of these nutritional habits, is it any wonder that the Quolla are hyperaggressive and violent?

A physician can learn many things from his patients — if he listens. One of my patients has raised horses for 30 years. He told me that horses become mean and vicious when they are fed sugar. Horses, like humans, love sugar, but they're apt to nip their owners when they consume the sweet. I'm quite sure that much

[33]Dr. Portis's work is discussed on pages 110 - 127 of the book *Body, Mind, and Sugar* by E. M. Abrahamson, M.D., and A. W. Pezet. Published in 1951 by Avon Books, New York, N.Y.

[34]*Prevention*, March 1975, page 50.

[35]There are those who believe that Dr. Fredericks' observations have no validity. His critics differ from him in viewpoint and manner. They probably criticize him most for his outspoken beliefs, verbal manipulatory ability, "exaggerated claims," and successful books for the lay public. I suspect that his razor-sharp mind has been many years ahead of his time. In regard to asthma and hypoglycemia, whether he is right or not harks back to one's definition of hypoglycemia.

[36]"Aggression and Hypoglycemia Among the Quolla, A Study in Psychobiological Anthropology" by Ralph Bolton, Ph.D. Published in the *Journal, International Academy of Metabology, Inc.*, Vol. 3, No. 1, March 1974.

"biting" behavior among humans has the same origin. The tragedy is that many persons with irritable, aggressive, and sharp personalities never trace the source of their behavior to sugar in the diet.[37] This is the case, because sugar is so ever-present in the usual modern diet. It is, indeed, difficult to escape sugar consumption unless one conscientiously sets out to do so.

Parents who discover that their child turns into a hellion when he eats sweets are most diligent at sugar avoidance. They do not add sugar to food; they do not buy packaged and bottled beverages; they avoid pies, cakes, and canned soups; and they use fresh, raw fruit, seeds, nuts, and vegetables to replace ice cream and jello. Unfortunately, however, relatives, neighbors, and friendly bank tellers with lollipops frequently thwart their efforts by thrusting sweets upon the child. It is unfortunate, too, that those persons are not the ones who must live with the hellion for the next few days or carry him to the doctor for another sugar-induced infection!

Although multiple factors operate, in general, persons who avoid sugar are peace-minded and nonviolent. There may be a lesson there for world leaders. Do you suppose they'll listen?

The child with relative low blood sugar is sometimes of an artistic nature. He may live in a fantasy world and be known as a daydreamer. Although he may not realize it until he grows older, he may feel chronically ill most of the time. Because of this, he is apt to be susceptible to the addicting influence of substances such as sugar, other refined carbohydrates, alcohol, caffeine-containing beverages, or drugs. These items will often produce a feeling of temporary well-being that he never otherwise experiences.

It is not uncommon to find that the hypoglycemic patient has frequent stomachaches. The symptoms are quite often similar to those of peptic ulceration. X-ray studies of the upper bowel may show irritability of the stomach, duodenum, and pylorospasm. Although one might expect excess stomach acid with such symptoms, it is more common to find a diminished supply of stomach acid.

I suspect that hypoglycemia is frequently a manifestation of food allergy. This would perhaps explain the frequent occurrence

[37]The reader must again be reminded that a diet high in sugar is also likely to be one high in other refined carbohydrates, chemical additives, and overcooked food, and low in essential vitamins, minerals, enzymes, and fiber.

of elevated blood eosinophil levels in such patients. Some of the case studies presented at the end of this chapter describe a number of such patients.

The study of the New York Institute for Child Development, Inc., reported earlier in this section, found that 86% of hyperactive children had an elevation of blood eosinophils ("allergy cells") of more than 6%.[38]

A hypoglycemic condition may be indicated by trace-mineral imbalance in the body. An analysis of hair for minerals can determine this imbalance. According to the late Dr. John Miller (Ph.D.), Bio-Medical Data, Inc., Chicago, Illinois, the changes in hair which suggest hypoglycemia are

> a high calcium to potassium ratio,
> a high magnesium to potassium ratio,
> a low potassium to zinc ratio, and
> perhaps a high lead to manganese ratio.

Dr. Miller believed that when this type of mineral pattern is present, glucagon is not produced, resulting in hypoglycemia.[39] Other investigators indicate that hair patterns which show high calcium, low potassium, and high zinc may be characteristic of hypoglycemia. Dr. Carl C. Pfeiffer (Ph.D., M.D.) indicates that very high calcium and magnesium in the hair may be a sign of functional hypoglycemia.[40] Disturbed values of sodium and potassium also appear to be characteristic of hypoglycemia and/or adrenocortical insufficiency.

Dr. Charles Rudolph Jr. (D.O., Ph.D.) indicates that the cardinal indicator of hypoglycemia in hair analysis is the increased calcium to potassium ratio with a low level of chromium.[41] My experience suggests that low potassium with elevated calcium and low or high chromium may be characteristic of hypoglycemia.

[38]"Glucose Tolerance and Hyperkinesis" by L. Langseth and J. Dowd. Published in *Food and Toxicology* in 1978.

[39]Personal Communication with John Miller, Ph.D.

[40]"Observations on Trace and Toxic Elements in Hair and Serum" by Carl C. Pfeiffer, Ph.D., M.D. Published in *Journal of Orthomolecular Psychiatry*, Vol. 3, No. 4, Fourth Quarter of 1974.

[41]"Trace Element Patterning in Degenerative Diseases" by Charles J. Rudolph Jr., D.O., Ph.D. Published in the *Journal of the International Academy of Preventive Medicine*, Vol. IV, No. 1, July 1977, pages 9 - 31.

Figure 113A shows a hair-mineral pattern which is highly suggestive of the presence of hypoglycemia in the patient from whom the hair was taken. Figure 113B shows the glucose tolerance test of this patient, indicating the presence of hypoglycemia. Cases such as this in which the hair pattern "fits" with the blood-glucose levels are common. Research is now being carried on to further clarify the relationships of hair minerals to body chemistry.

Dr. Carl Pfeiffer has found the blood level of spermine to be a reliable indicator of the presence of hypoglycemia.[42] He uses the spermine blood levels instead of the GTT to diagnose hypoglycemia. Spermine is a long aliphatic amine (a polyamine) associated with RNA (ribonuceic acid).

According to Dr. Hal Huggins (D.D.S.),[43] the buildup of glycogen from glucose in the liver is encouraged by potassium more than by sodium, whereas glycogen breakdown to sugar is encouraged more by sodium than by potassium. Calcium and magnesium levels in the body have a great deal to do with glycogen formation and breakdown. Thus, we may encounter blood-sugar disturbances when sodium, potassium, calcium, and magnesium levels are imbalanced.

In my office practice, I have been impressed with the improvements in mental and physical function that occur when elevated levels of hair calcium have been returned to normal. In some cases, they reflect a relative deficiency of magnesium in the diet. At other times, calcium elevation in the hair appears to indicate endocrine imbalance, such as hypothalamic dysfunction, low pituitary, low thyroid, and/or low adrenocortical function.

Diuretics are agents which encourage the elimination of fluid from the body through the kidneys. As fluids leave the body, various minerals are often carried out with the liquid. Probably the commonest diuretic used in our society is caffeine. Caffeine is contained in carbonated beverages, coffee, and tea.

Sodium is eliminated from the body under the influence of

[42]A Clinical Nutrition Tape by Carl Pfeiffer, M.D., Ph.D., International College of Applied Nutrition, 1975. Supplied by Tapette Corp., 7221 Garden Grove Boulevard, Garden Grove, Calif. 92641.

[43]Hal Huggins, D.D.S., Analytico Laboratories, 100 East Cheyenne Road, Colorado Springs, Colo. 80906.

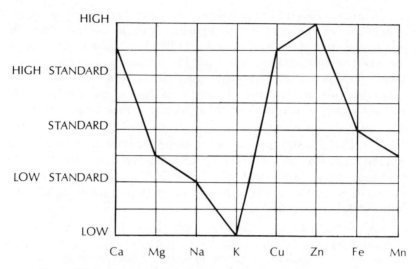

Figure 113A. Graph of mineral elements in hair showing the relative proportions of one mineral to another

This patient has high calcium to potassium, high magnesium to potassium, and low potassium to zinc ratios. This hair pattern is highly suggestive of a hypoglycemic condition in the patient from whom the hair was obtained.

Ca:	calcium	Cu:	copper
Mg:	magnesium	Zn:	zinc
Na:	sodium	Fe:	iron
K:	potassium	Mn:	manganese

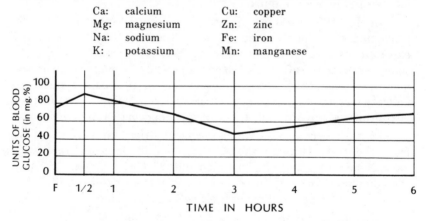

Figure 113B. GTT of same patient whose hair-mineral pattern is shown in Figure 113A

The blood sugar rises very little after a glucose load and then drops to a level of 47 mg.% at 3 hours. The individual reported symptoms of weakness, dizziness, tremor, and nausea from 2 to 4 hours.

This kind of functional hyperinsulinism precipitated by sugar is commonly associated with the kind of hair-mineral pattern shown for this individual.

diuretics. If sodium is depleted below the normal level, interference with the intestinal absorption of glucose may occur.

Potassium is also one of the body minerals which is eliminated in the urine when diuretics are taken. As indicated, potassium is needed in the body when carbohydrate in the diet is changed into glucose and thence into glycogen. If potassium is deficient, glycogen may not be formed, and the individual becomes hypoglycemic because of a lack of glucose storage reserve in the liver.

Zinc excretion may also be increased by diuretics.

The use of diuretics generally becomes unnecessary when diet is optimized and when other aspects of lifestyle are improved. Vitamin B_6, pantothenic acid, and vitamin E are substances that may exert a fluid-eliminating effect in certain individuals.

Seven minerals are especially important for proper glucose homeostasis. They are sodium, potassium, calcium, magnesium, zinc, chromium, and manganese. All minerals, of course, are important in their closely interwoven effects upon enzyme functions and other vital processes in the body. Our understanding of body biochemistry is limited but progressing.

The American population is almost uniformly low in manganese, according to the results of the trace-mineral analysis of hair.[44] This may be one of the reasons why hypoglycemia appears to be so prevalent today in the American population.

What is the link between manganese deficiency and hypoglycemia?

Gluconeogenesis is the formation of blood glucose from protein. Gluconeogenesis meets the needs of the body for glucose when carbohydrate is not available in sufficient amounts in the diet. Manganese is essential in the enzyme reactions involved in gluconeogenesis (see footnote 44). If manganese is deficient, gluconeogenesis may not proceed and hypoglycemia may occur. Also, manganese is involved in the proper working of the thyroid gland. Disordered thyroid function may impair the absorption of glucose from the gut.

Excessive use of milk and cheese has been blamed for

[44]"The Total Metabolic Approach," an audio-tape featuring Lloyd Horton, D.C., and Paul Eck, D.N., D.C., January 1977. Supplied by Western Academy of Biological Sciences, 115 South 38th Street, Tacoma, Wash. 98408.

hypoglycemia in some patients. This may be due to milk allergy or to the high lactose (milk sugar) content of milk and milk products. Lactose is a disaccharide (double sugar). When digested in the gut, it is broken down into its component sugars: glucose and galactose. When absorbed into the body, galactose is changed to glucose for use in the metabolic energy pool. The end result of consuming a large quantity of milk or cheese is an increased glucose load. (See Chapter 15 for further discussion of milk in the diet.)

In Jamaica, hypoglycemia is caused by eating certain unripe fruits which cause vomiting and low blood sugar. This is due to isovaleric acid toxicity. In our country, eating of isovaleric acid-related substances can bring on hypoglycemia. These blood-sugar-lowering chemicals may be present in fats that are somewhat rancid, as well as in canned fruit juice, nitrate-cured meats, soy sauce, some wines and bourbon, the volatile fatty acids of bovine muscle and liver, and tobacco smoke.[45]

The present state of our culture with its enormously high intake of sugar and other refined carbohydrate; with its sadly underexercised population; with its high caffeine, alcohol, and tobacco use; and with its high stress-loads undoubtedly sets the stage for disturbances of metabolism. We are able to increasingly recognize large numbers of functional hypoglycemics and sugar-sensitive individuals. Could there be that many? What kind of public health problem do we have?

We live in a world of crime and increasing personal disrespect for one another. The politican is quick to promise a fight for law and order. The world of business has too often become: Get mine and forget about you. "Progressive" urbanization brings with it urban din, rape, robbery, and physical violence. Verbal abuse heaps upon the physical battering we do to one another.

Although the problem has multiple roots, I present these thoughts again:

1. I have observed that cranky persons are usually nutritionally deficient sweet-eaters.

[45]See the article "Mineral Deficiencies in Chronic Disease" by John J. Miller, Ph.D. Published in *Journal of the International Academy of Preventive Medicine*, Vol. 1, No. 1, Spring 1974, pages 95 and 96.

2. A child can change from a hellion to an angel when he gives up sugar.

3. Candy bar wrappers were found at the spot where Lee Harvey Oswald presumably shot President Kennedy.

4. Albert Speer reports that Adolf Hitler was a sugar-glutton.

5. The vicious Quolla Tribe in South America has a high consumption of candy.

6. We nurse the nips on our hands from sugar-fed horses.

Sugar intake and all that goes along with it — nutritional inadequacy and body chemical imbalance — appear to be vital underlying factors in the production of unwanted social behavior.

What kind of a lesson is that for responsible adults? Can we build a better world based on personal respect for others? Is it worthwhile to pursue optimal nutrition and improved ways of living?

In Chapter 15 this matter is elaborated.

CLASSIFICATION

Now let us examine the various types of hypoglycemia in more detail. Low blood sugar can be divided into *organic* varieties and *functional* conditions. Organic conditions are those in which there is a recognizable anatomic cause. Functional conditions are those in which the chemical function of the body is off, but there is no known anatomic reason for the disorder. Hypoglycemia may also be categorized as *fasting*, or *nonfasting* (fed). In general, organically caused low blood sugar is of the fasting variety, and functional conditions are associated with low blood sugar after eating. Hypoglycemia could also be described as *long-lasting* (chronic) or *transient* (acute). Table 4 lists the causes of hypoglycemia according to Dr. Jerome W. Conn (M.D.).

Table 5 gives a classification of hypoglycemia which has been presented by Dr. Robert Pollett (M.D.).

TABLE 4

THE CAUSES OF HYPOGLYCEMIA

I. ORGANIC: RECOGNIZABLE ANATOMICAL LESIONS:
    A. Pancreatic islet beta-cell tumors.
    B. Epitheloid tumors derived from foregut anlage.
    C. Massive extrapancreatic neoplasms.
    D. Diffuse liver disease.
    E. Adrenocortical hypofunction.

II. FUNCTIONAL: NO RECOGNIZABLE ANATOMICAL LESION:
    A. *Impaired hepatic output of glucose:*
        1. Specific hepatic enzyme defects.
        2. Ethanol and poor nutrition.
        3. Hypoglycin ingestion.
        4. Ketotic hypoglycemia of childhood.
    B. *Stimulation of insulin release:*
        1. Rapid absorption of glucose following gastric surgery.
        2. Reactive hypoglycemia in early mild diabetes mellitus.
        3. Leucine hypersensitivity of infancy and childhood.
        4. Transient neonatal hypoglycemia in infants of diabetic mothers.
        5. Erythroblastosis fetalis.
    C. *Drug-induced hypoglycemia:*
        1. Insulin administration.
        2. Hypoglycemia sulfonylureas.
        3. Miscellaneous.
    D. *Undetermined etiology:*
        1. "Functional" hypoglycemia.
        2. Transient hypoglycemia in the newborn of low birth weight.
        3. Idiopathic hypoglycemia of infancy and childhood.
        4. Infantile gigantism with macroglossia, visceromegaly, omphalocele, and microcephaly.

SOURCE: Abstracted from *Current Concepts of Spontaneous Hypoglycemia* by Jerome W. Conn, M.D., and Summer Pek, M.D., Division of Endocrinology and Metabolism, The University of Michigan, Ann Arbor, Michigan 48104, printed as a *Scope Monograph* by Upjohn Company, Kalamazoo, Michigan, and used with the permission of the authors and the Upjohn Company.

TABLE 5

HYPOGLYCEMIA*
(Categorized according to the results of glucose tolerance test
and immunoreactive insulin levels)

REACTIVE (POSTABSORPTIVE)

Induced by glucose, galactose, fructose, or leucine. Apt to have central nervous system symptoms as well as autonomic nervous system (adrenergic) symptoms due to rapidity of fall.

A. Early phase
1. 1½ - 3 hours. Insulin inappropriately elevated but secretion is timed correctly.
   a. postgastric surgery
   b. alimentary
2. 2 - 4 hours. Unexplained sensitivity to insulin. Insulin levels not inappropriately elevated.
   a. functional
B. Late phase
1. 3 - 5 hours. Insulin inappropriately elevated, and its secretion is also inappropriately timed; that is, too much, too late.
   a. diabetes mellitus
   b. obesity
C. Variable
1. Occasionally an insulinoma will present this way.

FASTING

Normal individual should not become hypoglycemic even after fasting for 72 hours.

Apt to have only central nervous system symptoms due to relatively slow fall.

A. Decreased glucose production
1. Liver disorders
   a. glycogen storage diseases
   b. ketotic hypoglycemia (deficient liver enzymes interfering with gluconeogenesis)
   c. acquired massive liver disease (e.g., hepatitis of pregnancy)
2. Endocrine deficiency states
   a. ACTH
   b. adrenal cortex
   c. glucagon
   d. catecholamines

*Hypoglycemia is defined as a blood-sugar level of less than 40 mg. %.

TABLE 5 — *Continued*

FASTING — *Continued*

B. Increased glucose utilization
   1. Insulin overproduction
      a. offspring of diabetic mothers
      b. erythoblastosis fetalis
      c. visceral hypertrophy
      d. pancreatic islet hyperplasia or tumor
   2. Noninsulin mediated
      a. growth hormone deficiency
      b. extra-pancreatic neoplasms; 75% are mesenchymal or liver tumors; hypoglycemia associated with markedly elevated levels of nonsuppressible, insulinlike activity (NSILA)

DRUGS & TOXIC STATES

A. Insulin administration
B. Sulfonylureas
C. Ethanol (alcohol)
D. Miscellaneous

SOURCE: Based on table 10-4, page 639, in *Textbook of Endocrinology*, Fifth Edition, edited by Robert H. Williams, M.D., published in 1974 by W.B. Saunders Co., Philadelphia, Pa., and modified in accordance with lecture comments of Robert J. Pollet, M.D., Assistant Professor of Medicine, University of South Florida, College of Medicine, Tampa, Florida. Lecture presented at the American Medical Tennis Association, Sarasota, Florida, January 1975. This classification is used with the permission of the textbook publishers and Dr. Pollet.

In organic hypoglycemia, there is a demonstrable anatomic cause for the low blood sugar. Various tumors such as pancreatic insulinoma; carcinoid; multiple endocrine adenomas; fibroids and sarcoids; as well as liver, pituitary, and adrenocortical hypofunction may be included in this category. In functional conditions, there is no demonstrable anatomic lesion, but there is abnormal function of an organ, a system, or an enyzme. Nearly everyone agrees that functional hypoglycemia accounts for the vast majority of hypoglycemia. Functional hypoglycemia is twice as common in females as it is in males.

Reactive hypoglycemia refers to those conditions in which the low blood sugar is due to a dietary constitutent such as glucose, galactose, fructose, or leucine.

A classification of hypoglycemia which I have assembled will now be given. It includes hypoglycemia due to amino acids, hormones, liver dysfunction, and other forms.

HYPOGLYCEMIA AND AMINO ACIDS

Proteins in the diet are made up of amino acids. When protein is mentioned, one can think of amino acids, because digestion in the gut splits protein into its constitutent amino acids for absorption into the body.

When carbohydrate is eaten, glucose is the end product of digestion, and the rise in blood glucose stimulates the secretion of insulin. No glucagon secretion occurs.

When a piece of meat is eaten, the amino acids in the meat stimulate the secretion of insulin and glucagon. This is due to the variety of amino acids which make up the protein.

The amino acid arginine is a most potent stimulator of insulin as well as glucagon. Alanine increases glucagon but not insulin.

Leucine stimulates insulin but not glucagon, perhaps in part accounting for the clinical disorder of low blood sugar due to leucine in the diet.

Blood levels of total protein, albumin, and globulin can be readily measured along with blood glucose. When protein deficit is found or when the evidence points to incomplete or inadequate protein digestion, remedial steps should be taken to correct the situation. Some hypoglycemic individuals improve greatly when impaired digestion is adequately treated.

Leucine Sensitivity

In some children, eating a meal which is high in protein may bring on hypoglycemia. Other amino acids have also been observed to produce hypoglycemia. The diagnosis of leucine sensitivity may be suspected when a child becomes lethargic, pale, or when he has a convulsion following a high-protein meal. The diagnosis of leucine-induced hypoglycemia may be established by a specific tolerance test in which leucine is given to the child and blood-sugar values obtained thereafter. *It must be realized that leucine-induced hypoglycemia is a rare disorder and that most hypoglycemic states are thought to be improved by a high-protein diet.*

Maple Syrup Urine Disease

Maple syrup urine disease is a genetically determined inborn error of metabolism, involving the amino acids valine, leucine, and isoleucine. The odor of the urine of patients with this disorder resembles maple syrup. Low blood sugar in this disorder may be due to improper absorption of sugar, hyperinsulinism, or liver dysfunction. This is also a rare disorder.

HYPOGLYCEMIA AND HORMONES

Growth Hormone and/or ACTH Deficiency

The anterior pituitary gland secretes growth hormone as well as adrenocorticotropic hormone (ACTH). Both of these hormones have a blood-sugar-elevating effect. ACTH acts to do this by way of the adrenal cortex which it stimulates to secrete glucocorticoid hormones.

Both growth hormone and ACTH production are dependent upon stimulating hormones released from the hypothalamic portion of the brain.

If the anterior pituitary hormones or their stimulating factors are deficient, low blood sugar could occur.

Short stature (growth failure) is one indicator of growth-hormone deficiency.

It is likely that degrees of inadequate production of growth hormones and/or ACTH occur, accounting for hypoglycemia which develops under adverse conditions or excessive stress.

Thyroid Hormone Deficiency

The patient with hypothyroidism may fail to absorb sugar from the gut, or the absorption may be slower than normal. Administration of proper amounts of thyroid hormone corrects these problems.

Most traditional physicians, especially those with academic inclination, rely almost entirely on blood levels of thyroid hormones or related tests (T-4, T-3, T-3 uptake, etc.) to diagnose thyroid dysfunction. Most useful for this purpose is the measurement of thyroid stimulating hormone (TSH). When the blood-TSH level is higher than it should be, primary hypothyroidism is indicated.

Dr. Broda Barnes (M.D.) has championed the use of the basal body temperature measurement as an indicator of the need for thyroid treatment.[46] Dr. Barnes believes that a patient with a lower than normal basal body temperature may have hypothyroidism even though his blood thyroid tests are normal. My experience bears this out, although allergy, nutritional deficiency, hyposedentary lifestyle, low blood sugar, and adrenocortical insufficiency are other factors that need to be considered in the origin of low body temperature.

Glucagon Deficiency

Glucagon, the "anti-insulin" hormone, elevates blood sugar by breaking down glycogen, by opposing insulin release, and by inhibiting the uptake of glucose by cells. If glucagon is deficient, ineffective, or inhibited, insulin remains relatively unimpeded in its action, and low blood sugar comes about due to relative insulin excess. Much is yet to be learned about glucagon. It is possible that states of functional glucagon inadequacy are quite common. Certainly, the opposite state of affairs also occurs — that is, that glucagon, relatively unopposed, is associated with high blood sugar in states of insulin insufficiency or ineffectiveness (diabetes).

Glucagon deficiency can be diagnosed by giving an animo acid, arginine, to the patient intravenously. Arginine is a known stimulus for glucagon secretion. If blood sugar does not rise after arginine infusion, the patient is deficient in glucagon production.

Exercise increases the secretion of glucagon. In our greatly underexercised population, it is possible that a tendency to low blood sugar may be associated with a functional state of glucagon inadequacy.

[46]*Solved: The Riddle of Heart Attacks* by Broda O. Barnes, M.D., Ph.D., and Charlotte W. Barnes, A.M. Published in 1976 by Robinson Press, Inc., Fort Collins, Colo. 80522.

Glucocorticoid Deficiency

The outer part of the adrenal gland (the bark or cortex) secretes hydrocortisone, which is a major adrenal hormone (corticosteroid). Hydrocortisone is termed a *corticosteroid of the glucocorticoid type,* because this hormone has a profound effect on sugar (glucose) metabolism. The glucocorticoid hydrocortisone promotes increased levels of blood-sugar breakdown of glycogen. When this glucocorticoid is deficient, the blood-sugar lowering action of insulin remains relatively unopposed, and low blood sugar results. Addison's disease is the classic example of the pathological state in which adrenocortical deficiency is often associated with low blood sugar. Antibodies directed against adrenal tissue occur in many cases of Addison's disease, and this suggests an autoimmune (hypersensitivity, allergic) basis of the disease.

Congenital adrenal hyperplasia is another adrenal-gland disorder which may be associated with hypoglycemia. In this disease, which is an inborn error of metabolism, the adrenal gland produces an excess of hormones that masculinize the female and cause the male to have precocious development of secondary sexual characteristics. The offending hormones are produced instead of the essential cortisonelike hormones which profoundly influence glucose homeostasis.

It seems likely that lesser states of adrenocortical malfunction exist to explain the lesser degrees of hypoglycemia which are encountered in clinical practice as well as in children with learning and adaptation problems. When such individuals are temporarily treated with corticosteroids,[47] a favorable effect may be due to the effect of the corticosteroid on sugar balance or on an allergic condition in the person. The favorable effect of whole adrenal cortex extract on hypoglycemics as described by Dr. J. W. Tintera (M.D.)[48] suggests that hypoglycemics may be deficient in adrenal cortex secretory products. When we know more about the function of the

[47]See *Allergy, Brains, and Children Coping* by Ray C. Wunderlich Jr., M.D. Published in 1973 by Johnny Reads, Inc., Box 12834, St. Petersburg, Fla. 33733. The use of corticosteroids in treatment must be closely supervised by the physician. Corticosteroids should not be used when other treatments are effective.

[48]"The Hypoadrenocortical State and Its Management" by J. W. Tintera, M.D. Published in the *New York State Journal of Medicine,* Vol. 55, No. 33, July 1, 1955.

many trace hormones contained in the adrenal cortex, we will be more knowledgeable about low blood sugar and its treatment.

Recurrent vomiting, genital abnormalities, low blood pressure, a craving for salt, or increased blood levels of eosinophils may indicate adrenocortical insufficiency. Hypoactive adrenocortical function is discussed in some detail in Chapter 8 of this book.

Hypoglycemia with Adrenalin (Epiniphrine) Insufficiency

Whenever a lowered amount of sugar occurs in the blood of the normal person, he secretes increased amounts of adrenalin as a compensatory measure to elevate the blood sugar. (Adrenalin, as well as glucagon, does this by breaking down glycogen.) Rarely, a person may fail to secrete increased adrenalin in response to lowering of the blood sugar. It is not known whether this defect is in the adrenal medulla, the hypothalamus, or in the autonomic nervous system.

Insulin Excess

There is no question that hyperinsulinism is a common cause of hypoglycemia. The presence of excess insulin in the body may occur in the following instances.

1. *Administration of insulin by syringe:* The patient with diabetes mellitus experiences the effects of excess insulin when his blood sugar falls and he becomes sleepy, cold, tremulous, and incoherent. The diabetic who takes insulin injections takes care to always have sugar available to combat any possible insulin reaction from an overdose.

2. *Pancreatic (islet cell) tumor:* The insulinoma, an adenoma, is a benign tumor of the pancreas which involves the isles of Langerhans. It characteristically occurs in older children and adults. Attacks of hypoglycemia are apt to take place after exercise or prolonged fasting, unlike functional hypoglycemia. The fasting blood sugar is frequently low and very often may be below 60 mg.%. Prolonging a fast to 72 hours and having the subject exercise during the fast will bring out low blood sugar in the patient with insulinoma. The ratio of blood insulin to blood glucose is high in these patients and grows larger as fasting is prolonged. Proinsulin is usually elevated along with insulin. Occasionally

only proinsulin may be elevated. Intravenous administration of tolbutamide, leucine, or glucagon may be used to confirm the diagnosis of insulinoma. The intravenous glucagon test does not usually produce the severe hypoglycemia encountered with other provocative tests. Although these provocative tests are helpful in diagnosing insulinoma, the 72-hour fast with ratio of insulin to glucose is the cornerstone of diagnosis. Medical treatment of insulinoma is generally not effective, and surgical removal of the tumor is the treatment of choice. In pancreatic islet cell cancer, treatment with streptozotocin is utilized.

Dr. T. J. Merimee (M.D.) and Dr. J. E. Tyson (M.D.) at the University of Florida, Gainesville, Florida, have found that "normal" women exhibit absolute hypoglycemia when they are subjected to a 72-hour fast.[49] This could represent a source of confusion in the diagnosis of insulinoma.

In 40% of the subjects studied, the blood glucose fell to less than 45 mg./dl. Thirty-three percent had values as low as 30 to 35. Occasional values between 25 and 30 were recorded. Symptoms were not prominent and were believed to be due to ketosis rather than low blood sugar.

In these "normal" subjects, the plasma insulin concentration also declined during the fast and the ratio of insulin to glucose decreased in the vast majority.

In an attempt to find some way to distinguish this absolute hypoglycemia of "normal" women from pathologic hypoglycemia that accompanies insulinomas and some other organic lesions, Drs. Merimee and Tyson then compared normal women and insulinoma patients subjected to a 72-hour fast.[50]

Compared with "normal" women, the decline in blood sugar with fasting in insulinoma patients occurred earlier in the period of fasting and was more productive of hypoglycemic symptoms. Also, in insulinoma patients, the insulin to glucose ratios increased during the fast rather than decreased. According to the authors, "The ratio of insulin to glucose provided a clear cut distinction

[49]"Stabilization of Plasma Glucose During Fasting" by T. J. Merimee, M.D., and J. E. Tyson, M.D. Published in the *New England Journal of Medicine*, Vol. 291, 1974, page 1275.

[50]"Hypoglycemia in Man, pathologic and physiologic variants" by T. J. Merimee, M.D., and John E. Tyson, M.D. Published in *Diabetes*, Vol. 26, No. 3, March 1977, pages 161 - 165.

between the physiologic hypoglycemia of fasting and pathologic hypoglycemia."

3. *Hyperplasia of pancreatic islet cells:* In some individuals, there are substantially more (excess) insulin-secreting cells than in the usual person. These cells supply excess amounts of insulin, producing the findings of hypoglycemia. There is no specific diagnostic test for this state, but surgical removal of a portion of the pancreas may be curative.

It has been shown that insulin-secreting cells in the pancreas may be formed from noninsulin-secreting cells. This process (nesidioblastosis) may occur in response to prolonged stimulation of the pancreas by sugar.

4. *"Nervousness": Nervousness* is an inexact term. It is used here to mean *psychological imbalance, anxiety states, depression,* etc. It is probable that emotional instability can be the cause as well as the result of low blood sugar. It is known that nervous impulses can stimulate insulin secretion. The mechanism of this may be from the parasympathetic center in the hypothalamus through the right vagus nerve to the pancreas. This has been termed the *cephalic*[51] phase of insulin secretion.

Whenever hyperinsulinism occurs without a demonstrable organic cause, it is termed *functional hyperinsulinism.* Emotional ("nervous") hyperinsulinism may be a fairly common cause of functional hypoglycemia.

This variety of low blood sugar is also known as *reactive functional hypoglycemia.*[52] The blood glucose falls to low levels 2 - 4 hours after a meal; therefore this is a "fed" hypoglycemia. The fasting-blood-sugar levels are within the usually acceptable range of normal. A family history of diabetes is usually absent. These individuals are believed not to develop diabetes at a rate greater than the general population, but the accuracy of this statement remains to be seen.

5. *Repetitive sugar (sucrose) ingestion:* It is probable that the high dietary sugar load in our society results in hyperinsulinism in some individuals. This may be due to or associated with vitamin

[51]*Cephalic* means head.

[52]Terminology of Drs. Fajans and Floyd. See "Hypoglycemia: How to Manage a Complex Disease" by Stefan S. Fajans, M.D., and John C. Floyd Jr., M.D. Published in *Modern Medicine,* October 15, 1973.

and trace-metal imbalance. Hyperinsulinism has been previously discussed in this chapter and in Chapter 9.

6. *Diabetic reactive:* In this type of low blood sugar, the patient has a mild form of diabetes with elevation of blood sugar levels in the first 2 hours of the GTT, and hypoglycemia at the 3rd to 5th hours. Insulin release is initially delayed but later excessive to account for hypoglycemia (see Figure 114).

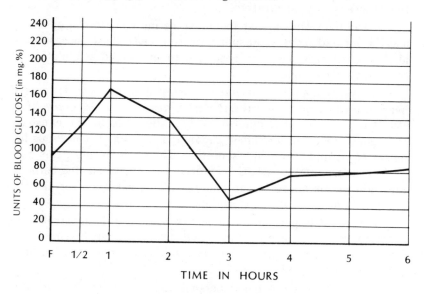

**Figure 114. Mildly diabetic glucose tolerance curve
with hypoglycemia (diabetic reactive type)**

This curve belongs to a 13-year-old boy who is overweight.

Notice the rather severe dip at 3 hours. The change from 170 at 1 hour to 47 at 3 hours is a profound fall. The somewhat elevated 2-hour level (40 mg.% more than the fasting level) is a characteristic of diabetes, as is the elevated peak of 170 at 1 hour. This curve might also be described as dysinsulinism.

7. *Alimentary hyperinsulinism:* This is usually, but not always, seen in persons who have had surgical removal of stomach tissue or who have undergone a surgical procedure to bypass the stomach. There is an accelerated absorption of sugar with peak blood-glucose levels at ½ - 1 hour on GTT. These early levels are quite high. The subsequent hypoglycemic drop occurs 1½ - 3 hours after glucose ingestion. It is possible for some individuals with intact stomachs to demonstrate the same type of glucose tolerance

curves. These individuals may have a particular type of gastro-intestinal function known as *dumping*, in which food rushes headlong out of the stomach into the intestine.

I suspect that alimentary hyperinsulinism is much more common than realized. It probably accounts for those persons with excessively rapid sugar absorption who experience a subsequent hypoglycemic dip at 2 to 4 hours and who experience a wide variety of symptoms. The person who consumes large quantities of sugar that is rapidly absorbed is, by virtue of input rather than by gut surgery, creating his own alimentary hyperinsulinism.

Figures 98 and 101 in Chapter 13 describe patients with very rapid absorption of glucose from the gut.

HYPOGLYCEMIA AND LIVER DYSFUNCTION

The liver plays a prominent role in maintaining normal blood glucose (see Chapters 3 and 4). Almost any interference with the function of the liver may lead to lowered blood sugar. There are three liver functions necessary to maintain normal blood-sugar levels. These are

1. storage of glycogen,
2. breakdown of glycogen to glucose, and
3. synthesis of glucose from amino acids (protein).

Injury to liver cells may interfere with any or all of these functions to produce hypoglycemia.

When the liver source of blood sugar becomes unavailable, the individual becomes entirely dependent upon his diet for minute-to-minute maintenance of adequate blood-sugar levels.

Since alcohol is a very rapidly absorbed form of carbohydrate, it may be that some alcoholics develop their drinking problem because of hypoglycemia. On the other hand, there is no question that alcoholics may develop hypoglycemia as a result of the toxic effect of alcohol upon the liver. Alcohol has a direct effect, interfering with the formation of glucose from amino acids (inhibition of gluconeogenesis). Administration of alcohol intravenously has been used as a measure of gluconeogenetic reserve.

It is well known that alcoholics frequently develop cirrhosis of the liver because of the chronic repetitive toxic effect of alcohol on this organ. Less well known is the fact that children can develop serious illness merely by ingesting alcohol from a concentrated source on a single occasion! Seizures, coma, and death have been reported. Because of this (and for other reasons) it is wise to prohibit children from guzzling the dregs of cocktails.

According to Dr. Ross Cameron (M.D.),[53] nearly 100% of alcoholics have hypoglycemia — when they are not drinking alcohol. In view of this, it is wise to order tests for hypoglycemia on the relatives of alcoholics. When this is done, according to Dr. Cameron, a higher than usual incidence of abnormal carbohydrate metabolism is found.

When one considers the profound harmful effects of alcohol — the most rapidly absorbed carbohydrate known — it seems rather foolish to permit this villainous drug to remain freely available in society. However, as long as refined sugar, other refined carbohydrates, hard drugs, and alcohol are abundantly available; as long as the moral fiber of society is weak; and as long as the population is unaware of the harmful effects of these substances; it is unlikely that such products will be turned aside. At some point, society may be able to properly educate the public to the dangers of these substances so persons do not consume them, even though they may remain available.[54] This book is a step in that direction.

Other poisons besides alcohol may likewise injure the liver — such agents as phosphorous, carbon tetrachloride, hydrazine, etc.

On the island of Jamaica and in other tropical habitats, ingestion of the unripe Ackee fruit may bring about a fatal condition with liver injury, vomiting, and severe hypoglycemia.

Infiltration of the liver by cancerous cells, scar tissue, cysts, or fat may also lead to hypoglycemia.

Acute or chronic inflammation of the liver (hepatitis) may also be associated with hypoglycemia but not usually until the final stages of the fatal disease.

Glycogen has been frequently mentioned in this book, and it has been noted that the liver is a major storehouse for this substance. Disorders which specifically affect glycogen metabolism

[53]Formerly with the Pinellas County Health Department, St. Petersburg, Fla. See Appendix C for further information about Dr. Cameron.

[54]Consider the power of the media to influence the buying habits and lifestyle of a population.

are known as *glycogen storage disease*. Hypoglycemia is associated with those types of glycogen storage disease which primarly affect liver glycogen. In these conditions, glycogen is either unable to be synthesized or unable to be broken down to glucose because of the congenital absence of specific enzymes.[55]

Galactosuria is an inborn error of metabolism in which the sugar galactose is improperly metabolized. Hypoglycemia is one manifestation of this disorder which is characterized by liver and spleen enlargement, failure to thrive, jaundice, cataracts, and mental retardation. Galactose is a component of lactose (milk sugar). Infants with this congenital disorder become ill when they consume milk with its galactose-containing milk sugar. A specific enzyme deficiency in the liver and in the red blood cells blocks normal metabolism of galactose. Because of this, a chemical substance accumulates, which has a deleterious effect on cell metabolism. If detected at birth or soon after, the harmful effects of the disorder can be prevented with a milk-free diet.

Fructosuria is a somewhat similar disorder, involving fructose (fruit sugar). Severe hypoglycemia and vomiting occur after an affected patient eats food containing fructose or sorbitol (converted to fructose in the body). This disorder of hereditary fructose intolerance is, like galactosuria, due to the absence of a specific enzyme necessary for proper metabolism of the sugar — in this case, fructose. As a result of this metabolic block, a chemical substance accumulates which may interfere with normal glucose metabolism, producing hypoglycemia. Persons with fructosuria are in perfect health when fructose is absent from their diet. Symptoms develop when they eat fruit or table sugar which contains sucrose, a sugar which is broken down into fructose and glucose.

It is not uncommon to find "minor" abnormalities of liver function in patients who have symptoms of ill health. Persons who are not well frequently have disturbed values for LDH, SGOT, SGPT, alkaline phosphatase, or bilirubin tests. This is not a surprising finding in view of the central importance of the liver, the largest organ in the body, in the body metabolism. Abnormalities in liver function may of course be part and parcel of whatever is

[55]There are approximately 12 forms of glycogen storage disease. Von Gierke's, Forbes', Hers, and glycogen synthetase disorders are types associated with hypoglycemia.

causing poor health. Frequently, this is found to be subadequate nutrition with or without drug or chemical toxicity.

Many persons in poor health also have disturbances in carbohydrate metabolism best described as *functional hypoglycemia*. It may well be that "mild" liver dysfunction is linked with "mild" hypoglycemia. Certainly the two disappear together as persons clinically improve under treatment. The most commonly effective treatment is that which improves the dietary input and effectively nourishes the body.

When I attended medical school, I was taught that liver dysfunction is characterized by a certain set of abnormal liver function tests. Since that time, I have learned that these tests may all be normal in the presence of suboptimal liver function. The performance of the blood sugar in the late hours of the GTT or NTT is a sensitive indicator of one kind of liver function. When blood-sugar levels fall at 4 to 6 hours, liver function may be compromised.

OTHER FORMS OF HYPOGLYCEMIA

Ketotic Hypoglycemia

Ketotic hypoglycemia is a fairly common cause of low blood sugar in children. Some believe it to be the most common cause. This disorder is due to the lack of important liver enzymes in the early years of life. It commences between 1 and 5 years of age, improves around 8 years of age, and disappears by the time of puberty. Low-blood-sugar attacks are accompanied by acetone (a ketone substance) in the urine. Attacks may be precipitated by fasting or by eating a high-fat diet. Lack of glucose formation from protein seems to be the main problem (inadequate gluconeogenesis). Patients with ketotic hypoglycemia when fasting have low levels of alanine (an amino acid) in their blood. This amino acid is the principal one which is transformed to glucose by the liver. Giving these patients intravenous alanine takes away their low blood sugar.

The following eye defects may occur in this disorder: eye jerking (nystagmus), cataracts, crossed eyes, and degeneration of the vision nerve head. Testing a patient by having him fast for a 24-hour period is an excellent diagnostic measure for this disorder; but blood sugars must be carefully followed, and the test terminated when the diagnosis of low blood sugar is established. Urines should also be tested for acetone. This substance may appear in the urine before low-blood-sugar levels are apparent.

Drug Toxicity

Low blood sugar may be due to salicylate (aspirin) overuse, acetohexamide, d-amphetamine, d-ribose, and, of course, overdosage with insulin or hypoglycemic drugs.

Propranolol (Inderal: Ayerst Lab) is a drug known as a *beta-adrenergic blocking agent.* It interferes with the function of the sympathetic branch of the autonomic nervous system. It is used for heart disorders, particularly those of heart rhythm and angina pectoris. Propranolol has been associated with serious hypoglycemia, especially when the patient has been fasting.[56] Individuals who are prone to hypoglycemia, as well as diabetics receiving insulin or oral hypoglycemic drugs, should use this drug with great caution. Propranolol, because of its blockade of the sympathetic nervous system, may prevent the appearance of symptoms of hypoglycemia (sweating, rapid pulse, and blood pressure changes).

Malnutrition

Hypoglycemia can be a complication of kwashiorkor. It has also occurred in children with phenylketonuria (PKU) during treatment with low phenylalanine diets. On the other hand, it has been

[56]This would not be unexpected. Pancreatic insulin release is under parasympathetic control via the vagus nerve. Propanolol is an antisympathetic drug that would result in parasympathetic dominance.

Recent research has indicated that fasting decreases sympathetic activity. ("Fasting Zaps Involuntary Nerves," *Science News,* Vol. 112, July 2, 1977. This is a report of an article in *Science,* June 24, 1977, by James B. Young and Lewis Landsberg of Harvard Medical School.)

Fasting is the first physiological state identified that reduces norepinephrine turnover and hence decreases activity of the sympathetic nervous system. This raises the interesting question of whether supervised fasting might not be a therapeutic procedure of choice in many disorders associated with sympathetic overactivity or unusual sympathetic sensitivity. In a frenetically-paced world beset with adverse stress, where fight or flight are frequently not viable behavorial alternatives, the ancient historical practice of fasting (medically supervised) may emerge as a most helpful nontoxic therapy.

shown that a gross insulin deficiency may be characteristic of malnutrition. Probably mixed patterns of carbohydrate abnormality occur, depending on the nature of the malnutrition.

The commonest form of malnutrition in our society is the excess consumption of sugar, other refined carbohydrates, and caffeine. Such empty calorie vitamin-, mineral-, enzyme-, and fiber-poor foods take the place of more wholesome nourishment that provides the nutritional elements needed every moment of every day for proper physical and mental well-being. Malnutrition and nutritional imbalance as a major cause of carbohydrate metabolic disorders is the major theme of this book.

Nonpancreatic Tumors

Chest or abdominal growths, usually sarcomas or fibrosarcomas, have been associated with fasting (nonfed) hypoglycemia. The condition is often severe.

Exercise

Anyone who has participated in a long, strenuous athletic contest has experienced the weakness that arrives as the event wears on. Although there are numerous variables that may account for this,[57] certainly low blood sugar may be a prominent cause of exhaustion and faltering performance.

Analysis of glycogen in the liver has shown a 60% reduction after 1 hour of heavy exercise. Drinking a sugar-containing liquid drink can elevate a blood-sugar level within 5 to 10 minutes and thus lessen the demand on the liver.

Dehydration, the loss of body fluid, plays an important part in disrupting the efficiency of cellular function in prolonged exercise. To avoid this, replacement fluids should be taken frequently in small amounts. Dr. David L. Costill (Ph.D.)[58] recommends 7 to 10 ounces of fluid every 15 minutes for long-distance runners. He indicates that excess sugar delays the transit time of fluid into the body from the gut. Fluids containing less than

[57]"Muscular Exhaustion During Distance Running" by David L. Costill, Ph.D. Published in *The Physician and Sportsmedicine*, Vol. 2, No. 10, October 1974, pages 36 - 41.

[58]See footnote 57.

2.5 grams/100 ml. of water are suggested, and Dr. Costill notes that most commercially available drinks contain far too much sugar in relation to fluid.

My experience also suggests that far too much sugar is contained in beverages that are consumed by sports participants. If the usually available beverages are diluted one to one with water, a more favorable balance of fluid and sugar may be obtained.

Hypoglycemia of the Newborn

It is medically agreed that hypoglycemia in the newborn period may result in mental retardation, learning disability, neurological deficit, or behavioral abnormality. Physicians who care for newborns are therefore urged to detect and vigorously treat hypoglycemia when it occurs. Manifestations of low blood sugar in the newborn may be cyanosis, limpness, low body temperature, coma, tremors, seizures, poor feeding, and breathing difficulties.

A blood-sugar level below 30 mg.%[59] has been considered abnormal in the full-term newborn.[60] There is some trend now to consider a level less than 40 mg.% as abnormal. In the premature or the undersized infant, levels below 20 mg.% are considered abnormal.

It is not uncommon for the newborn with hypoglycemia to manifest other chemical abnormalities which indicate that his metabolic homeostasis (steady-state) is disrupted. For example, a lowered level of calcium in the blood may accompany hypoglycemia. Interestingly enough, this is exactly what I find in adults who are overstressed and undernourished.

Infants of diabetic mothers and infants with blood-group incompatabilities who are hypoglycemic are known to be so due to hyperinsulinism. Most other hypoglycemic newborns fail to

[59]Blood sugar may be measured by a method which detects true blood sugar (glucose oxidase) or by a method which detects other elements in addition to blood sugar (Folin-Wu). The latter method gives values which are 20 mg.% higher than the former. Blood-sugar values in this book are true blood-sugar values. In the newborn period, it is especially important to use the true-blood-glucose method, because other sugars may be present in the newborn from the maternal circulation which would affect the Folin-Wu Test.

[60]Full-term newborn: an infant born after a full, 40-week pregnancy.

manufacture glucose from protein and fat (inadequate gluco-neogenesis).

When measured in accordance with the preceding criteria, hypoglycemia is known to occur in 10% of full-term newborns whose weight is appropriate for their length of pregnancy. Some studies indicate an even higher incidence. It can thus be seen that this disorder is not at all a rarity. When the full-term infant's weight is low in relation to the length of his pregnancy (the small-for-date infant), hypoglycemia occurs in 25% of these infants. These small-for-date infants are the products of intra-auterine malnutrition which occurs for one reason or another. Their hypoglycemia may be due to a lack of glycogen stores in the liver.

True premature babies whose weights are consistent with their pregnancy dates have a 15% incidence of hypoglycemia. The premature baby whose weight is small in relation to his length of pregnancy has hypoglycemia 67% of the time! These figures show the great importance of adequate nutrition (as indicated by the weight of the infant) in preventing hypogly-cemia of the newborn. Interestingly enough, the same is true for children and adults; namely, the presence of hypoglycemia is directly related to nutritional disturbance.

At times, an abnormality of the placenta may be the reason for intrauterine malnutrition. In the majority of cases, however, dietary inadequacy on the part of the mother is probably a major factor.

There is a higher-than-usual risk of hypoglycemia in the baby born to a mother with toxemia.[61] Blood-glucose levels of 20 mg.% or lower have been reported in 24% of the babies born to toxemic mothers, and levels of 30 mg.% or below have been reported in 65% of such babies. Since toxemia in all probablility is a manifestation of nutritional inadequacy,[62] the increased rate of hypoglycemia in infants born to toxemic mothers is under-standable.

Hypoglycemia is most common in babies of toxemic

---

[61]Toxemia: a disorder of pregnancy, characterized by fluid retention, ele-vated blood pressure, or kidney disorder.

[62]See *Metabolic Toxemia of Late Pregnancy, A Disease of Malnutrition* by Thomas Brewer, M.D. Published in 1976 by Charles C. Thomas Co., Springfield, Ill.

mothers whose urinary estriol levels are abnormally low. Estriol is a female hormone, estrogen. Its presence in adequate amounts during pregnancy is a sign of an adequately functioning feto-placental unit.

The baby born to a diabetic mother is notoriously prone to have hypoglycemia on the first day of life. These babies also usually have a growth disturbance, being larger than average. It has long been recognized that oversized babies are born to mothers who may become diabetics in future years or who have a mild undetected form of diabetes.

A serious cause of hypoglycemia in the newborn is the administration of sulfonylurea drugs (oral hypoglycemic agents) to pregnant mothers with diabetes. These drugs have been shown to pass through the placenta and remain in the newborn for long periods, causing protracted and serious hypoglycemia. Because of this, the pregnant diabetic should not take the oral hypoglycemic drugs (chlopropamide, acetohexamide, tolbutamide).

Some babies who develop bloodstream infection have been noted to be hypoglycemic. It is not as yet known whether the low blood sugar is due to inadequate nutrition in a sick child, or whether the infection has a more specific effect on blood sugar. It is conceivable that the infectious process stresses the infant's adrenocortical reserve to the point that hypoglycemia develops.

When twins are born, the smaller twin has an increased incidence of hypoglycemia in the newborn period. Again, it appears that nutrition as indicated by the weight of the child probably plays a large part in preventing or producing the lowered blood sugar.

Infants who are unduly exposed to a cold environment in the early weeks of life may also develop hypoglycemia. Cold may act as a physical stress for the infant. If the child's antistress mechanisms are not adequately developed, hypoglycemia may result.

Beckwith's syndrome is a disorder in which the newborn infant is found to have a large tongue, excessive size, oversized visceral organs, and a large rupture of the navel. Hypoglycemia occurs in the majority of these cases and may be due to an overabundance of insulin-secreting pancreatic tissue.

Transient neonatal diabetes mellitus (TNDM) may be a related disorder due to immaturity of the insulin-releasing mechanism. TNDM has been associated with large tongue and intra-auterine growth failure.

Beckwith's syndrome and TNDM may be examples of the wide spectrum of endocrine-carbohydrate imbalance that may occur in the fetus and newborn. In later life, hypoglycemia, diabetes, and dysinsulinism may be examples of the same basic endocrine-carbohydrate imbalance.

Babies who have erythroblastosis (blood-group incompatibility) may become hypoglycemic. The more severe the blood disorder, the more likely it is that an infant will be hypoglycemic.

The administration of dextrose (glucose) to a mother in labor by means of an intravenous drip could be responsible for low blood sugar in her newborn. Most of the time when this occurs, it is associated with a stressful labor and delivery. Presumably the hypoglycemia occurs due to an exaggerated insulin reponse to the administered glucose. It should be kept in mind that glucose in the intravenous bottle is often derived from corn. An allergic or allergylike reaction to corn could be responsible for blood-sugar deviation.

There is considerable evidence that small-sized premature infants have a fragile carbohydrate metabolism. These babies may devleop elevated levels of blood sugar as well as abnormally low levels of blood sugar. Stress in the form of infection, hemorrhage, etc., may accompany the elevated blood sugar.

Idiopathic Hypoglycemia of Childhood

Idiopathic means of unknown origin. This "waste-basket" category is declining in incidence as more knowledge is obtained about other forms of low blood sugar.

ANOTHER CLASSIFICATION

Dr. Herbert B. Goldman (M.D.)[63] presented the following classification of hypoglycemia at the Third Annual Meeting of the International Academy of Metabology in 1974. This classification is that of a functionally oriented physician. It is based upon Dr.

[63]61 Forestdale Road, Rockville Center, N.Y. 11570. This material is used with Dr. Goldman's permission.

Goldman's clinical experience and the use of the 5-hour GTT.

Dr. Goldman lists four types of hypoglycemia based upon the results of the glucose tolerance test. The types and their relative frequencies are

1. low flat (18%),
2. functional (57%),
3. reactive (15%), and
4. high rise with plateau (10%).

Low Flat

In the low, flat type of hypoglycemia, the patients show less than 25 mg.% variation of blood glucose throughout the test. Despite the ingestion of the standard loading dose of glucose, the blood-sugar curve remains essentially flat.

Dr. Goldman believes that female patients with this type of curve frequently have estrogen deficiency. Lack of thyroid may be responsible for the low, flat curve. Poor absorption of glucose through the intestinal wall may account for the curve.

The insulin response may be normal in individuals with the curves, but hyperinsulinism may account for the flatness of the curve in some cases.

The disorder may be inherited or acquired, but it is easier to manage when it is acquired. In the family history, it is not unusual to find that one parent often has maturity-onset diabetes.

The low, flat type of hypoglycemia is thought to be a precursor of diabetes. It frequently develops into the reactive form of hypoglycemia. When a patient progresses to diabetes, he may go through a stage when his GTT curve will be normal! The patient's symptoms, however, will remain with him, indicating that his problem with carbohydrate metabolism is still present and, indeed, even progressive.

Functional

The patient with functional hypoglycemia (according to the Goldman classification) starts out with a GTT that appears to be entirely normal at ½ hour and 1 hour. A fall of blood sugar to low levels occurs at 3 to 4½ hours, and the sugar level has usually risen back up to normal levels at 5 hours.

In these patients, blood-insulin levels are not higher than normal, but the peak insulin levels are reached later than normal. In a normal GTT, insulin peaks by ½ hour; but in these patients, the insulin peaks are recorded at 1 hour or 2 hours. This functional hypoglycemic may be said to have insulin lag.

Note that this is the commonest form of hypoglycemia in Dr. Goldman's experience.

According to Dr. Guy L. Schless (M.D.), functional hypoglycemia represents latent diabetes mellitus which may develop into overt diabetes if it is not treated.[64] Dr. Schless indicates that functional hypoglycemia is present when the blood sugar falls below 50 mg.% between the second and fifth hours of the GTT. He indicates that a four-year study of untreated latent diabetics showed that 50% progressed to overt diabetes and 10% became normal. (The rest were unchanged.) In his experience with latent diabetics, treatment resulted in return to a normal GTT in 48%, an additional 34% were significantly improved, and fewer than 5% developed overt diabetes. (The rest were unchanged.) Dr. Schless's treatment included dietary measures and the use of the oral hypoglycemic drug phenformin.

Reactive

The hallmark of reactive hypoglycemia is a marked initial rise in the blood sugar. The GTT curve is termed prediabetic because of this high glucose within the first 2 hours. The decline is gradual over a few hours, or there may be a precipitous drop. As is the case with any hypoglycemia, the absolute level of sugar reached may not be as important as the rapidity and magnitude of the drop.

Reactive hypoglycemia frequently occurs after emotional stress, accumulated severe stress, dietary indiscretion, or alcoholism.

Blood-insulin levels in this condition are both higher and later than normal. The insulin level peaks between 2 and 4 hours. These patients have some failure in turning off the insulin secretion.

[64]"Functional Hypoglycemia Need Not Become Overt Diabetes" by Guy L. Schless, M.D. Published in *Consultant,* February 1973.

High Rise with Plateau

These patients have a sustained elevation of blood glucose followed by a marked drop. This picture often occurs in liver damage — for example, in alcoholism. Dr. Goldman believes that the blood-insulin levels in this category are essentially the same as in the reactive type.

DIET AND ADRENAL CORTEX

The careful reader of this chapter will appreciate the many metabolic disorders of which hypoglycemia may be a part. In spite of all these considerations, the usual case of hypoglycemia encountered in office practice can be considered to be due to improper diet (with "hyper"-insulinism) and/or to insufficiency of the adrenal cortex. In either case, the relative balance between insulin and adrenocortical hormones appears to be disturbed, with hypoglycemia the result.

Some scientists believe that the only true way to identify hypoglycemia is by measuring the blood levels of cortisol· (hydrocortisone). In true hypoglycemia, there should be a rise in cortisol blood levels. Since cortisol is believed to be the principal adrenocortical hormone regulating glucose in the body, this seems quite reasonable. My experience with the use of whole adrenocortical extract (ACE), however, leads me to believe that there is much more to hypoglycemia than cortisol levels. Subsequent years will tell the story.

PSEUDOHYPOGLYCEMIA[65]

If a blood specimen is allowed to stand after it is drawn from a patient, a low-blood-sugar report can result that does not reflect the actual blood-sugar level within the body. When whole blood is allowed to stand, the glucose concentration progressively decreases. A significant change can occur within an hour. The warmer the environment, the more rapid is the decline in sugar level.

[65]"Leukocytosis and Artifactual Hypoglycemia" by Thomas J. Goodenow, M.D., and William B. Malarkey, M.D. Published in *Journal of the American Medical Association*, Vol. 237, No. 18, May 2, 1977, pages 1961 and 1962.

This pseudohypoglycemia can be prevented by promptly separating the liquid portion of the blood (plasma or serum) from the blood cells. Delay in transporting or processing blood could cause this factitious depression of blood glucose. If glucose determination is not promptly performed on the same day that blood is drawn, use of a chemical additive or freezing the specimen can prevent glucose decline.

Well-run laboratories whose directors appreciate the importance of optimal blood-sugar levels will take steps to avoid such causes of artificial hypoglycemia.

Elevation of the white-blood-cell count to a level of more than 60,000 cells per cubic millimeter of blood has also been associated with a decline in blood-glucose values. Such white-cell elevations are found in patients with leukemia. For the vast majority of individuals, this factor is not operative.

TREATMENT

When an individual has low blood sugar, he needs sugar and he needs it fast! Dramatic relief of cerebral and autonomic nervous symptoms is obtained by oral or intravenous sugar. In fact, an individual who is comatose or convulsing may, on occasion, be given a load of glucose by intravenous injection as a therapeutic trial, to see if the condition improves when sugar is given. By the time a laboratory determination of blood sugar can be obtained, the patient's clinical response to the injection may inform the physician of the patient's hypoglycemia.

In severe pathologic or absolute hypoglycemia (blood levels of glucose below 40 mg.%), as well as in convulsive states associated with hypoglycemia, permanent brain damage may result. Dr. Dale R. Leichty (M.D.) at a meeting of the Wyoming State Medical Society indicated that neuroglycopenia (low blood sugar in the nervous system) can cause permanent brain damage very similar to that caused by oxygen deprivation in infants and young children.[66] Prompt treatment may preserve brain tissue.

The majority of individuals with hypoglycemia do not have serious acute illnesses from their disorder, but they have chronic lingering and recurrent "misery" of one sort or another. The treat-

[66]"Brain Damage Seen in Neuroglycopenia Similar to That in Oxygen Deprivation," a report in *Pediatric News*, Vol. 11, No. 5, May 1977, Page 11.

ment of their problem is not as urgent as the comatose or convulsing patient, but it is equally as important.

We wish to get sugar into the *acutely* ill hypoglycemic as quickly as possible. In the *chronically* ill, *milder* hypoglycemic, we wish to keep sugar away from him at all costs. In the dietary treatment of hypoglycemia, it is probably the absence of sugar in the diet that is the single most important factor. Sugar stimulates insulin release that drives the blood sugar down. Body chemical stability (homeostasis) is fostered by a diet free of refined sugar.

Because the hypoglycemic is usually addicted to sugar, the patient may temporarily feel worse when he stops eating sugar. Such withdrawal discomfort may occur within the first week or so after sugar removal and may recur from time to time in later weeks and months during periods of low biorhythm[67] and during other stressful times. This is especially so if the individual has deficiency of vitamins and minerals.

In some individuals, it is necessary to restrict not only dietary sugar from refined sugar sources, but also to restrict naturally occurring sugars and starches such as occur in fruit, milk, bread, cereals, etc. Such a highly restricted carbohydrate diet may be quite helpful, but should only be done under the supervision of a qualified physician. Dr. William H. Philpott (M.D.) has pointed out that *any* food may be responsible for blood-sugar deviation in a particular individual.[68]

The hypoglycemic individual should be encouraged to eat as often as he likes, so long as the proper foods are taken. Avoidance of hunger is a prime necessity if the individual is to succeed in staying away from sweets in our society. Often, a programmed eating structure involving six or eight small meals a day is helpful. In some cases of hypoglycemia, especially those with a flat GTT curve, it may be necesary for the patient to eat every two hours for a temporary period in order to restore body reserves of glycogen.

A full complement of vitamins, minerals, enzymes, and diges-

[67]See *Is This Your Day?* by George S. Thommen. Revised edition published in 1973 by Crown Publishers, Inc., New York, N.Y.

[68]"The Role of Allergy-Addiction in the Disease Process" by William H. Philpott, M.D. Published in *New Dynamics of Preventive Medicine: Medical Progress Through Innovation*, Vol. 5, 1977, pages 89 - 104, edited by Leon R. Pomeroy, Ph.D., International Academy of Preventive Medicine, 10409 Town and Country Way, Houston, Tex. 77024.

tive substances, when indicated, should be supplied to the hypoglycemic. The B vitamins may be particularly important. The patient's liver and adrenal glands should be well fed. Hair-mineral analysis, wisely interpreted, can be of assistance in guiding the physician as he tailors the patient's diet and nutritional supplements to make up for deficits as well as to rid the patient of certain mineral excesses.

Dr. Seale Harris (M.D.) in 1924 used a high-protein diet to treat his newly described hypoglycemic patients. Dr. Jerome W. Conn (M.D.) in 1936 demonstrated the effectiveness of the high-protein diet for hypoglycemia. The high-protein diet has remained the treatment of choice to the present time. Slowly metabolized protein furnishes blood sugar to the body in a gradual, nonflooding manner. Peaks and valleys do not occur, and thus insulin secretion due to abrupt rise in blood sugar does not occur.

This is all well and good when protein itself does not trigger hypoglycemia. As Dr. Philpott (M.D.) and others have emphasized, however, a particular food protein may be responsible for blood-sugar disturbance. If a protein is the cause of low blood sugar, then increasing the amount of that substance in the diet certainly does not improve the condition.

The benefit seen with a high-protein diet in hypoglycemia may occur because the individual was effectively getting an inadequate amount of protein in the first place. Inadequate quality or quantity of protein in the diet or faulty protein digestion in the gut could account for protein deficiency in the body.

The favorable effect of a high-protein diet may have other origins. Whenever protein is increased as a source of nourishment, some other food is eaten in smaller amounts. If protein replaces sugar, other refined carbohydrates, and starches in the diet, the increased protein may be given credit for improvement when in actuality the avoidance of refined carbohydrates was the effective therapeutic change.

Minerals in the body also change when a high-protein diet is instituted. The excretion of calcium in the urine is increased. Changes in the level of one mineral often result in shifts of others.

My experience indicates that a proper "junk"-free diet may not need to be a high-protein diet at all. Many hypoglycemics respond very well to a diet that is relatively low in protein, when all nutritional needs are being met. Furthermore, a high-protein diet may actually be injurious, and, over time, it may lead to degenerative diseases. As Dr. Paavo Airola (Ph.D.) has said, "Because something is good doesn't mean more is better! It is not protein but too much protein which is bad."[69] Dr. Airola as well as The American Natural Hygiene Society[70] favor predominantly vegetarian, somewhat low-protein diets.

The availability of pleasant-tasting, high quality, protein supplements assists many hypoglycemics in weaning themselves away from sweets. Nutritional supervision is mandatory. Chemical determinations of the blood levels of total protein, albumin, globulin, blood urea nitrogen, uric acid, cholesterol, and triglycerides assist the physician in adjusting the dietary levels of protein, fat, and complex carbohydrate.

For some persons, predigested protein has been helpful in nourishing the lean-body mass. Probably this is so because the substance "slips into" the body without the necessity of using "fouled up" digestive and gastrointestinal-excretory machinery. Physician supervision is mandatory.

Surgical removal of an insulin-secreting islet cell tumor cures hypoglycemia due to insulinoma. This is a rare occurrence. If the tumor is cancerous and nonoperable, the use of the antibiotic drug streptozotocin may be helpful. This drug is toxic to the beta cells of the pancreas, and may also injure kidney tissue.

Diazoxide increases blood glucose by decreasing the secretion of insulin and other mechanisms. Use of a diuretic with it may enhance the elevation of blood sugar and minimize sodium retention. Diazoxide may cause an increase in body hair as a side effect.

Adrenocortical hormones can be very helpful in the management of hypoglycemia. In states of adrenocortical hypofunction, they may be given as replacement therapy. There are a considerable number of physicians who "swear by" the use of adrenocortical extract. Standard medical dogma has declared the use of this substance "obsolete." Further research needs to be carried out to establish the therapeutic role of whole adrenal-

[69]*Hypoglycemia: A Better Approach* by Paavo Airola, Ph.D. Published by Health Plus, Publishers, P. O. Box 22001, Phoenix, Ariz. 85028.

[70]The American Natural Hygiene Society, 1920 West Irving Park Road, Chicago, Ill. 60613.

cortical extract (see Chapter 8).

Individuals whose hypoglycemia is caused by ingestion of some particular substance are "cured" by elimination of the offending substance. Accordingly, persons with fructose intolerance, galactose intolerance, leucine hypersensitivity, etc., respond favorably to the elimination of fructose, galactose, leucine, etc., from their diet. The hypoglycemic patient must also assiduously avoid foods to which he is allergic in order to reduce the load of adverse stress on the body.

Very tiny microdoses of hormones have been used to bring about favorable changes in blood-sugar levels. Estrogen, testosterone, thyroid, whole pituitary, posterior pituitary, progresterone, insulin, and adrenal cortex extract may be used. The mechanism of the effect of these minute doses has not been worked out. It may have to do with the amplification cascade produced by adenylate cyclase, cylic AMP, and other chemicals within the cell. The effect of a hormone is amplified by at least 100 times at each of four enzyme reactions. Thus, a single molecule of a hormone could trigger the release of about 100 million glucose molecules! In other patients, more conventional doses of hormones (estrogen, thyroid, adrenocortical) may be needed. Some hypoglycemics require rather large doses of adrenocortical extract.

In individuals whose nervous state is accompanied by low blood sugar, it is, of course, wise to provide appropriate psychological counseling to reduce adverse tension. Various visual treatment measures may also alleviate unwanted stress. The program espoused by Dr. John McAmy (M.D.), Human Life Styling,[71] is a profound advance in the maintenance of health.

Experienced tennis players, as well as other athletes, know well the "unexplainable" slumps of poor performance that mysteriously plague them. Comments such as, "Oh well, you can't win all the time," or "Anyone can win on any certain day," or "Some days you just don't have it," attempt to ease the disappointment of poor athletic performance. Although many factors are involved, such as nutrients, training, sleep, nervous tension, aspects of home and job life, etc., a frequent explanation for swings in athletic performance may be related to blood sugar and glycogen balance in the body.

[71]Human Life Styling, P.O. Box 6585, St. Petersburg, Fla. 33736.

Vigorous exercise depletes the body stores of glycogen. Appropriate training, however, makes one more resistant to glycogen depletion. Glycogen can be restored by eating appropriate foods. A diet high in natural carbohydrates in the first 12 hours following exercise appears to be beneficial in restoring glycogen supplies and warding off subsequent relative or absolute low blood sugar with exercise. A period of relative rest for 2 days after severe muscular exertion may be needed for restoration of glycogen supply in muscles.

In some persons, the ingestion of protein or refined sugar can bring about fatigue and decrease in concentration which interfere with athletic effort. In yet other persons, even the sugar that is present in fruits may be responsible for erratic performance. Some athletes believe they do well with honey.

Young athletes are particularly unaware that dietary factors may be very important in determining the result of their efforts on the athletic field, arena, or court. The boundless enthusiasm and energy of youth seem so great that many individuals work right through the depressive effects of certain foods on athletic performance. Nevertheless, when a young athlete becomes aware of the improved play that may result from a selected diet, he, too, may experience more stable athletic performance.

The last word has yet to be said in regard to low blood sugar and its treatment. The reader should listen cautiously to those who appear to possess the ultimate truth. A particular understanding, a certain theory, or a way of thinking may be very helpful to those in need, yet not agree with scientific understanding.

Most health-related persons in our society genuinely wish to help others who are not well. Rigid beliefs and fixed adherence to ideas which have been taught may limit these individuals in rendering service to others. It is hoped that this book will serve to unbend minds and stimulate positive thinking. As a result, one may forge a more productive and happy life for himself and thus be better able to help his fellow man.

HYPOGLYCEMIC PATTERNS

The following GTT and NTT curves showing hypoglycemic patterns are representative of those seen not infrequently in persons who have problems in adaptation to school, job, or home life. They represent individuals who have symptoms and disorders of ill health. Their health problems may be primarily biochemical, allergic, mental, social, nutritional, degenerative, or eductional, but commonly their physiological problems, when corrected, lead to improved mental and social performance.

Many of these cases have had elevated levels of eosinophils ("allergy cells") in their blood, indicating an allergy, low adreno-cortical function, or both.[72] I increasingly suspect that a diet which contains large amounts of sugar; other refined carbohy-drates; and artificial colors, flavors, and preservatives tends to be associated with a higher-than-normal level of eosinophils in the blood. Allergy to sugar itself may be responsible for increased levels of eosinophils in the blood, although this has not been proven.

Figures 115 through 131 describe patients with these hypo-glycemic patterns as determined by standard glucose or natural tolerance tests. These patients demonstrate the whole spectrum of hypoglycemia. They have not been limited to the narrow definition of absolute hypoglycemia. The comments in the legends of the figures deal further with this matter of absolute versus relative hypoglycemia and the interpretation thereof.

[72]Examination of bowel movement specimens in every case has eliminated intestinal parasites as a cause of the elevated level of eosinophils.

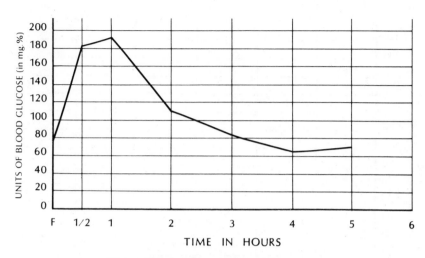

Figure 115. A glucose tolerance test curve

An 11-year-old retarded boy in special education classes.

Prediabetic or diabetic GTT due to excessive rise of blood sugar after glucose load. The dip at 4 and 5 hours to levels below the fasting level suggests additional hypoglycemia, although some investigators would indicate that this is not so. This pattern is fairly characteristic of the prediabetic or mild diabetic patient. It also resembles the pattern known as dysinsulinism. Insulin release in this patient is probably delayed (too late).

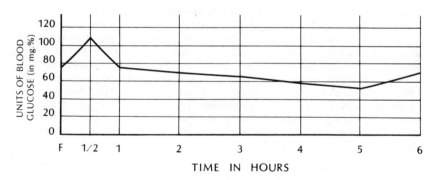

Figure 116. A glucose tolerance test curve

An 11-month-old boy with a disease known as cystic fibrosis and with allergic bronchial asthma.

GTT is generally somewhat low. Note the steady decline in blood sugar to the peak low at the 5th hour. Poor glucose absorption is a possiblity as well as hyperinsulinism. At a later date in this child's life, the GTT might become diabetic.

An optimal GTT curve is believed to have a peak glucose level at 1 hour rather than 15 or 30 minutes.

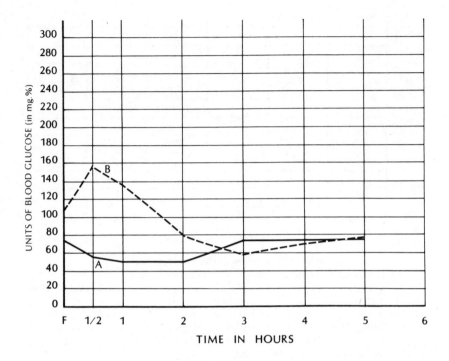

Figure 117. Glucose tolerance test curves

A 16-year-old slow-learning girl with marked fatigue. The lower curve (curve A) is the initial one taken in June. Note that her blood sugar *fell* after drinking an oral glucose load, and it remained low until the 3rd hour of the test! This undoubtedly represents hyperinsulinism — that is, excessive insulin secretion or an increased sensitivity of the body to normal amounts of secreted insulin with or without delayed absorption. The upper curve (curve B) is the repeat test done in August after 10 weeks of sugar elimination from the diet. Note the dramatic change in the configuration of the curve. The girl's fatigue was much less prominent at that time, and there were suggestions that her school work was improving.

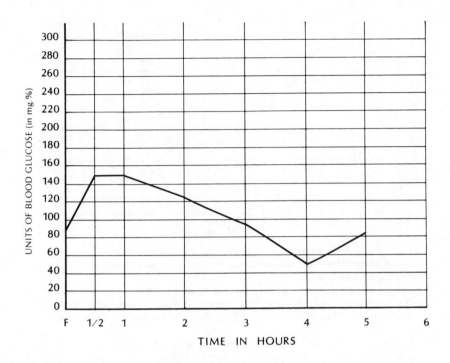

Figure 118. A glucose tolerance test curve

A 12-year-old 7th grade girl with obesity and chronic, severe, allergic disorder of the nose. Symptoms of nasal congestion, sneezing, drainage, and itching were more severe just before mealtimes.

The GTT curve shows a dip at 4 hours to 50 mg.%. Fasting level is 90 mg.%. The child complained of hunger, sleepiness, increased nasal symptoms, and boredom at 4 hours.

If one uses the criterion that blood sugar must drop below 40 or 45 mg.%, this child does not have hypoglycemia. Nevertheless, she has some condition or state associated with symptoms and a blood sugar of 50 mg.%. Many clinical physicians would term this relative, fed, reactive, or functional hypoglycemia. At any rate, it usually responds to dietary improvement. It also represents dysinsulinism.

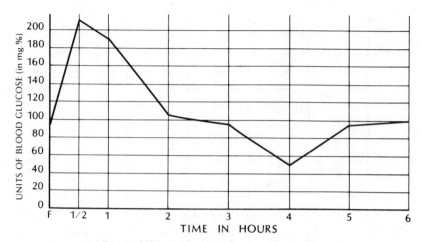

Figure 119A. A glucose tolerance test curve

A 60-year-old woman with postnasal drip, chronic sore throat, headaches, and chronic subnormal temperature.

GTT is typical of dysinsulinism — that is, abnormally high in the early phase and quite low in the late phase. This curve also fits the description of those described as hypoglycemia with mild diabetes as described by Drs. Fajans and Conn.

This individual was helped to become symptom-free only after allergic desensitization. Dietary improvement along with vitamin-mineral supplementation was ineffective until allergic treatment was given. The GTT will be repeated at a later date.

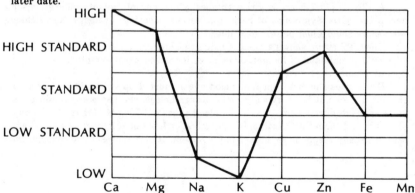

Figure 119B. A hair-mineral pattern

This is the hair-mineral pattern of the 60-year-old female whose GTT is shown in Figure 119A.

The low potassium (K) and high zinc (Zn) along with high calcium (Ca) are suggestive of hypoglycemia. Low sodium (Na) and potassium, along with somewhat low manganese (Mn) and elevated calcium and magnesium (Mg), is a pattern seen very often in maladapting individuals.

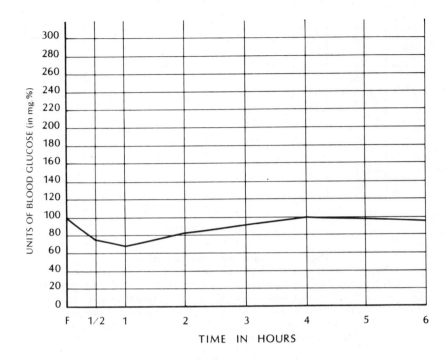

Figure 120. A natural tolerance test curve

This is the natural tolerance test of an 18-year-old boy with fatigue, constipation, and failure to gain weight. He had been treated by allergy shots for several years without relief of recurrent respiratory infections.

His breakfast for the test consisted of orange juice, 1 fried egg, Special K cereal with 2 teaspoons of sugar, ½ pint of milk, 2 slices of buttered toast with apple jelly, and a Danish pastry roll.

Notice the prompt decline in blood-sugar values from a fasting value of 100 to a low of 68 at 1 hour.

This is a form of hypoglycemia precipitated by food and very probably associated with excessive insulin release or enhanced susceptibility to normal insulin release (hyperinsulinism).

Dietary improvement with appropriate vitamin and mineral supplements improved this boy's fatigue and constipation. He was also found to have poor digestion of his food and was successfully treated with digestive enzymes. His recurrent respiratory infections disappeared.

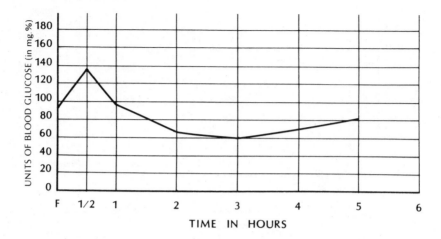

Figure 121. A glucose tolerance test curve

A 9-year-old boy, hyperactive, disruptive in class, with allergic nose.

The dip from 2 to 3 hours represents a reactive, fed, relative hypoglycemia. The case for this is strengthened if the patient has symptoms of disability during these hours. The insulin release in response to sugar ingestion is probably excessive, or the body's response to normal amounts of insulin is excessive. Frequently, adrenocortical function is lower than optimal.

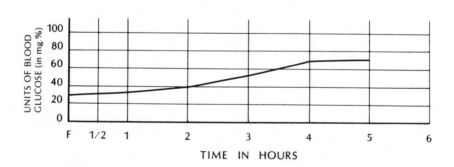

Figure 122. A glucose tolerance test curve

A 9-year-old boy with severe behavorial problem, chronic nasal allergy, abnormal brain wave, neurological dysfunction, and craving for sweets.

Severe hypoglycemia in response to GTT. Unfortunately, this boy failed to return for followup treatment.

Is there a relationship between neurological dysfunction and hypoglycemia? Certainly, low blood sugar of this degree can interfere with brain function. Is it possible that neurological dysfunction may produce hypoglycemia? The answer is probably yes, but the frequency of such occurrence is unknown.

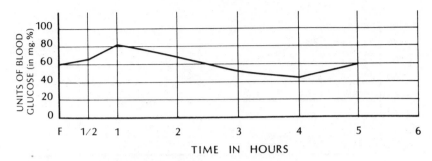

Figure 123. A glucose tolerance test curve

A 16-year-old girl diagnosed as mentally retarded. I.Q. 75. Third grade reading level. Normal brain wave. Consumes much candy.

Curve generally depressed with low rise after glucose load and rather impressive lowered levels of blood sugar (44 mg.% at 4 hours).

Her mental functioning improved after dietary treatment. A gain of 15 I.Q. points was recorded 6 months later.

It is possible to improve a person's I.Q. score by various means of intervention. Changing the diet is one way to improve mental function in some individuals.

This patient did not have a followup GTT done.

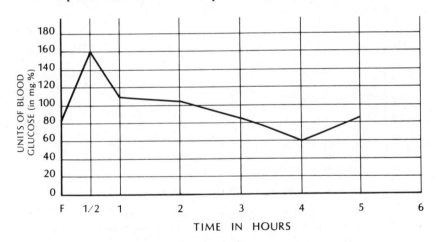

Figure 124. A glucose tolerance test curve

A 7-year-old boy with chronic runny nose, cough, shortness of breath, and fecal soiling. Emotional problems. Craving for sweets.

This GTT curve would be called normal by some. It could also be interpreted as hypoglycemia at 4 hours with the level of 60 mg.% being 23 mg.% less than the fasting level of 83 mg.%. It may be desirable to have the sugar level reach its peak later than it has in this curve, to have the peak slightly lower, and for the 4-hour dip not to drop so low. After restriction of sugar in the diet for several weeks or months these desirable changes often occur.

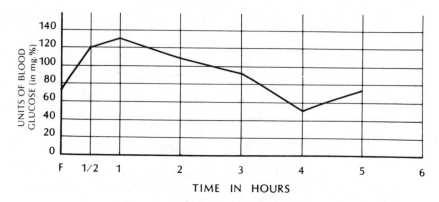

Figure 125. A glucose tolerance test curve

A 15-year-old boy with hay fever and marked fatigue.

GTT shows a large drop in the 4-hour value to 50 mg.%.

Is this normal or does it have significance in regard to this patient's physical and mental health? After dietary improvement, this degree of decline in the 4-hour dip is often seen to diminish. Many individuals are less fatigued when sugar and other refined carbohydrates are removed from the diet.

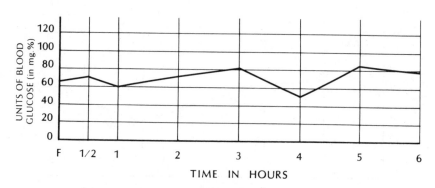

Figure 126. A glucose tolerance test curve

A 13-year-old 7th grader with learning disability, social withdrawal, and allergies to foods.

GTT is flat and has a dip at 4 hours to a sugar level of 49 mg.%. This type of curve is often seen in children whose diets are high in refined carbohydrates. The patient with this type of curve frequently also has social and/or learning difficulties.

Alleviation of social and educational stress has been followed by significantly changed GTT's. It occurs more readily when dietary change is also made. Children on restricted diets because of food allergies may become nutritionally deficient. In my experience, it is rarely necessary to use restriction of basic foods for long periods, although the amounts may need to be curtailed.

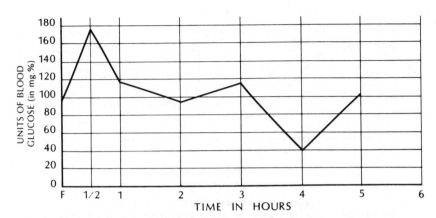

Figure 127. A glucose tolerance test curve

A 15-year-old somewhat mentally dull boy with bleeding gums and dissolution of bone around the teeth. He had been a heavy sweet-eater from infancy.

The GTT curve is erratic, with excessive rise and dip to 40 mg.%. This patient reported no symptoms at the 4th hour. He may have grown accustomed to his pattern of chronic swings from high to low blood sugar. His mental dullness, perhaps due to low blood sugar, may impair his ability to sense and report symptoms of low blood sugar.

Some individuals can become ensnared in a cycle of low blood sugar, eating sugary foods, additional low blood sugar, mental dullness, poor diet, and so on. Blood calcium, phosphorous, and dental health are intimately related to dietary consumption of sweets and other refined carbohydrates, as well as to the adequacy of the diet in regard to vitamins and minerals.

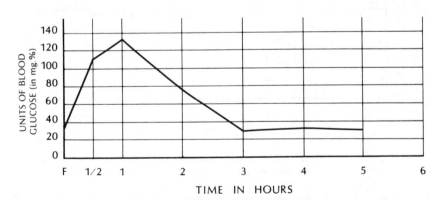

Figure 128. A glucose tolerance test curve

A 10-month-old girl. Asthma and neuromotor retardation of development.

Abnormally low fasting blood sugar and dip to severely low levels.

Investigation at a medical center is being conducted to attempt to define the cause of the hypoglycemia and the developmental delay.

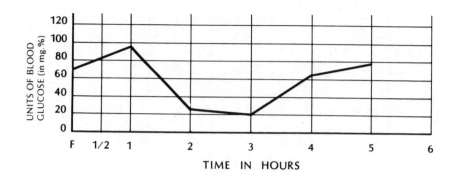

Figure 129. A glucose tolerance test curve

This is the glucose tolerance curve of a 49-year-old alcoholic woman. Notice the somewhat small initial rise in sugar level, followed by a severe, rapid decline to 25 mg.% at 2 hours and a further decline to 20 mg.% at 3 hours. The recovery of the blood glucose to higher levels is probably due to glycogen breakdown in the liver. This type of curve is quite characteristic of the alcoholic who is not drinking. The curve suggests hyperinsulinism rather than liver damage as the cause of the hypoglycemia. This patient's liver function tests were entirely normal.

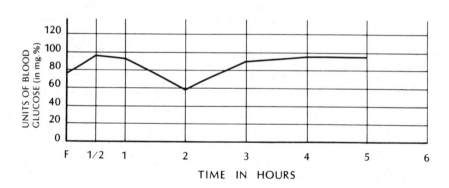

Figure 130. A glucose tolerance test curve

An 8½-year-old slow-learning girl with temper tantrums and feelings of being picked on by others.

Glucose tolerance test shows very little rise, early dip at 2 hours, and a generally flat curve.

Dietary improvement and megavitamin therapy were ineffective in management, but psychological counseling for the family resulted in alleviation of the problems. After counseling was completed, the parents noted that the child's behavior was "unsteady" when she ate colas, chips, candy, or cookies.

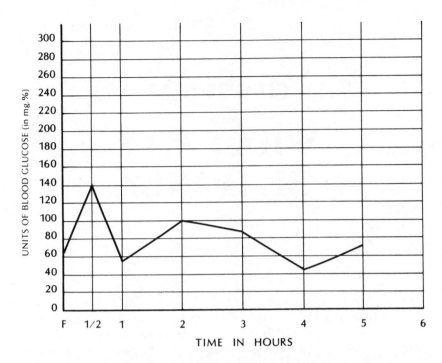

Figure 131. A glucose tolerance test curve

An 8-year-old boy with psychomotor seizures, learning disability, and stomaches. History of a traumatic birth.

Saw-tooth GTT curve. Spike-drop at 1 hour; low sugar of 43 mg.% at 4 hours. Certainly this lad is experiencing a markedly unstable glucose homeostasis. Low blood sugar is a known precipitating factor for seizures.

His stomachaches disappeared and his seizures lessened in frequency when he was placed on a high-protein diet with frequent feeding every 2 hours while awake.

In some patients with early morning seizures, a high protein snack at 11 p.m. to midnight may alleviate the seizures. Many seizure patients have nutritional disturbances and blood-sugar deviations that contribute to their convulsive problem.

CHAPTER 15

WHAT CAN BE DONE?

In a society that consumes sugar as a way of life — a sugar-permeated society — it is difficult to make headway in reducing sugar intake. Anyone who has witnessed ants marching to spilled jelly on the kitchen floor knows the powerful attraction of sweet. When sugar is injected into the amniotic fluid surrounding a human fetus, the little tyke gulps down the sweeetened fluid, swallowing it at a faster rate than he does the unsweetened fluid. Sugar has an addicting effect that is not easy to break.

In this last chapter, I intend to repeat some subject matter presented earlier in the book. I do so, in some cases, because of the importance of the information. In other cases, I will repeat in order to assist the reader in deprogramming himself from sweet. The profound ingraining of sweet in our culture makes it difficult for even the most dedicated person to change his dietary ways.

It can be speculated that our liking for sweet at one time may have contributed to our survival. Sweet-toothed early forerunners of man may have been drawn to the sweet and relatively harmless fruits of the jungle and away from bitter poisonous plants. At the present time, however, with the ready availability of refined sugar in our culture, our sweet tooth threatens rather than protects.

Preceding chapters have pointed out that refined sugar is an unnatural food. Indeed, sucrose can be considered an undesirable foreign substance when separated from the leaf, root, or stem where it is naturally found.

The present high cost of sugar in our economy (when this was written) may assist in reducing the sugar habit; however, I'm not going to hold my breath and count on it. The economic problem only challenges man's ingenuity. He presently seeks a way to alter the production of concentrated orange juice to provide refined sugar from citrus. Such is man's sugar habit — and the economic gains to be derived from it.

Nevertheless, despite the sugar addiction of modern man, the antisugar campaign is growing stronger. The research data of Dr. John Yudkin (M.D.) in England, the writings and lectures of Dr. Carlton Fredericks (Ph.D.), the *Journal of the International College of Applied Nutrition*[1] and journals of other similar groups, the lectures and writing of Dr. John McAmy (M.D.),[2] and the many popular books about low blood sugar are having an appreciable impact in educating our society about the possible dangers of sugar in the diet. Moreover, the growing numbers of physicians who practice preventive medicine[3] are improving the health of present and future generations through nutritional education as well as improving the lifestyles of their patients.

For the individual patient, much can be accomplished. States of low blood sugar and elevated blood sugar can be improved. There is preliminary clinical evidence that some states of chemical diabetes can be normalized. The case studies given in Figures 132 through 135 are illustrative of the progress that can be made.

It should be understood, of course, that some individuals with profound genetic influence or damage to the endocrine system may develop diabetes and other disorders despite the best preventive health programs.

---

[1]P.O. Box 386, LaHabra, Calif. 90631.

[2]*Human Life Styling* by John C. McAmy, M.D., and James Pressley. Published in 1975 by Harper & Row, Publishers, Inc., New York, N.Y.

[3]Many physicians who practice preventive medicine are members of the International Academy of Preventive Medicine, 871 Frostwood Drive, Houston, Tex. 77024.

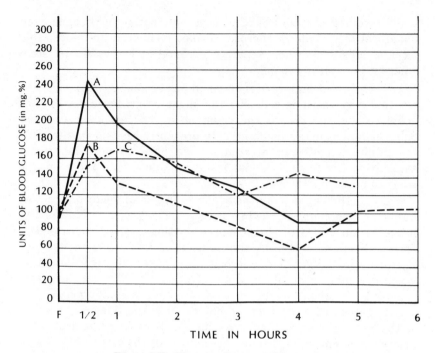

Figure 132. Glucose tolerance test curves

Randy, a 12-year-old boy, was seen because of chronic headache, irritability, and academic difficulties.

Three GTT's are shown over the course of 1 year. They are labeled curves A, B, and C. Curve A shows an early and abnormally high peak of blood glucose. After the first test, Curve A, the boy was treated with dietary restriction of sugar and the administration of appropriate vitamin and mineral supplements. All of his symptoms improved. Note the progressive lowering of peak sugar values in GTT B, done 5 months later, and in GTT C, done 1 year later. Also note that the peak values are reached at ½ hour in Curve A and Curve B; but in Curve C, the highest sugar value is reached at 1 hour. These changes (that is, the decline in peak values and the shift in peak time to the right) may be indicative of an improvement in the usage of glucose by the body. In Curve B there is a 4-hour dip to 69 mg.%, but this has disappeared in the subsequent GTT, Curve C. It is expected that further improvement in symptoms and GTT and a slight reduction in fasting glucose level will occur as the boy's diet and his body mineral balance are further improved.

This type of change toward normal in diabetic or prediabetic GTT's is often obtained when nutritional treatment and appropriate vitamin-mineral therapy is carefully administered by a judicious therapist.

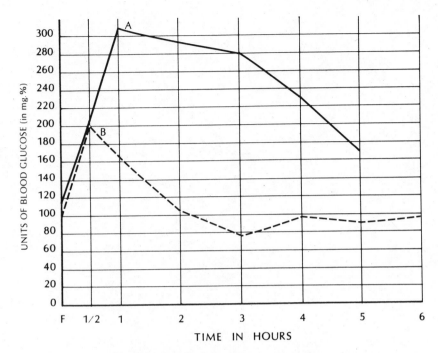

Figure 133A. Glucose tolerance test curves

An 11-year-old boy with severe hyperactivity, very much impaired attention span, rocking behavior, speech problem, and autistic traits. Surgery had been performed for crossed eyes. Severe nasal allergy was accompanied by an eosinophil blood level of 12% and a strongly positive nasal smear for eosinophils.

Curve A is the first GTT done in May 1973. It shows a definitely diabetic curve. One year later, the GTT (Curve B) has shown marked improvement and has returned considerably toward normal. Dietary improvement, vitamin and mineral supplements, allergy treatment, and a supervised exercise program had been carried on between May 1973 and May 1974. It is expected that the GTT will continue to show further improvement. Chromium supplementation is now being given in view of a low level of chromium in the patient's hair-mineral analysis, and the known association of chromium deficiency with diabetes.

**Figure 133B. Human figure drawing done by the 11-year-old boy
whose GTT's were shown in Figure 133A**

This boy's drawing indicates a serious degree of developmental retardation. His figure is like that drawn by children of an early school age and not like that of a normally maturing 11-year-old.

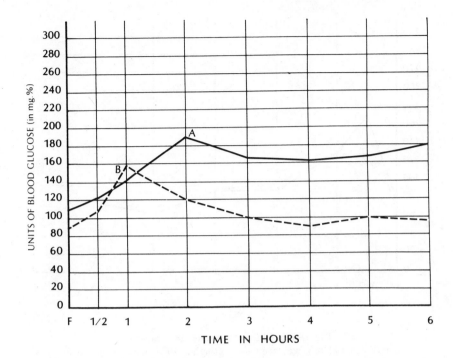

Figure 134. Glucose tolerance test curves

Rob, a 5-year-old boy, was tested with glucose tolerance tests because his mother was a diabetic. She wished to do whatever she could to prevent diabetes in her child.

The initial GTT, Curve A, was done in the spring of 1973. It shows a diabetic type of curve. The fasting blood sugar is suspiciously high, and the subsequent values rise too high and remain elevated throughout the test. Since this boy had no clinical symptoms of diabetes, no sugar or acetone in the urine, and was a "well child" in outward appearance, he could be termed a chemical diabetic.

The boy's mother was instructed to have him avoid all dietary sugar, to avoid all refined cereal and bread products, and to take vitamin and mineral products appropriate for him.

One year later, the GTT, Curve B, was obtained. It shows marked improvement and can be considered essentially normal. The fasting blood sugar is much lower, there is no excessive peak, and the sugar values return to normal within 2 - 3 hours after a glucose load.

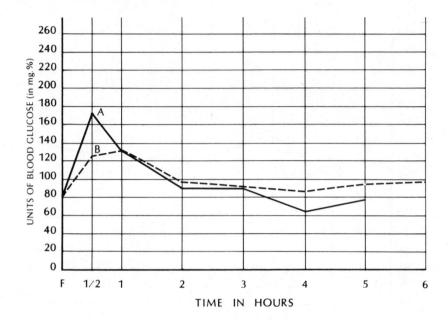

Figure 135. Glucose tolerance test curves

This 12-year-old boy was seen in September 1973 because of learning disability, fatigue, and poor appetite. His blood count showed 7% eosinophils ("allergy cells").

The initial GTT, Curve A, shows borderline elevation of the peak glucose level and a glucose level of 65 mg.% at 4 hours. After receiving allergy treatment and improved diet, 5 months later the GTT, Curve B, was obtained. It shows a lower peak level and somewhat higher levels at 2, 3, 4, and 5 hours. The boy had "come alive" according to his teachers and was learning at a greatly improved rate. His appetite was normal and his fatigue had vanished.

It is likely in this case that the boy was experiencing allergic fatigue, an example of the allergic tension-fatigue syndrome. The changes in GTT may be due to chance, but the regularity of this change, in my experience, indicates that it is due to dietary change and perhaps relief of the allergic process.

When one removes sugar from his diet, improved health is apt to result. However, because we wish to promote proper nutrition in regard to *all* nutrients, the following suggestions for dietary improvement are included in this book. These paragraphs on nutritional improvement are based on a booklet which I have written[4] and which is used in my practice.

[4]*Improving Your Diet* by Ray C. Wunderlich Jr., M.D. Published in 1976 by Johnny Reads, Inc., Box 12834, St. Petersburg, Fla. 33733.

IMPROVING YOUR DIET

It is becoming more and more difficult to obtain from our diet the nutrients that our bodies need for the best physical and mental function. These nutrients are vitamins, minerals, enzymes, and other food substances. Even the traditional "balanced" American diet (who has it?) may not be adequate to nourish our bodies and minds. Refining; processing; the storage, packaging, and transporting of food; food preparation; cooking; and the monotonous dietary habits of our convenience-oriented society too often encourage the development of nutritional problems.

The United States of America is the breadbasket of the world, yet food is often produced in quantity with insufficient attention to quality. Overworking of soils and "artificial" fertilizers can produce nutrient depletion in foods before they ever reach our kitchens. Wheat grown on the soils of today may contain on the average substantially less protein than wheat grown in earlier years.[5] Food additives may enhance the flavor, appearance, and shelf life of foods, but these additives have been linked with hyperactivity and decreased attention span in children. The long-term safety of additives, singly and in combination, has not been shown.

In the United States, the intake of soft drinks has more than doubled since 1960. In the early 1900's, almost 40% of our calorie intake came from fruits, vegetables, and grains. Today, only a little more than 20% of calories are derived from these foods. Instead, more is coming from sweets and fats.[6]

Myocardial infarction (commonly known as heart attack) was not recorded before 1900. Since that time, heart attacks and the milling of flour (processing, refining) have become everyday occurrences.

Pollution in today's environment increases the body's need for some nutrients and adds toxic elements such as lead, mercury, nickel, and cadmium to our surroundings. In addition, some individuals have special health problems that interfere with the di-

[5]Personal Communication with Standard Process Laboratories, Inc., Division of Vitamin Products Company, 2023 West Wisconsin Avenue, Milwaukee, Wis. 53201.

[6]"Preventive Nutrition — The Real Challenge" by Esther Winterfeldt, Ph.D. Published in *Food and Nutrition News*, Vol. 49, No. 1, October - November 1977. Published by National Live Stock and Meat Board, 444 North Michigan Avenue, Chicago, Ill. 60611.

gestion and absorption of food, or that magnify the needs of the individual for certain food substances, such as vitamins. Then, too, **it is likely that we all have increasing difficulty in digesting, absorbing, and utilizing foodstuffs as we grow older.** As nutrient intake declines, our bodies are likely to take on a higher toxic load of noxious environmental poisons.

In the paragraphs that follow are some suggestions that have proven helpful for improvement of the diet and, hence, for improvement of one's health. In individual cases, these dietary suggestions may not be appropriate. In such cases, the advice of a competent nutritionally oriented physician should be followed.

Fresh, Raw, and Varied Foods

Whenever possible, use food that is *fresh*, such as *fresh* vegetables, *fresh* fruit, etc. When this is not possible, use frozen food. Canned food is the last choice. *Your* soup made from fresh ingredients can be tastier and more wholesome than that purchased in a can. It *is* worth the time and effort. It *is* worth planning and experimenting until the soup comes out right.

Man is the only animal that cooks his food. Cooking destroys enzymes and other nutrients. It is best to include some raw foods in each meal. The more the better. Think of raw foods as counterbalancing the cooked foods in each meal. If raw foods can supply digestive power, then the body will not have to work to do so.

Don't forget that raw salads can consist of much more than lettuce and tomato. Avocado, raw potatoes, celery, watercress, romaine, carrots, cauliflower, collards, spinach, squash, parsley, beets, and bean sprouts can be used. For salad dressing, use none at all, lemon or lime, or a small amount of vinegar and cold-processed unsaturated vegetable oil (safflower, sesame, corn, etc.).

Cooked food is generally undesirable, but overcooked food is especially dangerous for nutrient supply. Unfortunately, many persons in our society eat largely in restaurants or cafeterias that provide overcooked food. Cooking and especially overcooking may increase fermentation and putrefaction in the digestive tract, because the raw enzymes in food are destroyed.

Vegetables are particularly vulnerable to heat. Their natural enzymes and vitamins are readily destroyed, and mineral losses may be substantial in the cooking process. A vegetable that is crisp — that is, one that is uncooked or lightly cooked — has more nutritive value than its limpid cooked or overcooked counterpart. It is advisable to eat vegetables raw or lightly cooked by steaming or preparation in a wok.

Protein fibers are toughened by high temperature, and amino acids may be destroyed by heat. Cooking meat, however, does destroy some bad things along with the good things. Parasites that infest meat are killed by high temperature. Trichinosis is avoided by eating pork that is well-done. Many persons choose to avoid parasites and the need for cooking meat by not eating meat at all.[7]

It has been said that one should eat only that which will spoil and that one should eat it before it spoils. This is excellent advice, but it is hard to carry out in today's society, because nutritional habits are geared to convenience. Nevertheless, one should attempt to obtain food "right from the farm" whenever possible. Food grown in soils that have been overcharged with highly soluble chemical fertilizers is likely to be inferior in quality to food grown on organic soils. He who grows his own food is doubly blessed, first by the opportunity to directly toil for his "bread," and secondly by his harvest.

Vary the diet. Try to avoid eating in a fixed pattern. *Dietary monotony leads to nutritional imbalance and food allergy.* Man probably developed his body over millions of years by taking advantage of the foods that were available during certain seasons of the year. Use fruits, vegetables, and berries that are in season. Avoid the excessive use of any one food, and vary menus whenever possible.

Fresh fruits and vegetables contain pectin, a substance found within the cell walls of plants, that is helpful in the maintenance of health. Pectin reduces the absorption of toxic substances such as mercury, lead, and strontium 90 from the gut. It substantially contributes to the lowering of copper when that element is found to be elevated in hair analysis. It may also lower the blood cholesterol. An antibacterial effect against staph and strep germs has also been attributed to pectin.

At all times, serve what you know will give your family good nutrition rather than what is expedient or popular. Do not be ex-

[7]Each person is unique. Many individuals flourish without red meat in the diet. Others become weak and malnourished. A nutritionally oriented physician should be consulted for advice.

travagant with food, but it is wise to cut financial corners in areas other than diet. Proper nutrition will pay for itself in better health with fewer doctor, pharmacy, and hospital bills. I believe the day is coming when health insurance programs will offer markedly reduced premiums to those who practice healthful living.

Too often in our society, our pattern of food intake is dictated by school, job, convenience food outlets, advertising, or the company that we keep. All efforts should be made to provide proper nutritional input despite daily activities that make it easy to ingest improper or inadequate foods. When deemed worthwhile, an optimal diet can be obtained every day of our lives. The key is sufficient planning, and planning means that some energies are devoted ahead of time to the procurement of proper food. Shopping several times a week may be necessary in order to obtain the fresh grains, vegetables, and fruits that are the cornerstone of an optimal diet. Provide a model of excellence in nutrition for your family. If you don't, who will? Believe me, it's worth the effort. You can depend on that.

Whole Rather Than Parts

When man developed, he had no extractors, juicers, mills, and separators. His body grew accustomed to eating whole foods. For this reason, it is probably wise, as a general rule, to eat whole foods rather than parts of foods. Fractionated foods abound in our present society and undoubtedly contribute to ill health.

If milk is to be used, whole, unprocessed milk straight from a healthy goat or cow is desired. Since each of us cannot be an excellent dairy farmer, we may have to compromise on this goal. Pasteurization preserves milk, but homogenization and skimming are man-developed processes that may not be particularly healthful. Pasteurization and homogenization are discussed in the next section ("What About Milk?").

Skimmed milk has lost more than fat. The manganese, molybdenum, selenium, chromium, and vitamins A, D, and E are reduced or absent in skimmed milk compared with whole milk. Skimmed milk is also low in cholesterol and linoleic acid.

There is a high consumption of sweet (sugared) drinks in our society. One sweet drink leads to a desire for other sweet drinks. Sugar and liquid, in one form or another, are obtained in all. Sometimes the sweet drink is man-made; at other times, it is a naturally sweet fruit juice to which sugar has been added; and at other times, it may merely be unadulterated fruit juice. Whichever it is, such drinks are indeed satisfying and delicious. They appeal to our need for instant satisfaction. They slake our thirst temporarily while they are creating our thirst as time goes by. It is quite easy to become enamored of such sweet drinks, especially in hot climates. Such drinks surround us. They are commercially available. We see others with them. To become part of the crowd, we swill the sweet with the rest.

For the person who takes stock of himself, however, there is an alternative. One can eat fruit in its whole, natural package and obtain natural sweet and liquid that way. When one eats an orange instead of drinking orange juice, he is doing several things:

1. obtaining the freshest possible food untouched by heat or mechanical contrivances that must be used to extract juice,
2. taking more time to eat the fruit than would be expended in drinking juice,
3. avoiding additives that may be inserted into juice,
4. avoiding the necessity of having the substance to be eaten contained in a container such as a can or a plastic or paper carton, and
5. probably consuming less of the sweet substance than he would consume were it in the form of juice.

For these reasons, and perhaps others, fruit should be eaten whole, as intact fruit, and not as juice, a mere part of the fruit.

When animal foods are eaten, it would probably be desirable to eat the whole animal. In other words, muscle, bone marrow, organs, etc., should probably be consumed. Except for desperate, hungry hunters on safari, this is not likely to be done. Then, too, there is the problem of toxic chemical accumulation in the bodies of animals subjected to our "enlightened" culture. Such toxic chemicals are often concentrated in the visceral organs, particularly the liver. For these reasons, meat eaters will probably stick with muscle flesh for their daily fare. It is notable that a whole line of helpful nutritional supplements are available that contain the internal organs of animals reared in a pesticide-free environment.

In some health food stores, one can obtain flesh and organ meats from animals that are reared without exposure to toxic chemicals.

When meats are eaten, it is advisable to use whole, intact flesh. Grinding meat does not improve digestibility, and it may hasten spoilage. When ground meat is used, it should be eaten as soon as possible after it is ground. Processed meats and meat mixtures such as meat loaf, weiners, sausage, chili, hash, and luncheon meats (bologna, salami, canned meat preparations, liverwurst, etc.) should be avoided.

Consideration should be given to removing all commercially produced land-animal meat from the diet.[8] Statistics suggest that those who do so are healthier than those who don't.[9]

Prepared, refined cereals, regardless of how many vitamins are added, are not whole foods. They are far removed from the original live foods from which they were derived. It is the height of presumption (man is known for his inflated ego) to believe that we can restore a food to its entire set of natural values by the man-made process of enrichment. What about all the bulk and trace minerals? Enzymes? Intangible life energy forces inherent in living matter? What about the relationships of substances within the food? Of course, we do not understand the answers to those questions. Man is in his infancy when it comes to his knowledge of nutrition.

Refined cereals are inferior to unrefined whole-grain products. We know that a more complete package of minerals is available in the whole grain; vitamins and enzymes, too. Processed cereals also usually contain preservative chemicals that might have an adverse effect on body function. In general, it is wise to avoid the use of all refined cereals. This includes the readily available dry breakfast cereals that come wrapped in wax paper and are attractively packaged in cardboard boxes.

[8]Before making any major dietary change, the advice of a nutritionally oriented physician should be sought.

[9]"On Your Risk of Stomach Cancer from Untreated Beef . . . and how the natural antioxidants, vitamins C and E and selenium, help protect you!" by Raymond J. Shamberger, Ph.D. Published in *Executive Health*, Vol. XIV, No. 12, September 1978. Published by Executive Publications, Pickfair Building, Rancho Santa Fe, Calif. 92067.

Whole grains such as oats, rye, wheat, buckwheat, barley, flax, and millet may be used to make breakfast cereals or breads. They are an excellent source of nutrients for those who tolerate them. Whole grains can be purchased from a health food store. Soaking whole grains for varying periods of time in fresh water will soften them so they can be eaten raw or lightly cooked. Oatmeal can be prepared by pouring hot water on it first thing in the morning and eating it later. Grains may also be sprouted. Soaking or sprouting begins the process of germination, thus activating the metabolic machinery of the grain seed for active living.

Easy Whole-Food Recipes by Elinor Wunderlich, R.N., contains complete directions for sprouting. Published by Johnny Reads, Inc., Box 12834, St. Petersburg, Florida 33733.

For those who do not have access to whole, fresh grains, wheat germ can be a valuable addition to a meal. Wheat germ can be used as a cereal; mixed with other cereals; or used as an ingredient in casseroles, bread, soups, muffins, hamburger, etc. Soybean lecithin granules, soy grits, pure natural bran, and brewer's yeast powder can be added to cereal or other foods.

Some individuals will not be able to tolerate certain grains in the diet. Some do better with a completely grain-free diet. If grains are believed to be creating a health problem, the advice of a nutritionally oriented physician should be sought and followed.

If a fat supplement is to be taken, use butter (right from the farm if you can get it) rather than margarine. Butter contains considerable amounts of chromium, manganese, cobalt, copper, and molybdenum. Remember, though, it is pure fat — saturated, at that, and loaded with calories. However, butter does provide essential fatty acids ("vitamin" F) that are important nutritional factors. It also assists in the absorption of the fat soluble vitamins: vitamins A, D, E, and K. If butter is used, try to obtain sweet cream butter without added food coloring or salt.

When the diet is sufficiently varied and includes fresh, whole foods such as grains, seeds, and nuts, the use of a fat supplement is probably unnecessary. Some persons are fat-intolerant due to gastrointestinal, liver, or pancreatic dysfunction. For those individuals, excess fat in the diet may only lead to distressing symptoms, fatty stools, mineral loss, etc. Nearly all Americans need less fat and oil in their diets.

What About Milk?

Cow's milk is a very nourishing food that provides many of the food elements that are needed in the daily diet. Milk is an excellent source of high-grade biological protein. When other sources of nutrients are not available, milk can be used to supply them. The potential drawbacks of milk, however, make it preferable that one obtain the bulk of his nutrients from sources other than milk, unless he is directed otherwise by his physician. It is suggested that the reader obtain and study the book *Don't Drink Your Milk! The Frightening New Medical Facts About the World's Most Overrated Nutrient* by Dr. Frank A. Oski (M.D.) with John D. Bell. (Published in 1977 by Wyden Books.) Dr. Oski is Professor and Chairman of the Department of Pediatrics at the State University of New York (Upstate Medical Center). He has coauthored 3 medical texts and more than 300 scientific articles.

Beyond infancy, milk is not an essential component of the diet for a person whose diet is otherwise adequate. Remember, the calf, for whom milk is intended, drinks milk for only a brief part of its life.

One wonders about such things as the relationship of milk drinking to cancer. Adelle Davis, America's aspostle of good health to millions, died of cancer after a lifetime of drinking a quart of milk a day. It has been said that the cancer rate is low in countries where milk consumption is low and vice versa.[10] We need more information than we have available to us about milk, the varieties of milk, the constitutents of milk, the additives in milk, and the impact of milk on human function and health.

Primitive cows secreted barely enough milk to suckle their young, but today's dairy cow, shaped by agricultural science, frequently yields 20,000 pounds of milk in 10 months. Is this favorable? Desirable? What effect does enhancing the milk supply have upon the milk and upon those who drink the milk? In a way, the modern Holstein dairy cow has been transformed into a freak of nature. (Perhaps this is inevitable in a society such as ours that glorifies the breast!)

---

[10]"Is Milk Cancer's Ally? A Leading Question Deliberately Spoken to Provoke an Investigation" by Walter Clare Martin. Published in *Coronet*, March 1937.

The disaccharide lactose is the carbohydrate component of milk. When it is digested, lactose produces the monosaccharide sugars glucose and galactose. These sugars can be rapidly absorbed and may contribute significantly to the elevation of blood sugar. Because that which goes up must come down, milk may not be helpful to the hypoglycemic patient whose metabolism does not efficiently handle such blood-sugar increases. The hypoglycemic patient who thinks he is doing the right thing by eating snacks of cheese and milk may not be helping his basic condition.

Milk is also a significant source of the amino acid leucine, and leucine may be responsible for inducing low blood sugar in some individuals.

The galactose of milk may lower the permeability of the intestinal wall to potassium, but not to sodium. Milk is high in sodium. A large intake of milk may encourage sodium-potassium imbalance in the body, with potassium deficiency and its attendant muscular weakness, fatigue, etc. Those individuals with high blood pressure or a family history of high blood pressure would do well to limit their sodium intake.

Galactose in milk may encourage cataract formation in the eyes of persons who lack the enzyme to properly metabolize this substance.

Calcium deficiency is not a rare problem in our society.[11] Muscle cramps, behavioral irritability, fatigue, and nervousness are symptoms of calcium deficiency. In severe cases, twitching, spasms in the hands and feet, numbness, tingling, or frank convulsive seizures may occur.

Milk is often used to supply calcium in the diet. In fact, it is often relied upon as the principal source of this vital mineral. There is some question about the availability to the body of calcium in milk and especially in pasteurized milk. Heat is said to destroy the enzyme alkaline phosphatase that is necessary for calcium absorption.

Dr. Frank Oski (M.D.) points out in his book that cow's milk contains 1,200 milligrams of calcium per quart and that human milk contains only 300 milligrams per quart. Yet the human infant

[11]"New Reports on the Dangers of Too Little Calcium in Your Diet, What It May Do to Your Bones, to Your Teeth — and, Possibly, Your Heart!" Published in *Executive Health*, Vol. 14, No. 2, November 1977. Published by Executive Publications, Pickfair Building, Rancho Sante Fe, Calif. 92067.

receiving human milk actually absorbs more calcium into his body than does the infant who receives cow's milk.

The advisability of using milk as the chief dietary source of calcium is suspect. Despite widespread consumption of milk in the populace, problems with calcium metabolism appear to be increasingly common. A number of reasons for this may exist.

Cow's milk differs significantly from human milk in its ratio of calcium to phosphorous. The ratio is 1.2 to 1 in cow's milk and somewhat greater than 2 to 1 in human milk. Relatively speaking, cow's milk is phosphorous-rich and human milk is phosphorous-poor. (Meat, poultry, and fish are also exceedingly rich in phosphorous compared with calcium.) Phosphorous can combine with calcium in the gut and thus deny calcium to the body. A condition known as hypocalcemia of the newborn is seen in some infants who consume cow's milk formulas. The condition is prevented or alleviated by switching to formulas with a more favorable calcium/phosphorous ratio. Such formulas resemble human milk.

I believe that milk in the diet of the older child and adult may contribute to disturbed calcium metabolism just as it may in the young infant.

Persons who drink large quantities of milk or who eat much cheese (derived from milk) are often found to have excess unavailable calcium as measured by hair analysis. This is frequently associated with imbalance in other body minerals, gastrointestinal dysfunction, and hormonal disorders. This suggests that there are problems associated with milk intake other than inadequate calcium absorption.

Persons who obtain their dietary calcium from food sources other than milk[12] often appear to be healthier and have fewer problems with calcium metabolism.

Milk is an excellent source of dietary protein. However, I suspect that many individuals who obtain their dietary protein needs from nonmilk sources may be healthier than those who rely

---

[12]Nonmilk foods containing high amounts of calcium are as follows (the foods are named in descending order of their calcium content): sesame seeds; blackstrap molasses; oysters; shrimp; turnip greens; collards; almonds; soybeans; kale; mustard greens; brewer's yeast; watercress; crab; parsley; figs; Brazil nuts; pinto, navy, and kidney beans; lobster; oranges; pistachio nuts; broccoli; egg yolk; wheat bran; beet greens; buckwheat; sunflower seeds; chard.

upon milk for this nutrient.[13] Milk protein contains casein, and casein contributes to arteriosclerosis (hardening of the arteries).

The cow is not as discriminating in her diet as you and I might be. She eats food that may contain pesticides and other extraneous material. She may have been treated with antibiotics. Other substances may have been added to her milk. She lives largely on grasses and grains, and these plants are known to cause frequent allergic reactions in many people.

Perhaps it is understandable that persons who obtain their nutrients from milk do not have as favorable a health record as those who acquire their nutrients from a wide ranging supply of nonmilk foods. Such is my experience.[14]

Milk may be taken in large quantities by an individual because he finds that it makes him feel better. This may be a conscious act or it may just subtly develop as a dietary habit. Some persons are milk-a-holics and appear to be addicted to milk.

The person who is nutritionally deficient will, indeed, find in milk the nutrients that he may need. The important question is whether milk is the best source with the lowest risk for him.

When large amounts of milk are taken for any reason, one may be filled up so that other nourishing foods are not eaten. Foods other than milk may more safely meet one's nutritional needs. It seems preferable to eat foods created for people than to drink a liquid nutrient made for the suckling calf. When the diet contains large amounts of green, leafy vegetables and nuts and seeds, a goodly supply of calcium is provided. Eggs, shellfish, salmon, and sardines are other good sources of calcium and protein from the animal kingdom. Beyond infancy, for most of us, milk need not be relied upon as a prime dietary nutrient.

There may be hormonal substances (growth hormone, for example) in milk which can have effects on growth and health that may not be desirable. Milk, it must remembered, is the product of a female animal, the cow; and as such, it contains a significant

[13]Nonmilk foods that supply high-quality protein are: eggs, fish, shrimp, red meat flesh, chicken, and turkey. Brewer's yeast, soybeans, nuts, whole-grain cereals, and seeds also supply protein, but it is thought to be less biologically complete.

[14]In some cases, persons are able to get along best by consuming milk and milk products. Human beings are unique individuals, and their function must be considered on an individual basis.

amount of the growth-promoting substances that nature intended for the suckling calf.

Roughly one quarter of the population in the United States (70 percent of the black population and 80 to 97 percent of Ashkenazic Jews and some orientals) lack the intestinal enzyme necessary to digest milk sugar (lactose).[15] Gastrointestinal symptoms such as pain, gaseousness, indigestion, diarrhea, acid stools, and failure to gain weight may occur. These symptoms should be completely relieved by a diet free of milk sugar.

Most allergists would agree that cow's milk is one of the most common foods responsible for allergy. Allergy to milk, milk-containing products, or foods derived from milk may exist as masked food allergy which is not easily recognized. Milk, for example, may have been responsible for bouts of wheezing in infancy, and it may have been recognized as the cause of intermittent wheezing later on in life. Milk may not, however, be recognized or considered as a cause of an individual's chronic sinusitis (or other disorders). Why? Because both milk intake and the sinusitis are present nearly every day. In such cases, it is necessary to "unmask" the allergy by withholding milk from the diet for 1 to 3 weeks. A subsequent challenge with milk may be used to try to precipitate the symptoms again.

Allergy to milk frequently coexists with sensitivity to beef. This is not strange since both milk and beef are derived from the same animal, the cow.

When one considers what cow's milk is and how much of it we consume, the high frequency of milk allergy becomes more understandable. Milk is certainly a protein load from an animal species rather distantly related to man. It is almost universally used to feed humans early in life. Even breast-fed (best-fed) infants in our culture "graduate" to cow's milk after they are weaned. As one grows older, his intake of liquid milk *may* decline, but he is then exposed to a culture that cooks in butter, slaps butter on bread and vegetables, and eats considerable cheese. In addition, milk is used in sauces, soups, and cremes; and it is placed in various food products as milk solids.

Cow's milk is a secretion of the cow's breasts. I know by clinical experience that the milk of human breasts may contain

[15]*Scope Manual on Nutrition.* Published in 1970 and 1972 by the Upjohn Company, Kalamazoo, Mich.

foods or parts of foods that the mother takes into her body. So it is with the cow. The grass that the cow eats, the fungus on the grass, the dust and pesticides in the cow's feed, and the chemicals in the cow's invironment may all make their way into the milk of ole Bessie.

Allergy to milk may be an immune reaction to the protein antigens of milk. It may also be various other allergylike reactions to the extraneous substances therein.

Milk is notoriously poor in iron content. The iron that it does contain is relatively unavailable, possibly because of the high phosphate content of cow's milk. Individuals in whom milk makes up a large part of the caloric intake often develop iron deficiency. Irritability, fatigue, loss of appetite, and susceptibility to infection may be noted. Growing children and females who have menstrual periods have increased requirements for iron. Iron deficiency, at times, can be subtle and not reflected in lowered hemoglobin values and red-blood-cell count.

The low copper content of milk may also be implicated in the origin of "milk anemia."

The element manganese is also low in milk, and it appears that manganese deficiency is common in most Americans. Manganese may also be related to the adequacy of iron in the body.

A deficiency of chromium is widespread in the American population. Milk (as well as sugar) may well have much to do with this deficiency. Milk has a high content of phosphorous. In addition to combining with iron, the phosphate in milk may combine with chromium. Chromium phosphate or chromium hydroxide are insoluble and pass out of the body with the stool.

Milk as we usually purchase it in our society is pasteurized milk. Pasteurization involves heating milk to reduce its bacterial content. This process of heating destroys helpful raw enzymes and tends to destroy the amino acid lysine. As a result, milk protein becomes of less biological value when pasteurized. Pasteurization is, of course, an excellent safeguard against bacterial contamination. Nevertheless, pasteurized milk must be considered "dead food," because of its exposure to heat and somewhat long storage.

Cow's milk as we usually obtain it is also homogenized. It is practically impossible, today, to obtain commercial milk that is not homogenized. There is some evidence to suggest that homogenization of cow's milk may be harmful. According to Dr. Kurt A. Oster (M.D.), homogenization allows the enzyme xanthine oxidase in milk to be included in the small, fat globules that are absorbed into the body rather than digested.[16] Xanthine oxidase in the body is reported to break down the elasticity of the walls of blood vessels. Heart attacks and other blood vessel disease could be the result. Certainly, one wishes that this thesis could be affirmed or categorically refuted. Until such time that objective conclusions can be reached, however, it may be wise to avoid the use of homogenized milk. Perhaps someday, a milk that is free of xanthine oxidase may be available. Interestingly enough, human breast milk is free of xanthine oxidase.

Whenever milk is used, it would be wise to use certified raw milk if it is available. Goat's milk is excellent but, again, large quantities are unnecessary if the diet is otherwise adequate.

Cottage cheese is a good source of animal protein for those persons who are not allergic to cow's milk products. Unfortunately, chemical additives may be used in its preparation. The low-fat product is preferred.

Plain low-fat yogurt can be a desirable food item when it is fresh, free of preservatives, and contains a full complement of living bacteria. A little blackstrap molasses or fesh fruit can initially be added to blunt the sour yogurt taste until one's taste becomes adjusted. Pineapple packed in its own juice may be used for this purpose. The mix of plain yogurt and molasses or fruit is not a desirable one from the standpoint of food combining. Nevertheless, the combination may be used as a temporary expedient to enable a person to move away from less desirable to more desirable eating patterns.

Plain yogurt may be used as a topping on strawberries, potatoes, and other vegetables. Before long, one may see plain yogurt as the health equivalent of whipped cream and sour cream!

Commercial varieties of yogurt with fruit added may contain considerable amounts of sugar and are best avoided.

When yogurt is made by the natural process using the lactobacillus organism, milk sugar (lactose) is almost completely broken down into its component sugars, glucose and galactose. Be-

---

[16]"Homogenized Milk Seen as Damaging Heart," a report of research by Kurt A. Oster, M.D., in *Newsletter, International Academy of Clinical Nutrition*, September 1974.

cause of this, persons who are intolerant of milk sugar may be able to tolerate yogurt.

Buttermilk contains lactobacilli and, like yogurt, often aids the digestive process or soothes an irritated gastrointestinal tract. Buttermilk is also a good source of potassium.

Fresh, natural cheeses (usually white in color) may be permissible, but they should not be allowed to make up a large part of the diet. Processed cheese "foods" should be avoided.

As a general rule, it is wise not to rely upon skimmed (fat-free milk) for holding down weight unless one can be absolutely certain that the diet is otherwise adequate in regard to fats and oils, minerals, etc.

Protein

The entire span of human existence is a drama in which protein is the director. Hormones, enzymes, organs, glands, nerves, blood vessels, blood cells, albumin in plasma, and even a major part of bones and joints are made of protein.

In the body, fat is stored as fat and carbohydrate is stored as glycogen in the liver and muscles. *Protein, however, has no storage depot.* The structural and functional integrity of the body is dependent on the daily input of an adequate quantity of high-grade biological protein. High-grade protein is that which supplies the 8 essential amino acids in liberal amounts and correct proportions.[17] Foods that provide high-grade biological protein are eggs, milk, cheese, meat, fish, and poultry.

Whenever the diet is deficient in protein, tissue breakdown occurs (catabolism), and the individual is said to be in negative nitrogen balance. Because the heart muscle has high priority for protein, other organs may be forced to give up protein in order to keep this vital life-pump operative. Bones, joints, glands, etc., may degenerate in order to sustain life.

Persons who are haphazard about their diets may lapse into protein deficiency. Such persons leave the welfare of their bodies to chance. Frequently, other nutrients are also in short supply.

---

[17]Essential amino acids are those that cannot be manufactured in the body. They are leucine, isoleucine, phenylalanine, tryptophane, methionine, valine, lysine, and threonine.

A full supply of nutrients is needed every day of a person's life. As the years pile up and an individual ages, it becomes even more important to have an optimal daily supply of protein and other nutrients. With advancing age, there is increasing difficulty in digesting and absorbing food in the gut. The accumulated errors of living make it more and more difficult for an individual to obtain what he needs, as the years go by.

It may be necessary for an individual to do some experimentation with his own diet in order to find the correct balance between protein foods and others. I firmly believe that there are individuals whose requirements are far different than those of others.

In some cases, a high-protein diet will lead to excess mucous in the respiratory tract; mental, gastrointestinal, joint, or skin disturbances; etc. In such cases, decreasing the amount of meat eaten and increasing the vegetables in the diet can be a powerful stroke for health.

High-protein diets are usually high in meats, eggs, cheese, fowl, or milk. The use of fresh fish is very often overlooked as a positive health measure. All of these foods are usually eaten after exposure to heat (cooking). They lack chlorophyll and raw enzymes. When beneficial effects take place after a low-protein diet is instituted, the benefit may be due to an increase in the raw foods in the diet.

All health authorities are in agreement that an adequate quantity of protein is essential in the daily diet. There are many opinions, however, about what constitutes an adequate quantity of protein. For one view, consult the pamphlet *Protein, How Much Is Enough?* by Dr. Jean Mayer (M.D.).[18] Dr. Mayer's booklet presents a viewpoint prevalent among nutritionists.

According to the recommendation of Dr. Mayer, an adult nonpregnant, nonnursing woman should have at least 46 grams of protein per day. An adult male should have at least 56 grams per day. Illness and stress are said to increase the protein need.

Richard Talbot has written a small pamphlet in which he emphasizes the importance of an adequate quantity of high-grade protein. In this pamphlet, *How to Count Grams of Protein . . . and*

[18]Published in 1975 by Newspaperbooks, 800 3rd Avenue, New York, N.Y. 10022.

how to select proper protein,[19] Mr. Talbot recommends 75 grams of protein per day for adults.

The low-protein point of view is presented in the booklet *Proteins in Your Diet* by Dr. Alec Burton (N.D., D.O., D.C.).[20] Dr. Burton indicates that nuts represent the best source of protein for humans. The reasons he gives for this are that nuts

1. are eaten raw,
2. do not contain toxic end-products of metabolism,
3. are generally fresher than meat and fish, and
4. are not subject to undesirable practices in growing.

Most vegetables are low in protein. In addition, vegetable proteins are said to be incomplete proteins because they lack a full supply of the essential amino acids. There is no question, however, that a vegetarian diet can adequately meet protein needs if it is sufficiently varied and properly combined. This, however, is somewhat difficult to carry out unless one is willing and able to devote considerable time to procuring food and planning meals.

Vegetarians usually have far lower blood-cholesterol values than meat eaters. This may be because they do not eat high-cholesterol animal foods, but it may also be due to their practice of consuming soybeans. Soybeans are one of the best sources of vegetable protein, and their consumption is an excellent cholesterol-lowering practice. Fermented soy food known as Tempeh[21] is a tasty product that could become as popular in our country as yogurt has become. Soybean curd, Tofu, is also popular.

Some persons believe that improved health follows a strictly vegetarian intake (veganism). This may well be the case for some persons under some circumstances. Religious groups that are predominantly vegetarian do have a better health record than the population at large. For other persons, however, veganism could be a health disaster. When veganism is healthful for some persons, it may be because those individuals have faulty gastric, pancreatic,

[19]Published by the Life Therapy Institute Press, Life Therapy Institute, P.O. Box 227, North Palm Springs, Calif. 92250.

[20]Published by Natural Hygiene Health Centre, P. O. Box 85, Pymble, N.S.W. 2073, Australia.

[21]See "Tempeh, A New Health-Food Opportunity" by Robert Rodale. Published in *Prevention,* June 1977, pages 25 - 32.

or intestinal digestion of animal foods. A shift to purely vegetarian foods alleviates their health problems — at least temporarily — but so might effective medical care.

Progress in understanding digestive disorders has been somewhat slow, hence it is understandable that veganism may be followed. After all, if that practice does the job in creating a symptom-free individual, it may continue to be the best alternative for a person in poor health or who does not feel well. Recent advances in preventive medicine and metabology, however, enable many persons with faulty gastrointestinal function to participate in less restricted diets and to maintain good health.

In these days, there is a new danger for the vegetarian who relies on food that he purchases in the market. There is considerable belief that the vegetables of today "ain't what they used to be." The vegetarian of today may become severely deficient in vitamins and minerals (especially the latter), because he is consuming food that itself is deficient in nutrients. Overworked soils and the use of "artificial" fertilizers that produce a large-volume crop may seriously jeopardize the quality of food upon which the vegetarian depends. The answer? Organic gardening in one's backyard or joining a food co-op that supplies organic food in quantity on a regular basis.

The practice of veganism has at times become associated with religious and psychic beliefs. For some, veganism is a complete way of life rather than a style of eating. Belief is a very powerful force. Belief in a particular system often is sufficient to make that system work. For some vegans, the belief that it will keep them healthy may be sufficient to do just that — at least for a time.

Proteins of animal origin are said to be complete proteins because they contain all of the essential amino acids. Vitamin B_{12} is the one vitamin that is apt to be in low supply in the diet of the strict vegetarian. Animal foods supply vitamin B_{12} in goodly amounts. It is also readily available as a vitamin supplement.

Many vegetarians decide to incorporate eggs, milk, or milk products in their diet in order to more readily meet their nutrient needs. Such persons are then known as ovo-lacto-vegetarians. Other persons decide to exclude red meat and milk from their diet, obtaining their animal protein from fish and/or eggs.

Most persons appear to be more vigorous when they incorporate some animal protein in their diet each day. At the present time, desirable animal sources of protein appear to be fish from noncontaminated waters; wild game; privately reared cattle, lamb, and pigs that have not been fed pesticide-contaminated foods and that have not been exposed to hormones, antibiotics, and other chemicals; fowl reared on the ground and fed natural foods; fertile eggs from such fowl; goat's milk and goat's milk cheese without additives; certified raw (nonpasteurized) milk from cows that are reared in a natural lifestyle; and perhaps plain yogurt, cottage cheese, and natural, nonprocessed, white-colored cheese.

Whether it's healthy for them or not, most American citizens of today are confirmed meat-eaters. The hamburger is America's "gift" to the world. Let us look more closely at this substance, beef, that occupies such a prominent place in the American diet.

The modern beef animal is as bad off as his milk-producing mate. His natural lifestyle has given way to enhanced production methods. His job is to grow as fast as he can so he can be slaughtered and make room for more of his kind. The modern beef animal has often been subjected to forced feeding, hormones, antibiotics, restricted quarters, and frightening death experiences. The health of the animal whose flesh is presented to the public is unknown. In addition, the saturated fat in meat, the high phosphorous load, the frequent frying of meat with its fat, and the frequent use of salt and other condiments on meat may not be optimal health factors — especially when continued on a long-term basis. For these and other reasons, some persons avoid or curtail their intake of beef. The same comment may apply to other animals that are prepared for the commercial market.

Limited quantities of chicken, turkey, fish, shellfish, lamb, pork, beef, glandular meats, and eggs may be included in the diet of a flesh eater to avoid excessive intake of animal food from any one source. When fresh fish is not available, frozen fish may be used as a less-desirable second choice.

Because shellfish are filter feeders that commonly live in contaminated waters, some persons choose to avoid them. It is doubtful that one can go through life avoiding all risk, although gross contamination would certainly make shellfish avoidance desirable.

Cooking meat destroys parasites that may be contained within animal flesh. Cooking protein, however, may increase amino acid linkages that are resistant to enzymatic digestion and increase putrefaction in the digestive tract. Since the eating of raw meat does not seem to be a particularly viable alternative in our culture, some persons avoid the flesh of animals altogether for that reason.

A common problem is to find that adequate amounts of protein are present in a person's diet but unavailable to the body. This points up the high frequency of digestive and assimilative disorders in our society. Predigested protein can be of help to such persons when this material is appropriately used to nourish the lean body mass under the supervision of a physician.[22] Whenever possible, however, digestive/absorptive problems should be remedied.

When protein is inadequate in the diet, it is commonly because sugar and other refined carbohydrates are eaten in excess. In other cases, overzealous vegetarians may become protein deficient and malnourished in other ways because of rigid adherence to diets that are supposedly health-giving but which may be unsuited to their metabolic types.

While underprivileged countries struggle to provide their citizens with sufficient protein, excessive protein intake is not uncommon in the affluent American society. The average American male is the heaviest meat-eater in the world. He eats his weight in meat each year, about 172 pounds of it. Besides being wasteful and expensive, it is likely that this excessive protein intake is not beneficial to health. Man is an omnivore but he probably did not evolve with such a steady load of dietary animal flesh. (Also, when he did feast on animal food, he ate the entire animal, not limiting himself, as we usually do, to muscle flesh.)

Elevated levels of uric acid and urea nitrogen in the blood are tests that may signify excess protein in the diet. A diet high in animal protein is high in acid and phosphorous. It may intensify deficits (for example, minerals) that stem from inadequate vegetable food in the diet.

Considerable improvement in a variety of health departures has been seen in persons who have switched from high- to low-protein diets — and vice versa! The conclusion is inescapable that one

[22]Mega Am Pro, a liquid made by Saron Pharmacal Corporation, 1640 Central Avenue, St. Petersburg, Fla. 33712, is one such product. Chick-Amine and Beef-Amine are tasty, predigested, protein powders (soups) made by the same company.

man's meat may be another man's poison. The advice of a nutritionally oriented physician should be sought if there is any question about the correct diet for an individual. We badly need additional nutritional research carried out on persons of varying metabolic types.

Starches and Sugars

Carbohydrates in the diet should be obtained from natural sources of starch such as potatoes, sweet potatoes, brown rice, barley, beans, lima beans, corn and other grains, nuts, and vegetables. Potatoes should be served raw, lightly baked, lightly boiled, or steamed. If starch or calories need to be limited, the center of the potato can be left uneaten. Whole-grain brown rice should be used rather than white rice.

All refined carbohydrates should be eliminated from the diet. Refined carbohydrates include white bread, crackers, chips, cookies, pies, cakes, donuts, processed cereals, and that addicting demon — sugar. Do not be fooled by the words *enriched*, *super*, and the like. A superdonut is a devitalized product. Enriched white bread is still basically wheat stripped of most goodness.

Although bread has been described as the staff of life, one can get along quite well without it. Eating bread is, however, one of those gastronomic pleasures that may be acceptable. Since bread contains fat and carbohydrate, it may need to be limited in the diet of the overweight or the carbohydrate-sensitive individual. Bread is also a cooked food and thus would not be desirable for those who wish to observe a predominantly raw-food diet.

Most bread, as it is now prepared commercially, contains about 8% sugar as well as preservatives to lengthen its shelf life. Consumer demands are, however, beginning to produce results. Improved breads such as stone-ground whole wheat with no additives is available in most supermarkets. Honey is appearing as a sweetener to replace sugar. Unfortunately, the person who is interested in improving his diet is penalized by having to pay higher prices for these "specialty products."

Most health food stores provide a number of different breads made from whole grains. Wheat, buckwheat, oat, rye, millet, and 7-

grain breads may be available. Homemade bread made with molasses can be a delight.

Be aware that all risen breads contain baker's yeast. This yeast can be responsible for the production of allergic symptoms in some individuals.

Use 100% whole-grain bread, cereals, and other starches whenever they are used. Macaroni, spaghetti, lasagna, and similar foods can be made from whole-grain materials usually available in a health food store. Vegetable macaroni may be obtained. The frequency of use of such foods depends upon an individual's tolerance to carbohydrate in the diet. Most persons do well on a diet containing considerable complex carbohydrate. Some persons, however, are intolerant of carbohydrate even though it is not refined.

At the present time, cereals represent 26% of the total calorie intake in the United States. This may be excessive for some of us. Vegetable intake probably suffers in the diet of those persons who are excessive cereal eaters. If cereals are truly whole grain and fresh, and not allowed to crowd out other desirable foods, they can be a major positive contribution to health. After all, grains have been with us since the development of civilization in the middle east, India, Egypt, and Central America.

On the other hand, man was man for millions of years before he domesticated himself and began to raise grains for regular dietary consumption. Man may be in a state of incomplete adaptation to grains. Evidence for this is seen in the many patients with ill health related to eating grains. Sensitivity to wheat and corn are common. Behavioral irritability, psychosis, lethargy, hyperactivity, diarrhea, joint symptoms, rash, and malabsorption of food (celiac syndrome) are some of the health disturbances associated with grain sensitivity or intolerance. When wheat or corn are offenders, other grains such as barley, oats, rice, flax, rye, and millet may be tolerated. In some persons, it is necessary to avoid all grains. In gluten enteropathy, all grains except rice and corn must be excluded.

Sugar is the ultimate in refined carbohydrates, since it supplies calories with no nutrients. The consumption of sugar in our society is extremely high. Sugar is added to many, many foods to enhance their palatability. A more palatable food will be more likely to be purchased again and again. It seems quite simple.

Sales are stimulated by the addition of sugar. An addicted public "demands" sweetened food products, cash registers continue to ring, and the quality of life and health inexorably decline. Dr. Ross Hume Hall (Ph.D.) has cogently examined in his writings the issues of food production, food fabrication, and the well-being of our citizenry.[23]

In Chapter 1, a list of foods and their sugar content is given. Label reading is a must if one is to escape the sugar onslaught. Dinner wafers, canned soups, salad dressings, catsup, seafood sauce, some commercially available bulk laxatives, and many other items are well-laced with sugar.

There is a high correlation between sugar intake, heart attacks, functional hypoglycemia, diabetes mellitus, and other illnesses. Elevated levels of triglyceride fat in the blood frequently fall to normal when sugar is removed from the diet.

Most persons who remove sugar from their diets experience a feeling of well-being after they have successfully altered their dietary habits. Bonus nutritional effects take place because the nonsugar eater improves his diet in other ways. A chain of favorable health measures is often invoked, and it starts with riddance of sweet.

One beneficial effect is weight reduction. Two reasons for this may be operative. Sugar, an osmotically active substance, holds water. Less sugar means less water retention and less weight.

Sugar ingestion sets off the release of insulin in the body, and insulin stimulates the formation of fat. No sugar means less insulin and less fat synthesis.

I have emphasized the improved personality that may follow sugar elimination. "The Sugar-Related Irritability Syndrome" is the name of a paper that I am now preparing for publication.

Some individuals who renounce sugar in their diets develop an uncanny ability to detect the presence of sugar in foods. Their bodies tell them they ate sugar when restaurants don't. Such persons notice some untoward symptom and relate it back to what they ate. They soon discover the power of sugar to produce unwanted changes in their bodies. One alert observer, for example, notices a painful sensitivity of the gums that comes on within a

[23]*Food For Nought: The Decline in Nutrition* by Ross Hume Hall, Ph.D. Published in 1974 by the Medical Department, Harper & Row, Publishers, Inc., 2350 Virginia Avenue, Hagerstown, Md. 21740.

few hours after sugar and lasts about 2 days. Another individual has as his index symptom, tenderness in the terminal joint of the index finger of the left hand. Others note backache, urinary irritation, full sinuses, sore throat, etc. A common symptom is irritability. A lousy temperament is rather characteristic of some persons who eat sugar. It is unfortunate that all persons do not develop symptoms when they eat sugar. Actually, many persons do, but they just never become aware that sugar is the causative culprit.

Dr. William G. Crook (M.D.) has called attention to sugar (sucrose) as a cause of hyperactivity and minimal brain dysfunction in children.[24] I see this problem very frequently in my practice. Some children cannot even tolerate the natural sugar contained in fruits.

For these and other reasons, one's goal should be the complete elimination of refined sugar (sucrose) from the family's diet. If this can be accomplished abruptly, so much the better. Certain individuals who crave sweets and who are eating large amounts of sugar in their diet, however, will show distinct withdrawal symptoms when sugar is abruptly removed from their diet. This appears similar to the withdrawal discomfort seen in alcohol and narcotic addiction. Dr. Crook indicates that the patient may worsen for five days or so after the elimination of sugar before improvement occurs. Such patients, like addicts in the withdrawal phase, may be very uncomfortable. As a result, they may reincorporate sugar in their diets and not go off it again. It may thus appear to the individual concerned that sugar is a necessary substance for him. In persons who experience these reactions, it is wise to provide extra psychological support during the time of sugar withdrawal. Eating frequently and consuming nuts and seeds, vegetables, whole grains, and fruit may be helpful. A plentiful supply of protein tailored for the individual's needs is also important. Fish, chicken, turkey, eggs, plain yogurt, cheese, and cottage cheese may be used.

Taking large doses of vitamin C and the B vitamins may minimize distress associated with sugar withdrawal. Balancing the mineral composition of the body may also be important in minimiz-

[24]*Can Your Child Read? Is He Hyperactive?* by William G. Crook, M.D. Published in 1975 by Pedicenter Press, P.O. Box 3116 (present name and address: Professional Books, Box 3494), Jackson, Tenn. 38301.

ing discomfort. The temporary use of tranquilizers may be necessary but should be avoided if possible. Finally, it may be wise to set up a program of gradual reduction in sugar intake over several weeks, months, or years, always with the goal of eventual complete removal of refined sugar from the diet. As previously stated, it may be necessary to reduce the dietary intake of natural sugars as well. In some cases, a reduction of total carbohydrate in the diet is necessary; but often, natural whole (complex) carbohydrates in the diet may be allowed or encouraged. For some persons, this means that potatoes, sweet potatoes, corn, lima beans, etc., may be eaten. The advice of a nutritionally oriented physician should be followed.

One way to abruptly withdraw sugar from the diet is to fast for a period of 48 hours or longer. Fasting, however, should never be undertaken without the advice and supervision of a physician. The use of appropriate protein supplements, with potassium and other appropriate minerals and vitamins (under physician supervision) may successfully break the back of the sugar habit and ease the pangs of transition to a no-sugar diet.

It goes without saying, of course, that one should surround himself with a facilitating environment. It requires superhuman dedication and effort to resist sweet when persons around an individual are consuming ice cream cones, colas, and chocolate sundaes. For this reason, that job in the ice cream parlor should be dropped. Many individuals will see the wisdom of including the entire family in a program of dietary improvement. When the person in the kitchen does not have the cooperation of the remainder of the family, she may still accomplish her goal by very gradually altering buying and cooking habits. Because a child obtains sweets out of the home is no reason to serve them in the home. Over months and years, children often lose their yearning for sweets when the diet in the home is consistently nonsweet.

A person's taste changes when he no longer eats sweets. He learns to appreciate smells, tastes, and flavors that he formerly avoided or did not recognize. Food acquires new meaning, because it gains a qualitative significance. When sugar is subsequently taken, it tastes very sweet; and many times the eating of the forbidden sweet will bring about symptoms of fatigue, irritability, depression, sinus irritation, sore throat, etc.

The improvement in taste discrimination may include a new-found appreciation for the taste of drinking water. Individuals who talk about the delightful taste of pure spring water may have a sensory awareness that is denied to others who partake of sweet.

I believe that human beings were not created to consume re- fined sugar, and I believe that human beings were not created to consume chemicals known as sugar substitutes. The use of sugar substitutes such as saccharine or cyclamate misses the point that one can profit by *not* reinforcing the taste of sweet. There are enough sweet foods in nature to meet our needs. As a matter of fact, a person is often in need of guidance away from these sweet foods of nature. The fruit-a-holic or the avid juice-drinker can have overweight, arthritis, functional hypoglycemia, recurrent infec- tions, rash, and other disturbances relieved by avoidance or re- duction of fruit or juices. The same may apply to honey.

It is wise to avoid the use of all chemical sugar substitutes. The energies used in fretting about their loss should be positively turned into nutritional education. One should become a busy bee, intent on building the best possible program of nourishment input. In this program, the sweetest things around an individual should be fruits (if they are tolerated), a compassionate nutritionally oriented physician, and the loved ones that constitute a person's family.

Ice cream contains about 16% sugar and many other addi- tives. It should be avoided. An ice cream made of "natural" food substances defeats its purpose when it uses pure cane sugar, the epitome of processed foods and a rather dangerous substance, in my view. Ice cream from the health food store made with honey and perhaps carob and containing no additives may be permissible in limited quantities. Remember, however, the comments made in this book about milk and the need for most Americans to consume a diet lower in fats than they are presently eating.

Be sure to study the list of sugar-containing foods given in Chapter 1 and read the fine print on labels to find hidden sources of sugar. At times, it is conceivable that the nutritional value of a food may be more important than the fact that it contains sugar. In almost all cases, however, the same or better nutritional value can be had in a food that is not sweetened — *if the individual wishes to take the time and trouble to obtain it!*

Canned fruits, when used, should have no sugar added. The label should indicate that canned fruit is packed in its own juice and is sugar-free. Canned fruit, of course, is not live food and is far less desirable than fresh fruit.

If juices are used, read the label on the packages carefully. It is amazing how often sugar creeps in as an added ingredient. One cannot stress this enough.

It may be wise to avoid the use of sweet foods with protein foods, as sugar may inhibit or retard the action of protein-digesting enzymes. Some examples of such sugar-protein combinations are: baked beans, ice cream, chocolate milk, desserts with meals, sugar-cured ham, milk shakes, and sugar in coffee with meals. Other examples are sweet fruits eaten with milk, eggs, meat, or cheese. A consideration of this one factor alone begins to show us that the high frequency of digestive disorders in our society may not be accidental.

Sugar may also impede the effectiveness of the starch-digesting enzyme in the saliva of the mouth. For this reason, it may be wise to eat sweet natural foods such as fruit alone whenever possible. This is especially important for those who are ill, weak, or who have gastrointestinal problems. As Peter Reuter, a natural hygienist, has pointed out in a personal communication, improper digestion puts an extra drain-load on body energy. A person who is trying to combat illness and regain energy can ill afford to waste energy in alimentary digestion that could be accomplished with less effort when optimal food-combining is carried out.

One may choose to use honey or molasses when any sweetener is used. Dark-colored (crude) honey and blackstrap molasses are preferable. Molasses contains considerable amounts of chromium, manganese, copper, zinc, molybdenum, and some selenium. Blackstrap molasses contains generous amounts of calcium, iron, phosphorous, B vitamins, and a large amount of potassium. Perhaps the most effective use of molasses is as a sweetener and nutrient for persons who are in transition from a high-sweet diet to an optimal sugar-free diet.

Small amounts of well-diluted honey, molasses, or rice bran syrup may be used with whole-grain pancakes, French toast made with whole-grain bread, and hot cereal. It should be possible, however, for an individual to grow to the point that no sweetener at all is used. Whole-grain pancakes are delicious when one's taste is not dulled by sweet and when his mind-set envisions this as a desirable food in its own right without sweet. Dehydrated banana flakes may also be used as a sweetening agent when such is deemed necessary. Sweet fruits such as bananas, raisins, and dates may come to represent the ultimate in sweetness for the individual who has renounced refined sugar.

Drinks and Additives

Bottles and cans strewn along American highways indicate the widespread littering desecration of our land. These same bottles and cans also reflect the widespread internal pollution that besets our society. Every can along the highway means another load of sugar, sugar substitute, or alcoholic drink has been taken by a fellow citizen. Pop-tops, cartons, six-packs, and cases make it easy to obtain and transport sweetened beverages or allied drinks that contain many chemical additives. They do little for one's nourishment but may do much to his health.

The soft drink habit in the youth of our culture is widespread and deplorable. Many commercially available drinks contain artificial colors, artificial flavors, artificial preservatives, and sweeteners in substantial amounts. They should be avoided.

Some individuals believe they have done enough when they avoid sugar-containing beverages. Many of the so-called diet drinks are sugar-free, but they usually contain a sugar substitute, and they often contain caffeine. The end result of ingesting caffeine may be variations in blood sugar similar to those caused by eating sugar. Sugar substitutes and caffeine should be assiduously avoided. Every effort should be made to avoid reinforcing the sweet habit.

When sugary substances and refined carbohydrates are eliminated from one's diet, an individual's thirst usually subsides. Because of this, the need for beverages in the diet is not great. Cool water, perhaps with a small amount of lemon or lime, is most appealing to the individual who is not consuming a large mount of sugar.

Tomato or mixed vegetable juice contains a low concentration of carbohydrate but appreciable amounts of sodium and potassium. Highly diluted, unsweetened grape juice may be used to quench the thirst in the warm summer months.

It is always preferable to obtain one's supply of liquid, before and after the completion of exercise, from fresh whole foods.

The major component of vegetables and fruits is water, filtered through the organic system of living plant cells at no extra charge to the consumer. Water-laden vegetables and fruits are attractively packaged by mother nature, but unfortunately they have no Madison Avenue representative.

A hydrating liquid substance known as E.R.G.[25] is available for the individual who loses excess fluid in association with exercise. Regular E.R.G. contains no sucrose and no fructose. Glucose in the product is absorbed directly into the blood from the stomach. E.R.G. also contains citric acid, sodium chloride, potassium chloride, vitamin C (12 mg. per 8 ounces), sodium bicarbonate, dibasic potassium phosphate, magnesium and calcium carbonates, natural citrus flavor, and U.S. certified color. The coloring agent is held to a minimum level of 7 parts per million or less.

Whenever possible, it is desirable to avoid food additives such as artificial coloring, flavoring, and preservatives. One should read labels and try to purchase foods which do not contain these substances. Preservatives are used to prevent food from spoiling by retarding the activity of microorganisms. This allows us to have food available when we want it, but if foods do not appeal to "bugs," they probably contain little nourishment for humans! Some children become hyperactive when they encounter certain food colors or flavors in their diets.

Most persons in our society have not developed the healthful habit of drinking water. This may be due to the ready availability in our society of sweetened drinks that crowd out any opportunity or need for water drinking. The taste of water from the usual water supply in the home may also leave much to be desired.

When pure spring or distilled water is used, the refreshing taste of the water may lead the individual to consume more of it. Taste for this water is developed as sugar is eliminated from the diet.

Water-softening units should not be used in connection with drinking water due to the excess sodium load they impart to the water.

When bottled water is used, the label should be checked for total hardness content (calcium carbonate). Present information

[25]E.R.G. is supplied by D.M.K., Inc., 5946 Wenrich Drive, San Diego, Calif. 92120.

indicates that a total hardness of greater than 100 parts per million is desirable.

Drinking untainted fresh water is a healthful habit. Even more healthful is to hydrate one's body by consuming fresh whole vegetables and fruits.

Fats and Oils

A large proportion of one's dietary fat should be obtained from plant sources. Plant foods, rather than processed oils and fats, are the best source of dietary fat. A diet that includes ample quantities of vegetables, whole-grain cereals, seeds, and nuts supplies an adequate quantity of essential fats. Safflower, sesame, sunflower, corn, and cottonseed oil are unsaturated oils that some may wish to use. These oils should have been prepared by the cold-press method, and they should contain no preservatives. Health food stores stock such oils. Since they become rancid easily, they must be used as fresh as possible.

Eggs provide additional nutrient fat. Considerable animal fat (saturated) is supplied in meat (including all-beef hot dogs), fowl, milk, cheese, and butter. It has been considered wise to keep the amount of saturated fats low in the diet, since it has been shown that the blood cholesterol drops when unsaturated fats are eaten. The matter of dietary cholesterol will be discussed in the next section, dealing with eggs.

Fats such as lard, shortening, and many oils are refined, hydrogenated, or preserved. They should be assiduously avoided.

It must be kept in mind that fats and oils provide more than twice as many calories per gram than carbohydrate and protein. Because of this, their quantity should be limited. Calories do count, especially when the quality of one's diet leaves much to be desired.

Heat is the enemy of all fats. Prolonged heating — as used in preparing French-fried potatoes in restaurants — is harmful, because the unsaturated fatty acids of even the best oils become saturated and rancid. For this reason, avoid the use of all fried foods whenever possible. Better yet, avoid added fat whenever possible.

If a fatty spread is to be used, it is probably wiser to use butter rather than margarine. Although margarine may contain polyunsaturated acids, hydrogenation of these acids has often been carried out. Many of the polyunsaturated acids change from their natural form to an altered form during the hydrogenation process. The altered form of fatty acids may be even more damaging to arteries than saturated fatty acids.

When the diet is otherwise complete, the use of butter is a concession to one's taste buds and culture rather than a dietary necessity.

Eggs

The first thing that comes to mind when one considers eggs is cholesterol. One can even purchase imitation eggs that have been fabricated and made available to avoid the high cholesterol input associated with eating eggs. Some discussion on cholesterol is warranted.

Cholesterol is a fatty alcohol. An elevated level of blood cholesterol is a risk factor for heart and blood-vessel disease. Eggs contain considerable cholesterol. Cholesterol in the diet may be related to high blood cholesterol, *but* this appears to depend upon many other factors. The intake of coffee, tea, dietary additives, sugar, and salt, as well as exercise habits and the amount of daily exposure to sunlight are some important variables. Chronic consumption of cooked and overcooked food; lack of raw enzymes; vitamin and mineral deficit or imbalance; lack of dietary fiber; overconsumption of meat, milk, and cheese; and mental stress are other factors that I believe to be more important than the single factor of cholesterol input from eggs.

Let us remember that our body hormones are synthesized from cholesterol. The liver has a great deal to do with cholesterol metabolism. When there is a thyroid deficit in the body, blood cholesterol levels rise. Sexual hormone disorders may be associated with disturbed cholesterol levels. Too little cholesterol in the blood may be an indication of the malabsorption of fat.

What is a desirable blood cholesterol level? No one knows for sure, but values in the range of 130 to 180 milligram percent appear to be optimal. The determination of high and low density lipoproteins assists in evaluating the significance of the blood-cholesterol level.[26]

[26]"New Look At Lipids — Why They're Not *All* Bad" by William P. Castelli, M.D., and Irving M. Levitas, M.D. Published in *Current Prescribing, The Journal of Practical Therapeutics*, June 1977.

Eggs are high in cholesterol, but there is evidence that the inclusion of cholesterol in *natural* form in the diet may not be harmful.[27] In animal experiments, atherosclerosis has been produced by feeding cholesterol; but in these experiments, crystalline cholesterol or heat-dried egg-yolk powder has been used. It appears that it is not cholesterol in the diet, but oxidized cholesterol products that produce atherosclerosis. Accordingly, it may be wise to avoid foods that contain dried egg yolk because of the altered cholesterol that they contain. Foods containing dried egg yolk are pastries, commercially baked bread, cake mixes, dried soups, and other convenience foods.

Fresh eggs may be used in the diet unless indicated otherwise by the physician. An excess of any food, including eggs, should be avoided. The blood values of cholesterol and perhaps those of high density lipoproteins may be used as a guide to the amount of eggs in the diet. It is imperative that the physician in charge consider the many other factors of importance in cholesterol metabolism as well as eggs. Being "hung-up" on the egg-cholesterol association has often gone along with a failure to consider other important aspects of nutrition, exercise, stress, etc.

When eggs are used, serve them soft boiled or poached. Many will wish to use them as a mixer in casseroles, French toast, etc., even though they may not wish to eat them alone on a regular basis.

Try to obtain eggs from farms in which the chickens are raised on the ground rather than in wire cages. Strong shells and dark colored yolks are desirable features in eggs and indicate favorable nutrition in the hens that produced them. Eggs are an excellent source of animal-complete protein and a storehouse of nutrients.

Purchase fertile eggs from the health food store or farm and use them in preference to nonfertile eggs. It is said that when "flu" virus is grown, fertile eggs are used to support the growth of the organism. If fertile eggs are better for the "flu" virus, they may well be better for us!

[27]"On the Much-Maligned Egg" by Mark D. Altschule, M.D. Published in *Executive Health*, Vol. X, No. 8, 1974.

Fiber

There is an increasing amount of evidence that indicates that fiber (roughage) in the daily diet is beneficial to health. The intake of a diet high in fiber (roughage) has been shown to assist in reducing blood fats, and in preventing obesity and heart disease. It also is preventive for gallbladder disease and gallstones, and appears to promote normal blood-sugar levels. A diet high in fiber may reduce the likelihood of developing diverticulitis and cancer of the bowel. Hemorrhoids and varicose veins are disorders associated with diets low in dietary fiber.

A high-fiber diet produces a bulkier stool, a shortened intestinal transit time, and fewer numbers of anerobic bacteria in the stool. Fiber also promotes the normal pattern of bowel activity in which the solid portion of the stool is excreted more rapidly than the fluid part.

Bran provides fiber but should be used as the 100% natural crude bran from wheat or rice. Other good sources of fiber are 100% whole grains (oatmeal, rye, wheat, etc.), prunes, carrots, brussel sprouts, potatoes, apples, plantains, peas, beans, peaches, figs, berries, alfalfa, and lentils.

In some persons, bran may be excessively irritating to the gastrointestinal tract. The right quantity in each person's diet may have to be arrived at by trial and error.

It is always preferable to eat whole, fresh foods rather than to rely upon some part of a food (bran, for example) for a specific purpose. Recalcitrant constipation that responds to bran, for example, would be likely to disappear more healthfully with a diet high in raw or lightly cooked vegetables, fruits, whole grains, and appropriate quantities of protein foods. In some individuals, however, it is necessary to use fractionated products in treatment.

Breakfast

The word *breakfast* indicates the function of this important but often overlooked meal. Breakfast does just what it says: It breaks the overnight fast. This important meal is usually not given the attention it deserves. It is often too easy to slip into the habit of not eating breakfast or just pouring something out of a box or a can. A breakfast consisting of a cup of coffee and a doughnut or sweetroll is an invitation to poor health, especially such conditions as nervousness, fatigue, weakness, vitamin and mineral imbalance, and "female complaints."

Breakfast can be a high-protein meal. Fish, eggs, or whole grains may be eaten. Wheat germ may be incorporated on toast, fruit, or in cereals, although it is preferable to use whole-wheat grains themselves. Soybean lecithin granules may be taken, although it is preferable to use whole soybeans as food. Brewer's yeast and nutritional yeasts are used by many persons as high-nutrient additives at breakfast. These can be extremely helpful. When high-quality foods are regularly consumed and digested, however, the use of such yeasts may not be needed.

Because it is difficult for many persons to obtain a regular supply of high-quality, fresh whole food, the use of wheat germ, soybean lecithin, and yeasts can be an excellent dietary practice. Whole bran, plain yogurt, and blackstrap molasses round out the list of food products that may be desirable. At no time, however, can one ignore the central importance of whole, fresh food properly combined, prepared, and served as the basis of health and vitality.

Raw, unsalted, and mixed seeds and nuts provide protein, unsaturated fat, and complex carbohydrate all in one package. They are also rich in vitamins and minerals. Seeds, nuts, and whole grains soaked in water for a day or two make a delicious vegetarian breakfast. They can be served "as is" or mixed with hot cereal.

When eggs are used, they should be fresh and cooked: soft-boiled or poached. Whole, natural carbohydrates such as potatoes, beans, lima beans, sweet potatoes, corn, and whole-grain rice are not objectionable unless starchy foods need to be limited. It is recognized that these foods are not usually considered to be breakfast foods, but there is no reason why they shouldn't be. Fresh milk (cow, goat, or soybean) may be used, but liquids should not be allowed to replace whole, fresh, solid foods in the diet. Some prefer to avoid all dairy foods.

Bananas, apples, pears, prunes, plums, dates, and figs are sweet fruits. Oranges contain considerable natural sugar. Melons, berries, and pineapple are apt to be less sweet in that they contain less sugar. Some individuals will not thrive as well when sweet fruits are eaten. Sweet fruits taken with starches or protein can interfere with digestion. For this reason, some believe it advisable to eat fruits alone whenever they are consumed. This appears to be less of a problem with papaya and pineapple, both of which contain digestive enzymes (papain and bromelain).

Some individuals, usually vegetarians, prefer fruits alone for breakfast. For others with different metabolism or different states of dietary development, this would be a disastrous practice. Wide swings of blood sugar with weakness, midmorning breakdown, and a generally poor day might follow.

Tomato and vegetable juices may be preferable to orange juice, because they contain significantly lower carbohydrate. A large glass of orange juice in the morning will certainly serve as an effective pick-me-up, but a number of individuals then suffer 2 to 4 hours later as their blood sugar declines. Eating a whole orange is always preferable to drinking its extracted juice.

Sometimes the reduction or elimination of all fruit in the diet is necessary to normalize blood sugar, alleviate clinical symptoms, or to stabilize body chemistry.

Hot cereals are generally more desirable than cold, processed cereals. Oatmeal (steel-cut), wheat, rice, millet, rye, barley, flax, and several cereal mixtures are available. One can buy whole grains in the health food store, soak them for a day or two, and cook them lightly for breakfast. Some prefer to eat them raw. Delicious mixtures using the varied cereals can be fashioned.

The granola cereals are a step in the right direction in that they contain nuts, seeds, and whole grains. Granola should be made in the home whenever possible, and the ingredients should be fresh. A health food store that has a rapid turnover of goods may be able to supply a desirable product. A small amount of blackstrap molasses is the preferred sweetener if any is used.

Ideally, breakfast should be eaten by a person because he is hungry. Automatic eating because one has arisen from sleep is less desirable than being mentally or physically active for an hour or two before breakfast. In some persons, at some stages of nutritive development, breakfast may be omitted as part of a total health-care plan. It is almost never advisable, however, to omit breakfast in order to get to work or school on time, or to starve oneself into weight loss.

Some of the common reasons for skipping breakfast are

eating a heavy meal the previous night at supper, consuming a heavy snack at bedtime, nibbling on starchy and sugary substances during the evening, and drinking cola drinks or milk during the evening. Some persons are only "half alive" in the early awakening hours or they suffer from a.m. indigestion or mucous collection in the sinuses, throat, or stomach. Each of these conditions calls for definitive investigation and management.

A most effective way of developing an appetite at breakfast is to reduce the quantity of the evening meal and to eliminate bedtime snacks. The gut is part of the body and needs a rest during the hours of sleep. It will perform its digestive, absorptive, and excretory functions more efficiently the next day when it is permitted to "recharge its batteries" at night.

Eating supper at an early hour may also be helpful. A snack thereafter is not objectionable if it does not encroach upon the bedtime hour. When the habit of eating breakfast regularly has been well established, there is much more leeway in the day's meal events.

Another effective way to develop an appetite for breakfast is to increase one's exertional effort. Cycling, running, swimming, walking, etc., may be effective appetite stimulants for this important meal.

Parents must remember that they are models for their children. If Mom omits breakfast, she cannot realistically expect her child to eagerly eat his. Cheery encouragement along with a good model of eating behavior on a parent's part can do wonders in developing the good breakfast habit in children.

A sleepy child (or a sleepy parent) will not be anxious to start the day with a hearty breakfast. Attention to bedtime hour the night before, elimination of evening television, and a firm, decisive approach with a child can do much to foster morning appetite. It is not fair to the child, parent, or school to allow a schoolchild to skimp on a meal as important as breakfast. The wise parent will use breakfast time to plan the day with a child, to discuss the daily news, or to interact in some other pleasant way with the child.

When an individual attains optimal nutritional balance and physical activity patterns, his needs may best be served by less than 3 meals a day. It such cases, breakfast may be omitted. For most persons, however, breakfast remains the cornerstone upon which the day is built.

Snacks

Man developed as a nibbling animal living in a natural style. Between-meal snacks need not be avoided if an individual is hungry for them. Rather than coffee and doughnuts, however, nuts and seeds, fruit, whole grains, raw vegetables, and tomato or vegetable juice should be used.

Although some of the following items are not strictly live foods, their use is preferable to candy, colas, and the like:

Brown rice wafers are fairly low in calories and go well with a little natural cheese. Lightly roasted soybeans are tasty and handy to have at hand. Yogurt chips are difficult to stop eating once started. Sunflower, sesame, and pumpkin seeds are highly nutritious and are enjoyed by most persons. Care should be taken to obtain unsalted products that are fresh and do not contain rancid oil. Peanut butter should be raw, unsalted, and nonhydrogenated. It can be obtained at health food stores where it can be freshly ground for the customer.

It is not unusual for some children to crave peanut butter. They may consume large amounts on a daily basis. If the peanut butter is raw, fresh, unsalted, and nonhydrogenated, and if it is not being eaten as a peanut butter and jelly sandwich, then it probably contributes to the child's nutritional well-being. Peanut butter contains protein and zinc in good supply. Both are needed for growth. When a child is not allergic to peanuts and when peanut butter is not allowed to crowd out other desirable foods, it is not objectionable.

Raw, unsalted nuts such as Brazil nuts, filberts (hazelnuts), almonds, cashews, walnuts, and pecans are usually available in health food stores. Pistachio nuts (plain, not dyed) are loaded with valuable nutrients, if one can afford to buy them. Nuts are generally meaty and quite filling, even when relatively few are eaten. This kind of snack "stays with you" and does not leave one with a hungry, empty, or sleepy feeling that so often follows a sugary or other refined carbohydrate snack.

It is wise to consume a *mixture* of nuts and seeds and their butters whenever possible, because their makeup of nutrient elements varies. One kind supplies something that another doesn't. Nuts and seeds, like grains, can be soaked in water to activate enzymes therein and to soften them for chewing. At times they are better tolerated when they are soaked than when they are eaten dry.

Salt

Unless salt losses are pathologically excessive,[28] no supplemental salt should be needed. Ordinarily, salt in the organic form should be obtained in ample amounts from the vegetables contained in the diet.

Unfortunately, however, many persons are beset by stress and have disturbed mineral balance, sometimes with functional adrenocortical insufficiency. These individuals are frequently of the low-blood-pressure type and may require supplemental salt (sodium chloride) in their diets for adequate or optimal function. Sea salt may be considered for use in these cases.

Individuals who crave salt may have acquired their habit by mimicry of others who live with them or with whom they associate. Others who crave salt may be deficient in body minerals. It appears that mineral deficiency should be satisfied by the provision of nutrient minerals such as calcium, magnesium, zinc, manganese, copper, chromium, selenium, etc. If these are not forthcoming in the oral intake, then salt may be used by the body as a filler or poor second choice. It is as though the body would prefer correct mineral balance but will go along with salt to fill mineral gaps. Analysis of hair can provide information about the long-term status of sodium and other minerals in the body. Blood analysis gives a picture of the sodium balance from hour to hour.

Persons living in parts of the country known as goiter belts may need to consume iodized salt, sea salt, or kelp on a regular basis to prevent thyroid dysfunction.

The physician may advise the restriction of salt as part of the management of specific health problems such as edema or congestive heart failure. At times, the use of diuretics and dietary salt restriction can be responsible for severe depletion of sodium, potassium, or perhaps other minerals. This is, of course, the object of diuretic therapy: to get rid of salt and water. Most physicians are well versed in the potential dangers of diuretic therapy.

[28]Patients with cystic fibrosis, salt-losing adrenal disorders, and pathological adrenocortical insufficiency usually require extra salt in their daily intake to sustain life. Stress in these individuals may produce life-threatening shock associated with severe salt deficit.

It is wise, however, to practice the most diligent preventive medicine so diuretics need never be used. This means careful attention to diet, exercise, stress reduction, and environment at the earliest possible time in life.

Do not salt a food item just because "it tastes better" or because a saltcellar happens to be nearby. As one grows accustomed to the taste of natural food, salt will not be missed. If one suspects that his salt craving is biologically based, then special tests on blood, urine, or hair should be carried out.

For those who do better with some salt in the diet, several forms are available. Iodized salt, sea salt, kelp powder, and vegetable concentrate powders may be used. The use of the last three would be more in keeping with the philosophy of using whole products rather than processed, refined, or fragmented products.

Other Nutritive Items

Brewer's yeast is a remarkably nutritious substance, and it may be included in the diet if it is tolerated. Brewer's yeast provides chromium in the organic form. This form is thought to exist as an entity known as glucose tolerance factor (GTF).

Chromium is in short supply in most Americans, and its deficiency has been linked with blood-sugar disturbances and atherosclerosis. Chromium deficiency, however, may not be universal. Some of the most disturbed body chemistry patterns appear to be associated with chromium excess in the body as measured by hair analysis. One should not automatically take brewer's yeast to supply chromium unless he is advised by a physician to do so.

Brewer's yeast is also a good source of the trace mineral, selenium. Selenium is an important factor in vitamin E metabolism, apparently enhancing the activity of vitamin E in the body. Deficiency of selenium in the body is a common occurrence, but one should not automatically take brewer's yeast (or selenium supplements) unless he is advised to do so by a physician.

Use of brewer's yeast powder in various foods or juices on a daily basis can add greatly to the B vitamin intake. B vitamins in foods are water soluble and heat labile, hence they are often in need of supplementation in the diet. Brewer's yeast powder in amounts up to 2 tablespoons per day may be used for this purpose.

Unfortunately, the powder has an objectionable taste even when debittered. Tablets of brewer's yeast may be used. A person should start with a few tablets once or twice a day and gradually increase until best results are obtained.

There is evidence that brewer's yeast lowers blood cholesterol and triglyceride and that it also has a favorable effect on blood-sugar values. Some persons experience a decided increase in energy and a mental calming effect when they take brewer's yeast.

People seem to fall into 2 distinct groups: those who can take brewer's yeast and those who cannot. Side effects from yeast are by no means rare. Nausea, gas, cramps, diarrhea, rash, and other symptoms may occur. Undesirable symptoms may disappear with long-term low dosage or slow, gradual dosage increase.

The individual with a well-functioning gut and metabolism who has integrated his mind and body and who attends to daily exercise needs may have no need for brewer's yeast or other supplements. If he consumes a "junk food" diet, however, it is likely in time that supplements will promote improved function and health.

By this time, the reader of this book realizes that I favor the incorporation of nuts and seeds in the diet, unless, of course, they are not tolerated. Choose a wide variety. Eat them raw and preferably soaked overnight in water to activate the natural food substances involved in germination.

Sunflower and sesame sprouts, and sprouts from alfalfa, mung beans, azuki beans, and lentils are delightful by themselves, in salads, or stir-fried with vegetables. Sprouts provide raw living food brimming with natural enzymes and vitamins.

Blackstrap molasses and plain yogurt have been staples in the health movement for years. Various other forms of acidified milk are also available.

Other nutritious foods are millet, buckwheat, lima beans, raspberries, elderberries, undyed pistachio nuts, and macadamia nuts.

Unless a physician has indicated otherwise, it is wise to avoid eating unless one is hungry. It is *not* necessary to eat 3 meals a day unless one functions best by doing so. Fortunate is the person who can listen to his body and act according to its needs.

Summary of Dietary Improvements

It is suggested that the diet for nearly all human beings consist of a predominance of fresh, raw, whole foods. This would include vegetables, fruits, seeds and nuts, and grains if tolerance permits. Tomato and vegetable juices may also be used.

When other aspects of the dietary are optimal, it appears that the limited use of eggs is a positive health measure for most persons, although one should follow his physician's advice on this matter.

Some persons will wish to incorporate milk, plain yogurt, cottage cheese, and pure natural cheese in the diet. Some will insist that these be goat products, believing that the incidence of allergy to goat's milk is considerably less than that from cow's milk.

Fresh fish from noncontaminated waters appears to be a universally desirable addition to the diet unless one has an allergy to fish or unless he has the fixed mental belief that no animal products should be eaten.

Yard-raised poultry and additive-free red meat — fleshy and glandular — are desirable for many persons, but some individuals possess the kind of metabolism that apparently functions well without such animal food.

Drinking water should consist of fresh or bottled mineral water that is characterized as hard water.[29] It should ideally be free of all additives and chemical contaminants.

Do not ordinarily use

1. refined sugar in any form,
2. processed grains (white flour, refined cereals, alcohol, grits, corn syrup, white rice),
3. processed meats ("luncheon meats"),
4. processed cheeses, including artificially colored cheeses and spreads,
5. hydrogenated (saturated) fats,
6. salt,
7. most bottled and canned drinks,

[29]However, the use of distilled water can be helpful in some circumstances.

8. sweet fruit juices (the whole fruit is preferred), and

9. caffeine (coffee, tea, "decaffeinated" coffees, diet drinks), chocolate, and cigarettes.

Dietary improvements enhance health over a long period of time. Instant improvements are not the nutritional way. Although not immediate, improved health due to better nourishment builds a strong inner core that leads a person to make better judgments.

Figure 136. A personal choice: options for body fuel
Each person has the responsibility to choose for himself the quality of fuel that he makes available to his body. The individual who eats refined, processed, incomplete food to the exclusion of more wholesome nourishment suffers the consequences in ill health or a diminished quality of living. The individual who eats in an informed fashion to provide his body with optimal nourishment is more likely to be disease-free and to enjoy a higher quality of living.

Dietary supplements may be needed to make up for longstanding depletions in nutritive substances. Vitamins, minerals, and enzymes may be vital factors in promoting and maintaining health.

The recommendations given in this chapter may not produce overnight cures; nor can they "change a sow's ear into a silk purse." Nevertheless, the beneficial effects of optimal nutrition are many and longlasting.

My wife, Elinor, has prepared a practical guide to sensible eating. Her book, *Easy Whole-Food Recipes* (For wholesome, healthful, and delicious meals — naturally), is available from Johnny Reads, Inc., Box 12834, St. Petersburg, Florida 33733.

The best food put into a sluggish, nonexercising body can only result in a fair degree of health and a pedestrian level of vigor. For optimal vitality, it is necessary to combine best food with a regular program of body building and movement fitness. Exercise programs should be graduated according to one's ability and should involve stretching, strengthening, and improvement of endurance. Recommended is the book *Keep Your Heart Running.*[30]

Most persons think of dietary factors when they consider blood fats. Recent research has found that exercise is profoundly related to the level of alpha and beta cholesterol in the blood. Regular exercise is an excellent way of building up one's level of high-density lipoproteins (alpha cholesterol). These are the "good guys."

The evidence to date indicates that these "good guys," the high-density lipoproteins, carry cholesterol in the blood and are associated with longevity and freedom from heart attack. Low-density lipoproteins (beta cholesterol, the "bad guys") are correlated with shorter life spans and a higher risk of heart attacks. Long-distance runners characteristically have a high level of the "good guys" and a low level of the "bad guys."

It is interesting to note that the regular exerciser sooner or later becomes interested in improving his diet. Also, the regular diet-improver often becomes a regular exerciser.

It is recognized that we live in an imperfect world. Soils, crops, fertilizers, and people need to be improved. Furthermore, ours is a culture that presently does not encourage enough the observance of dietary recommendations presented here. There may be considerable pressure from peers, TV, and social contacts to eat "junk food" or to exist on suboptimal nutrient input. The decision is up to each individual himself. One's own health as well as that of future generations depends on how wisely he makes decisions with regard to what he does and what he eats.

MEDICATIONS

When one follows an optimal nutritional program, the use of medication to treat illness can often become a thing of the past. It is always desirable to get along without medication if possible. Nevertheless, at times, medication is needed.

---

[30]*Keep Your Heart Running. A graduated total health and fitness program for people of all ages* by Paul J. Kiell, M.D. and Joseph S. Frelinguysen. Published in 1976 by Winchester Press, 205 East 42nd Street, New York, N.Y. 10017.

The many medicines that are taken by individuals in our society often contain sugar. This refers to both prescription and nonprescription medicines. A syrup, by definition, is a sugar-containing liquid. The many cough syrups that are available contribute to the sugar intake of the individual who wants to get well by taking the medication. At times, it seems that the sugar-containing cough syrups assist in alleviating the symptom of cough. In fact, honey has long been used as a folk remedy for coughs. In other individuals, the ingestion of sugar-containing syrups for colds and coughs appears to hasten the development of a bacterial complication of a cold such as infection in the throat, ears, bronchial tubes, or skin.

Sugar-free liquid preparations are available for use in treating colds, coughs, and allergic upper respiratory conditions. The Cerose preparations are made by Ives Laboratories, Inc., and the Conar preparations by Beecham-Massengill Pharmaceuticals. Dimetapp Elixir, made by A. H. Robins Company, is sugar free. An excellent sugar-free over-the-counter preparation for cough is Buckley's Mixture. It is a Canadian product distributed by W. K. Buckley, Inc., Cleveland, Ohio 44103.

In those cases when a medication is being taken daily or several times daily over a long period of time, it may be particularly important to use a sugar-free preparation. Persons who are being treated with large quantities of vitamins in a megavitamin therapy program should obtain sugar-free preparations.[31] Several manufacturers now provide sugar- and starch-free vitamin preparations. A physician who practices orthomolecular medicine,[32] will be able to recommend such products and supervise treatment when it is indicated.

[31]The following are sources known to the author: Willner Chemists, Inc., 330 Lexington Avenue, New York, N.Y. 10016; Organic Image Products, Inc., 708 Atlantic Shores Boulevard, Hallandale, Fla. 33009; Ortho-Vite Corporation, 6380 Wilshire Boulevard, Los Angeles, Calif. 90048; and North Nassau Dispensary, 1691 Northern Boulevard, Manhasset, N.Y. 11030. If the reader knows of other companies that provide such products, it is requested that the author be informed.

[32]Many physicians who practice orthomolecular medicine are members of the Academy of Orthomolecular Psychiatry, 1691 Northern Boulevard, Manhasset, N.Y. 11030.

SUGAR AND LEARNING

Children who do not learn well in school may have psychological, social, educational, or biological reasons for their learning problems. In my previous books I have pointed out the importance of brain dysfunction and allergy as factors in learning disability.[33]

There is no question that pathological (absolute) low blood sugar can interfere with the mental and physical abilities necessary to attend and perform in school. There is more of a question about states of relative or functional hypoglycemia. In my experience, children who have GTT's which are flat, low, saw-tooth, or highly irregular have more difficulty in school achievement than those children who have more regular GTT curves. The child who has symptoms or signs which appear when there is a significant change in blood-sugar level appears also more likely to have difficulty in adapting to the tasks of school than the child who is free of symptoms and signs. The child whose fasting blood sugar is in the 60's or 70's does not appear to learn as well as the child whose levels are higher. He also seems to be the child who is most vulnerable to adverse stress, whether it be academic, psychological, social, visual, infectious, allergic, or traumatic.

The student who consumes high quantities of refined carbohydrate, especially sucrose, is likely to experience underachievement or wide fluctuations in learning performance. When such a child's diet is improved, teachers often note improved attention span, reduced aimless behavior, enhanced motivation to learn, lessened irritability, and improved socialization. The child who learns well one day but is "out of it" the next is experiencing a disruption of homeostasis often due to nutritional imbalance.

Dr. Stephen Gyland (M.D.) had some children with learning disorders in his group of patients with functional hypoglycemia (see Chapter 14). These children all did well on the antihypoglycemia diet set forth by Dr. Gyland.

It is a common observation made by many teachers that children in the classroom are hyperactive, distractible, and rowdy

---

[33]*Kids, Brains, and Learning,* published in 1970, and *Allergy, Brains, and Children Coping,* published in 1973; both books published by Johnny Reads, Inc., Box 12834, St. Peteresburg, Fla. 33733.

for two or three days after Halloween. One of the principal reasons for this is the consumption of high sugar-containing "junk" from those trick-or-treat bags.

As Dr. William Crook (M.D.) has emphasized, sugar in the diet is a fairly common cause of hyperactivity and minimal brain dysfunction.[34] Dedicated teachers, counselors, psychologists, reading and learning specialists, and parents are giving their "all" to help children with minimal brain dysfunction attain the learning skills necessary for academic progress. Their dedicated efforts can be greatly enhanced when proper nutritional measures are simultaneously applied. The first step in supplying proper nutritional measures is elimination of sugar intake.

Dr. Sylvia Richardson (M.D.) has said, "If a child shows any abnormality in behavior or development, we cut down or eliminate sugar from his diet."[35] Undoubtedly, she, like many of us who see children with learning disorders, has found that a child generally has more stable, attentive behavior when he does not consume sugar (and all the other additives that go along with it).

When fasting blood-sugar values are found to be over 100 mg.%, an individual may be sleepier than he would otherwise be if his blood sugar were in the range of 80 to 90 mg.%. When these higher blood sugars are recorded, the individual can often improve his mental and physical performance by reduction of sugars, other refined carbohydrate, and perhaps fruit in the diet. The amount of reduction in these foods necessary to obtain an optimal blood-sugar level is highly variable. The optimal blood-sugar level, like the optimal diet, is one which permits the person to function most efficiently and with the least symptoms. This must be arrived at through trial and observation, but it is commonly found to be about 85.

I have found that many children show changes in their drawing and writing as their blood-sugar levels change. When regular error-free writing and creative nonconstricted drawings

[34]*Can Your Child Read? Is He Hyperactive?* by William G. Crook, M.D. Published in 1975 by Pedicenter Press, P.O. Box 3116 (present name and address: Professional Books, Box 3494), Jackson, Tenn. 38301.

[35]Dr. Richardson is a pediatrician, author, and director of the Center of Developmental Disorders, Cincinnati Medical College. The quotation is from the *ACLD Newsletter*, April/May 1975, which reported Dr. Richardson's remarks made at the 12th Annual ACLD Conference in New York City, February 26 to March 1, 1975.

are made, the blood sugar is usually found to be between 80 and 90 mg.%.

What then can we say about blood-sugar levels and the hyperactive child? It is true that absolute hypoglycemia, with blood-sugar levels of 40 mg.% or below, is very rarely encountered in the hyperactive child. My experience suggests, however, that other deviations from optimal blood-sugar levels are exceedingly common.

Most hyperactive children are consuming diets high in refined carbohydrates in which sugar and various chemical additives abound. Furthermore, these kids are usually "eating all the time." There may be several reasons for this.

The processed food that the children are eating lacks an adequate supply of vitamins, enzymes, minerals, and other nutrients. The body, in its inherent wisdom, may call for further eating in an attempt to obtain nutrients that are needed. The children may also be driven to eat by the homeostatic disruption that occurs when a high-sugar intake results in wide swings of blood sugar from high to low. In some cases, the highs never occur and the lows are dominant.

In a paper, L. Langseth and J. Dowd present data — including 6-hour glucose tolerance tests — on 261 hyperkinetic children who were treated at the New York Institute for Cihld Development from 1973 to 1976.[36] The children were predominantly white and middle class, and they ranged in age from 7 to 9 years.

When the GTT's were evaluated by standard medical criteria, 26% were found to be normal. *Seventy-four percent showed abnormality of some kind.* The predominant abnormality, accounting for 50% of the abnormal GTT's, was a low, flat curve. Fifteen percent evidenced abnormally high peaks with extremely rapid decline. Almost 11% had abnormally high peaks with slow recovery times. (In 15 of the 21 subjects, elevated blood cholesterol levels were found. Also, glucose was present in the urine during peak glucose levels in 13 out of 21 of these cases.) Another 11% of the abnormal curves showed a decline immediately after glucose in-

[36]This information is used with the permission of the authors and the New York Institute for Child Development, Inc., 205 Lexington Avenue, New York, N.Y. 10016. The paper was published in *Food and Toxicology* in 1978.

gestion with a slow rise thereafter and a terminal value higher than fasting values. Of the remaining 14% of abnormal curves, 8% had normal peaks with slow declines, and 6% had high late peaks or rapid declines.

Of particular importance in this study was the finding that the blood level of eosinophils ("allergy cells") was distinctly elevated (more than 6%) in 86% of the subjects! This may be evidence for allergy in these patients or it could be evidence for functional adrenocortical insufficiency or the presence of intestinal parasites. Parasites, however, have not been found in the usual case, by the customary techniques of stool analysis for ova, cysts, and parasites. My experience suggests that many factors may be operative in the hyperactive child who is commonly overstressed by social, academic, nutritional, and allergic factors.[37]

The most likely explanation for the elevated level of blood eosinophils is allergy to the substances that are being swallowed by the child. Sugar, other refined carbohydrates, and chemical additives are the most likely culprits. It seems, however, that the ingestion of "junk" food such as that, creates a nutritional deficiency state that favors the production of allergy to other foods. Most refined carbohydrates are devitalized grains to which additives may have been added. It is likely that the processing of grains makes them more allergenic to susceptible individuals.

It should also not be forgotten that various respiratory inhalants may land on the mucous membranes of the nose and mouth, thence to be swept backward for swallowing with mucus. Thus, pollens, dusts, danders, and molds may also be agents that might contribute to the allergic disorder.

My experience indicates that the patient improves (sometimes dramatically) and eosinophils and GTT's change toward normal when appropriate care is rendered. Treatment may involve dietary maneuvers, vitamin and mineral supplements, digestive enzymes and other gut aids, food eliminations, allergic desensitization, and, rarely, the use of medications.

I first noted the association of allergy and hyperactivity in

---

[37]All generalizations are dangerous since every hyperactive child is unique. Nevertheless, as the study of the New York Institute for Child Development suggests and as my experience bears out, hyperactive children as a group appear to be those who are particularly vulnerable to adverse nutritional and allergic factors.

1970 in my first book, *Kids, Brains, and Learning*,[38] from which I quote: "Many hyperkinetic children are allergic children by virtue of clinical history or by physical examination of the nasal mucous membranes. It is wise to remember that oranges, chocolate, eggs, wheat, or milk can be directly responsible for hyperactive behavior. There is nothing more tragic than the untreated allergic hyperactive, for proper allergic management will often result in disappearance of hyperactivity!"

Later, in 1973, in my second book, *Allergy, Brains, and Children Coping*,[39] I first emphasized the elevated eosinophils, the high-sugar diets, and other nutritive factors in the hyperactive child.

For a review of nutritional factors as possible contributary factors to learning disorders, see the paper by Harriet Thompson.[40] She indicates that nutritional counseling is needed for all children and especially the special child.

OVERWEIGHT

Overweight can be due to obesity — that is, an increase in the number or size of the fat cells in the body — or it can be due to excessive retention of fluid in the body. It is commonly due to both.

Obesity is characterized by a decreased responsiveness to insulin (insulin resistance). Insulin, on the other hand, is instrumental in stimulating the synthesis of fat in the body. Obesity is one of the principal factors responsible for the onset of diabetes in the adult.

Once established, obesity is a contributing factor in perpetuating a sedentary lifestyle. Underactivity, in turn, is conducive to the development of further obesity.

Eating refined carbohydrate foods, especially sugar, is intimately involved in the typically American problem of overweight. It is, however, not the only factor of importance.

[38]*Kids, Brains, and Learning* by Ray C. Wunderlich Jr., M.D. Published in 1970 by Johnny Reads, Inc., Box 12834, St. Petersburg, Fla. 33733.

[39]*Allergy, Brains, and Children Coping* by Ray C. Wunderlich Jr., M.D. Published in 1973 by Johnny Reads, Inc., Box 12834, St. Petersburg, Fla. 33733.

[40]"Malnutrition as a Possible Contributing Factor to Learning Disabilities" by Harriet Lieber Thompson. Published in *Journal of Learning Disabilities*, Vol. 4, No. 6, June/July 1971.

Let us examine the matter of water retention and sugar. Some individuals who eat large amounts of sugar grow larger in size due to water retention. In women, for example, the dress size may increase or decrease as they eat more or less sugar. The same applies to carbohydrates in general. Sugar is an osmotically active substance. This means that the molecules of sugar hold water wherever they are. If one's skin becomes full of sugar, as it often does when sugar-intake is high, then one's skin also becomes water-logged. The same is presumably true for all the body tissues. Cutting down on sugar in the diet can be responsible for slimming down by reduction in the water content of the tissues in the body. When sugar and other refined carbohydrates are eliminated from the adult's diet, a weight loss of about 7 pounds usually occurs.

It is also believed that sugar-water swelling in small blood vessels in the body can lead to decreased blood flow through various tissues, and perhaps even lead to damage in the walls of the blood vessels by interference with oxygen diffusion. Thick walls are nourished less readily by oxygen diffusion than thin walls. Sugar-associated oxygen deprivation may be followed by accumulation of fats in the inner wall of vessels, a condition known as atherosclerosis. Thus, it can be seen that repetitive and chronic sugar ingestion may be associated with narrowing and hardening of the arteries, the precursor of heart attacks, strokes, and other circulatory disturbances. Overweight, itself, places an increased load on the circulatory system of an individual. As all insurance companies know, from a statistical standpoint, there is ample reason to avoid overweight.

It is possible, of course, to grow fat because of the ingestion of excessive calories from other foods, such as fats. The combined elimination of food additives such as refined sugar and excess fats and oils goes a long way in helping the overweight individual to shed unwanted pounds. At the same time, it is imperative that the lean body mass — that is, the muscles, vital organs, bones, joints, connective tissues, and ligaments — be fed a full quota of nutrients.

In any discussion of overweight, additional factors must be explored because they interplay to determine the presence or absence of this important health problem. The problem of hormonal imbalance in the overweight is a complex one. The traditional medical viewpoint is that overweight 99% of the time is not due to a glandular problem. I suspect that this will be found to be erroneous in due time. Microimbalances in hypothalamic, pituitary, and gonadal regulatory hormones will probably be found to be frequently associated with the clinical problem of overweight.

But let us now devote our attention to physical activity and cultural programming.

Obesity is a disorder usually associated with physical under-activity in association with overnutrition and/or malnutrition. Since the American population is, by and large, a sedentary society, it is not so difficult to understand why so many Americans are overweight. For every participant in an athletic contest there are millions of persons sitting at home on a soft piece of furniture, passively enjoying the vicarious thrill of spectator sports.

But the average American is not only sitting and watching, he is also munching. He compounds the harmful effects of underactivity by stoking himself with unnecessary calories, usually from carbohydrate and fatty food sources. Dead foods such as beer, pretzels, chips, cookies, "buttered" popcorn, candy, ice cream, and other goodies flow down the red lane while muscles grow lax and bellies and hips protrude. Sugar provides its giant share of these offending calories.

The amazing thing is that the human body does as well as it does for so long a time without proper use and despite such gastronomical abuse.

It is my belief that the human body was meant to be used vigorously every single day of its life. The task of each individual who has a body is to find the proper amount and type of use for his state of inactivity. This is a delicate task which must be carefully managed to avoid excessive bursts of activity for which one's body is unprepared. Gradual increase in activity is desirable with long-term goals set up over months and years.

Dr. Kenneth Cooper's program of aerobic conditioning exercises is a great boon to mankind.[41] His program permits one to become the physically active person that he should be in a world of mechanical devices that constantly encourage him toward inac-

[41] *The New Aerobics* by Kenneth H. Cooper, M.D., M.P.H. Published in 1970 by Bantam Books, Inc., New York, N.Y.

tivity. When one sets aside an hour a day for personal physical conditioning, he can then use his auto, TV set, and electric washers to advantage. It is quite possible to live with labor-saving devices and still stay "in shape." Dr. Cooper has shown us how.

There is a close relationship between underactivity and eating. In general, the less activity one performs, the more he eats, or wants to eat. A normal appetite for food is apt to come about in the person whose physical activity is just right for his body.

Not only the quantity but the quality of food may change when one becomes physically active. Frequently one develops an improved mental viewpoint when he is in good physical condition. This favorable balance in mental and physical use of the body is reflected in a wiser choice of foods geared more directly to the needs of the individual himself.

Eating food is a repetitive act vital for survival in animal forms of life. Man's bodily function and structure emerged as we now know it through millions of years of struggle to survive. For much of this time, the body of man has had to get along with basically a scanty food supply. Periods of fasting were not infrequent. Primitive man stole what he could get to eat from the kill of other animals. Then he became a hunter and gatherer and ate well when he was skillful enough to catch his prey through physical activity. Physical activity and eating were, thus, basically linked together. It has only been since the agricultural revolution, a mere 5,000 years ago, that man has been able to count upon a constant food supply, food excess, and food availability without expenditure of vigorous physical activity. This state of affairs is very recent compared with the many years that man was a jackal, hunter, and direct food gatherer. It is very unlikely that primitive man was obese unless he became physically unable to move about.

Primitive man, like the forest animal or the fish in the sea, worked for his food. The situation today is that man has essentially no need at all to vigorously use his body in order to eat. The rigors of earning a living and buying groceries in modern society are no task at all for a body constructed to run, strain, hit, and to endure the hunt and the chase. Cultural programming today is out of phase with our biological experience in the formative period of human development.

The pernicious habit of eating more than we need on a regular basis is a derivative of man's leisure time and his ability to manufacture readily available foods of inferior quality. The refrigerator, the cardboard carton, the automobile, and the TV set make it easy for man to overconsume without exercise. Overeating and the consumption of "empty calories" is largely a disorder of "civilized" countries that possess labor-saving devices. In our "advanced" civilization, we have been culturally programmed for this way of life, thus we find it difficult to change.

As long as man's eating is split-off from the need to perform vigorous daily physical activity, it appears that overweight and other health problems may continue to plague us. Physical conditioning programs that attempt to put modern man "in shape" may be a solution. Attitudes and lifestyle changes come about as one lives to move. Primitive man probably had to walk, jog, or run many miles each day in order to gather the foods that were natural to his environment.

If we see to it that what we eat is basic whole food in its natural form, we will go far in preventing ill health. If we also see to it that we obtain daily, vigorous activity in a way that conditions our bodies for use, we will be stimulating the conditions under which our bodies developed, and we will be preventing "rust" from collecting in our muscles, joints, brains, and arteries.

The following statement appeared in an article in *Modern Medicine*[42] and is reprinted here (with the permission of this journal) because of its great importance to every member of our society.

Why exercise is beneficial

Strenuous exercise increases oxygen demand of the muscles, and whether the demand is met depends primarily on the adequacy of cardiac output. After several weeks of physical training, the healthy individual increases cardiac output to its maximum and thereby increases the rate of oxygen delivery to the tissues. This increased cardiac output is accomplished primarily through a larger stroke volume, and after a period of training, the stroke volume is increased for *all* levels of activity. The result is a lower heart rate for any given work load — and a lower heart rate at rest in the supine position. A slower heart rate yields a lower figure for heart rate times blood pressure (HR × BP), two major determinants of how rapidly the myocardium will require oxygen. A reduced myocardial oxygen requirement

[42]"How Much Exercise for Your Cardiac Patient?" by Morton E. Tavel, M.D. Published in *Modern Medicine*, June 1, 1975.

can lessen the heart's vulnerability to ischemic stresses and thereby reduce its susceptibility to infarction, sudden death, or angina pectoris.

Another benefit of exercise is the reduced oxygen requirement of trained skeletal muscles. After conditioning, an individual will have a decrease in skeletal muscle blood flow[9] and lactic acid production. Such changes probably result from morphologic and enzymatic adaptions in skeletal muscle.[10] After conditioning, therefore, the heart is required to pump less blood to the muscles for a given amount of oxygen required by exercise.

Physical conditioning also causes an interesting effect upon systemic blood pressure: After a conditioning period, hypertensive individuals show a reduction in systolic pressure amounting to an average of 13 mm Hg.[11] Blood pressure response to a given work load also is less after training,[12] while lower pressure itself helps reduce myocardial oxygen requirements. So conditioning also may be useful in the management of systemic hypertension.

Additional benefits to be gained from physical conditioning are a reduction of serum triglycerides,[11] a helpful adjunct to dietary weight reduction, and an increase in sense of well-being. For the cardiac patient, physical conditioning also helps psychologically to combat depression and the fear of invalidism. It really promotes a more positive outlook on life.

[9] Clausen, JP, Trap-Jensen J: Effects of training on the distribution of cardiac output in patients with coronary artery disease. Circulation 42:611, 1970

[10] Clausen JP: Effects of physical conditioning, a hypothesis concerning circulatory adjustment to exercise. Scand J Clin Lab Invest 24:305, 1969

[11] Bonanno JA, Lies JF: Effects of physical training on coronary risk factors. Am J Cardiol 33:760, 1974

[12] Naughton J, Shanbour K, Armstrong R, et al: Cardiovascular responses to exercise following myocardial infarction. Arch Intern Med 117:541, 1966

By development of concentrated effort, we must bring our cultural way of life into harmony with our biological heritage. When we do so, we can throw away the diet pills and let sugar cane grow peacefully as grass in the field, rather than harvesting and refining it to worsen man's ills.

WHAT LIES AHEAD?

Space research has shown us that tools and methods of investigation developed for one purpose may yield important information and practical benefits when applied in other directions. The momentum of nutritional research provided by the investigation of sugar will produce dividends that we now cannot realize.

In future years, it is likely that detailed, accurate analysis of blood, urine, hair, stool, and body tissues will guide our diets and supplements for optimal function. Intervention before disease becomes established will be the rule.

Preventive Medicine is the wave of the future. It must be so for humane reasons but also because the cost of traditional medicine as we now know it is self-defeating.

An inflated economy and the cost of health care in these times make it imperative that national political leaders address themselves to Preventive Medicine. Every reader of this book should contact the President of the United States and other key government officials to express his or her belief in the urgent need for Preventive Medicine. The more quickly we prevent, the less we have to treat. We wish to educate, not medicate.

The core of Preventive Medicine is nutrition — whole nutrition that considers soils, agricultural practices, fertilizers, transportation and marketing of food, additives, packaging, home preparation, digestion and absorption, and metabolism. Growing numbers of persons in the world, with a dwindling level of quality food, pose a serious problem for nations and, indeed, our whole globe.

Land, water, energy, and fertilizers are the basic elements in food production. None of these is in good supply today. Depleted soils upon which crops are raised produce depleted crops.

Our leaders must be chosen for their interest in providing efficient and adequate nutrition to the peoples of the world. How many political candidates today are seriously addressing themselves to these issues?

The elimination or marked reduction of sugar in the diet will not solve all the nutritional problems of mankind. It may not bring nations together for common purpose. It will, however, be a thunderclap of progress, a stroke which promises a longer and better life for those who avoid the excess of sweet, and might it not be easier in a sugar-free world to prevent an Adolf Hitler or a Lee Harvey Oswald?

A generation of mothers whose pregnancies are properly nourished, promises a future generation of children who require fewer psychologists, fewer reading specialists, fewer physicians, fewer judges and jailers, fewer weight-reducing diets, fewer hormones, fewer medications, and fewer hospitalizations. I suspect that such a generation would be more likely to attack and solve problems successfully and thus be more able to rear a family without guilt and fear.

Removal of refined sugar from the diet is the first step in the search for nutritional enlightenment.

Let us hasten to get on with the job of providing proper nutrition for every person in our society. In order to do so, the help of every reader is needed. It is hoped that each of you will become a "nutritional expert" and a purveyor of information for those around you who need to know more. To do less is to delay progress for our "advanced" society which may be stumbling headlong into its grave before its time.

APPENDIX A[1]

EVALUATION OF THE SIX-HOUR GLUCOSE TOLERANCE TEST

In the paragraphs that follow, I will present a method of inspection and evaluation of the standard six-hour glucose tolerance test performed after the administration of a standard glucose load. Use of this method or a similar approach can help insure that as much information as possible is obtained from the glucose tolerance test. Emphasis in this evaluation is placed on descriptive analysis of change rather than on fixed levels of normalcy or abnormality. A symptom checklist is provided for use during the test (see Figure 137).

It is also desirable to have the patient perform some standard mental and fine-motor task at the time of each blood drawing to assess the changes that may occur during the test. Writing a standard sentence (for example, "Now is the time for all good men to come to the aid of their country."), drawing a picture, and working arithmetic problems may be requested.

The results of the blood-sugar determinations should be plotted on a graph in which the sugar concentration in milligram percent is recorded on the vertical axis, and the time in hours is recorded on the horizontal axis. An example of such a form is shown in Figure 138.

[1]Appendix A is intended primarily for the use of physicians.

SYMPTOM CHECKLIST
FOR USE WITH GLUCOSE TOLERANCE TEST

SYMPTOM:	½ Hr.	1 Hr.	2 Hrs.	3 Hrs.	4 Hrs.	5 Hrs.	6 Hrs.
Nervousness							
Tremor							
Weakness							
Sleepiness							
Yawning							
Stomach ache							
Blurring of Vision							
Fainting							
Nausea or Vomiting							
Sweating							
Pallor							
Diarrhea							
Confusion							
Clumsiness							
Stuffy Nose							
Overactivity							
Other:							
COMMENTS:							

**Figure 137. Symptom Checklist for use with
glucose tolerance test**

The assistance of some responsible individual should be enlisted to fill out
this form as the GTT progresses. This individual could be the physician, a laboratory technician, a nurse, or in some cases the parent of a child. In my experience,
parents have usually been reliable in performing this role.

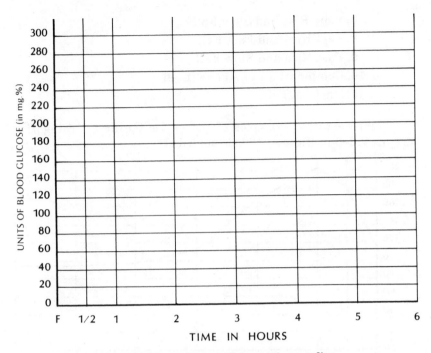

**Figure 138. Exemplary graph for use in recording
the results of GTT**

Blood-glucose values are plotted on the vertical axis, and units of time are plotted on the horizontal axis.

Next, look at the overall contour of the curve which has been plotted. Make a decision as to the outstanding characteristics of the curve. On the basis of these characteristics, assign the curve to one of the following categories:[2]

1. Diabetic (see Chapter 10)
2. Prediabetic (see Chapter 10)
3. Saw-Tooth
4. Flat
5. Hypoglycemic (see Chapter 14)
6. Dysinsulinism (high and low)

[2]A more detailed analysis of glucose tolerance curves with blood-insulin levels is in preparation.

7. Slow Rise and Slow Fall
8. Fast Rise and Fast Fall
9. Fast Rise and Slow Fall
10. Combinations of Above (1 - 9)
11. Unclassified

Representative examples of curves for the first 9 categories are provided in Figures 139 through 147.

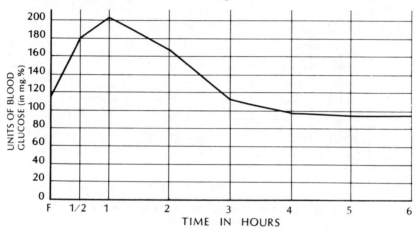

Figure 139. Diabetic GTT curve

The blood sugar rises excessively and remains elevated for too long a time.

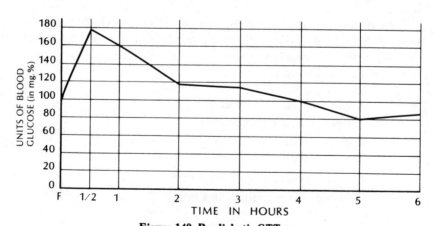

Figure 140. Prediabetic GTT curve

The changes are similar to those in the diabetic GTT, but they are less pronounced.

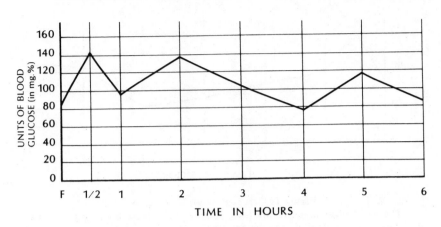

Figure 141. Saw-tooth GTT curve
The erratic up and down blood-sugar levels indicate unstable glucose homeo-stasis.

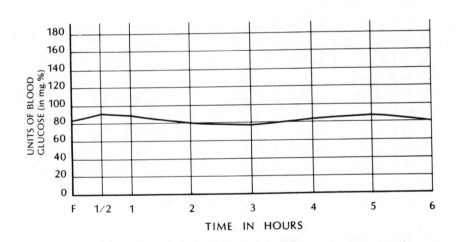

Figure 142. Flat GTT curve
Glucose is either not absorbed, or insulin activity is such to keep the glucose levels at a continually low level.

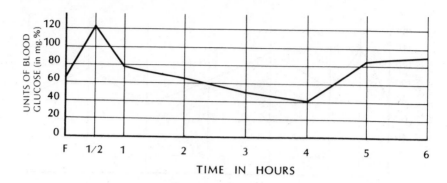

Figure 143. Hypoglycemic GTT curve

The blood sugar falls to a low level of 40 mg.% at the 4th hour. Many other varieties of GTT fall within the hypoglycemic spectrum, and these are presented in Chapter 14.

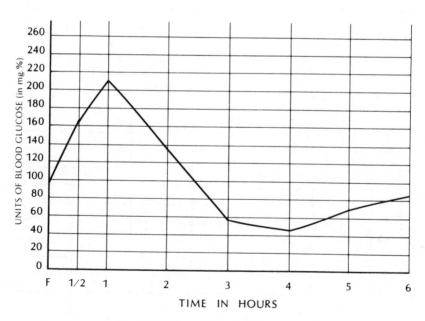

Figure 144. Dysinsulinism GTT curve

The combination of a diabetic or prediabetic curve in the early hours of the test with a hypoglycemic reaction in the later phase is termed *dysinsulinism*. It refers to the faulty timing of insulin release which is too much, too late. This pattern is also known as reactive hypoglycemia of early or mild diabetes.

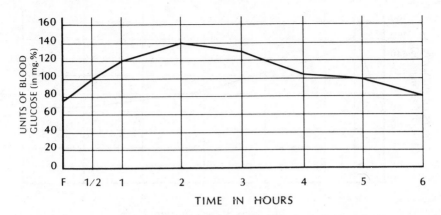

Figure 145. Slow rise and slow fall type of GTT curve

This pattern is less common than most other GTT patterns. It may represent a slow rate of glucose absorption and a slow blunted rate of insulin secretion. It could be a normal variant but more likely it is a phase in between some of the other stages of carbohydrate imbalance.

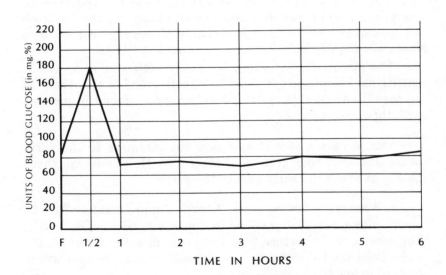

Figure 146. Fast rise and fast fall type of GTT curve

This is apparently due to an accelerated pace of glucose absorption and insulin release. Although this curve shows a glucose peak at ½ hour, some individuals have an even more rapid rise and fall (see Figures 98 and 101 in Chapter 13).

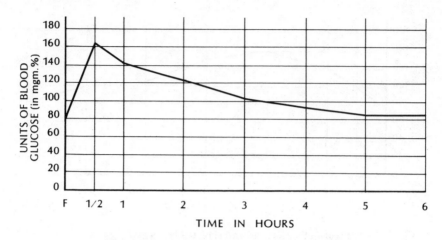

Figure 147. Fast rise and slow fall type of GTT curve
This is apparently due to rapid glucose absorption with a slow steady rate of insulin secretion or it is due to the juxtaposition of counterregulatory factors.

Particular comment is needed in regard to the type of curve that I describe as glucose decline. In this type of hypoglycemic curve, the blood sugar *declines* instead of rising after the loading dose of glucose or food. A variation of this type is the curve that rises slightly at ½ hour, then plunges at 1 hour.

The glucose-decline curve is usually indicative of a serious nutritional disorder. It is more commonly found in thin than fat persons. Considerable caffeine intake may be found in persons with this kind of curve.

The physician should now turn his attention to the early phase (the first hour or two) of the curve. He should ask these questions about the early part of the curve:

1. What is the fasting level? A level of approximately 75 to 95 mg.% with a midlevel of 85 mg.% appears to be most desirable. The greater the departure from this range in either direction, the more likely the individual is to be unhealthy and/or to have symptoms of ill health.

2. Does the curve rise, fall, or remain even? A rising curve can be thought of as one which increases 15 mg.% or more. A descending curve is one which falls 5 mg.% or more. An even

(flat) curve is one which does not rise or fall or which rises less than 15 mg.% in the first ½ to 1 hour.

3. If the curve rises, when is the peak reached? At ½, 1, or 2 hours? Most of the time the peak is expected at 1 hour. Peaks at ½ hour are very common but are probably evidence of disturbed carbohydrate homeostasis.

4. What is the height of the peak? If more than 160, it can be considered a high value. Between 140 - 160 is a moderate value. Below 120 can be considered low.

5. What is the amount of rise between the fasting and peak values? A flat rise can be thought of as less than 20 mg.% above fasting. A very low rise is 20 - 34 mg.%. A low rise is 35 - 44 mg.%. A moderate rise is 45 - 65 mg.%. A high rise is more than 65 mg.%, and a very high rise is more than 100 mg.%. It is believed that the one-hour level should be at least 50% greater than the fasting level.

6. What is the value at the second hour? Anything over 120 mg.% can be considered to be elevated. Above 150 is quite high. A level more than 8 mg.% above the fasting level but below 120 is suspiciously elevated. A level 8 mg.% or more below the fasting level is suspiciously low.

This is the end of the early phase.

The physician should now turn his attention to the last few hours of the curve, the late phase. Ask these questions:

First, about the descent:

1. Does the curve descend as a single straight line (no variance more than 5 mg.%), are there two descending slopes, or does the curve descend, rise, and then descend again? It is believed that a steady, slow, single descent is desirable.

2. Is the descent slow, moderate, or rapid? A slow descent takes more than 2 hours from the peak to the lowest subsequent point. A moderate descent takes 2 hours from the peak to the low point. A rapid descent takes less than 2 hours from peak to low point. A steep and precipitous drop is associated with greater adverse stress than a slower drop is.

3. What is the numerical value of the total fall from the peak

to the lowest subsequent point? A fall of 65 - 95 units is a high figure; more than 95 units is a very high figure. A drop of about 50 units is usually considered to be optimal. Dr. Robert Atkins (M.D.) indicates that 80 units is suspicious of hypoglycemia and/or diabetes and that 100 units is definitely abnormal.[3]

4. What is the amount of fall within the first 2 hours after the peak? Less than 35 mg.% may be considered a small drop; 35 - 65, a moderate drop; and more than 65, a large drop.

When the peak is higher than 160, a drop of at least 55 mg.% is usually desirable. When the peak value is below 130, even a moderate drop may be excessive because of blood-sugar change to relatively hypoglycemic levels.

5. Does the low point of the curve remain more than 8 mg.% above the fasting level? Does it reach the fasting level or within 8 mg.% above or below fasting level? Does it reach a point more than 8 mg.% below fasting? Does it reach a point more than 15 mg.% below fasting? If the low point is more than 15 mg.% below the fasting level, it is likely that this represents an excessive dip.

Next, about the dip:

(A dip is a lowered value, which occurs after the peak and which is followed by a rise of more than 3 mg.% on the next value.)

1. Is there no dip?
2. When does the dip occur? At 1, 2, 3, 4, 5, or 6 hours?
3. Is there a double dip? (A double dip is two troughs separated by a rise of at least 5 mg.%. Note that an initial dip in the early phase followed by a dip in the late phase is termed a double dip.)

A single small dip occurring at the fourth hour is thought to be normal. The earlier the dip, the more out of balance is the metabolism. A later dip may indicate liver dysfunction.

And finally, about the postdip curve:

[3]*Dr. Atkins' Superenergy Diet* by Robert C. Atkins, M.D., and Shirley Linde. Published in 1977 by Crown Publishers, Inc., New York, N.Y.

1. Is it flat? (varies up or down less than 10 mg.%)

2. Is it saw-tooth? (rise and fall of 10 mg.% or more)

3. Is it climbing? (rises 10 mg.% or more and remains steady or continues to climb)

4. Does it descend?

Usually this portion of the curve represents the hepatic (liver) stage of restoration of blood-sugar levels following the dip.

A careful review of the symptom checklist should be made. Symptoms noted during the test should be checked against the time that they occurred. It is not unusual for symptoms to occur at times when the blood sugar is *changing*. Most often the change is from a higher level to a lower level, but there are times when the symptoms occur when the blood-sugar curve rises.

The standard writing, drawing, or mental tasks should also be examined and correlated with the results of the GTT.

Individual patients seem to fall into those who freely develop symptoms with blood-sugar change, and those who do not readily develop symptoms even with larger degrees of blood-sugar change.

Last but not least, the reviewer of the GTT should be sure to know whether there has been any spillage of sugar in the urine. There should be no sugar in the urine at any time during the test. Its presence means that the kidney threshold for the excretion of sugar has been surpassed. Most often this is seen in diabetes mellitus. Acetone should also be absent from the urine.

This completes the evaluation of the glucose tolerance curve. If a physician conscientiously examines the curve according to this format, he will be in a better position to make his own judgments about a patient's glucose balance. He will also be able to compare sequential GTT curves in a more detailed fashion. As a result, he will be in a positon to more reliably interpret and meet the needs of his patient.

For further information about interpretation of the GTT, the interested physician is referred to the paper by Dr. Juan Wilson (M.D.) entitled, "Physiological and Psychological Implications of the Glucose Curve." This paper appeared in the *Journal, International*

Academy of Metabology, Inc., Vol. III, No. 1, 1972. (Published by International Academy of Metabology, Inc., 1000 East Walnut Street, Suite 247, Pasadena, California 91106.)

The author is indebted to Mrs. Karen Buell for her assistance in preparing and analyzing glucose tolerance tests and for assistance in organizing the material contained in Appendix A.

APPENDIX B

GLUCOSE TOLERANCE TEST CURVES

These glucose tolerance test (GTT) curves illustrate the wide variety of test results that are encountered in individuals with clinical problems.

If the reader is not familiar with the glucose tolerance test, Chapters 11 and 12 should be read first.

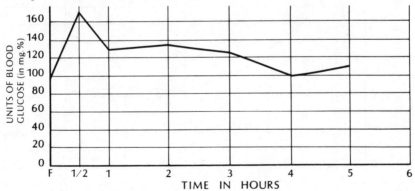

Figure 148. A glucose tolerance test curve

A 7½-year-old boy with recurrent middle ear infections and a resultant conductive hearing loss. His learning was below par, primarily due to reading disability.

The GTT curve shows a borderline peak elevation and some delay in returning to normal. Fasting blood sugars done on other occasions were high at 120mg.% and 115 mg.%. These findings indicate a prediabetic condition.

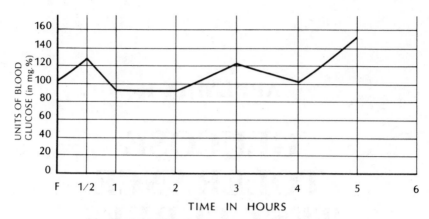

Figure 149. A glucose tolerance test curve

A 12-year-old boy with learning disability and social inadequacy. Also, mark-edly recurrent middle ear infections and sinusitis.

This GTT curve is saw-tooth. It climbs to an unusual height at 5 hours. The fasting level of 103 mg.% is also perhaps somewhat higher than optimal.

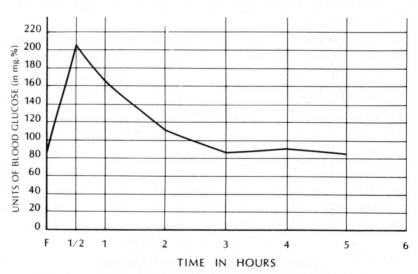

Figure 150. A glucose tolerance test curve

This 14-year-old boy was failing in school and was actively engaged in socio-pathic behavior.

The GTT curve shows a distinct diabeticlike initial response to oral glucose loading. The total drop after the peak is a large amount, and the amount of drop between 1 and 2 hours may also be excessive.

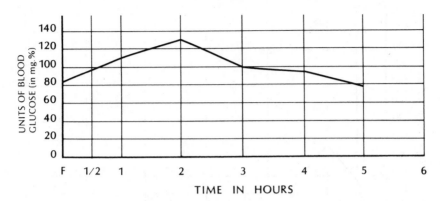

Figure 151. A glucose tolerance test curve

An 8-year-old hyperactive boy.

The GTT curve shows a slow rise and slow fall of blood sugar. Is this GTT normal for this child? Does this GTT reflect optimal metabolism for this individual? We need more research to provide the answers to these questions. The fasting blood sugar of 83 mg.% suggests that glucose metabolism may be at an optimum for this individual. On the other hand, the fasting blood-glucose value does not give complete information regarding carbohydrate metabolism.

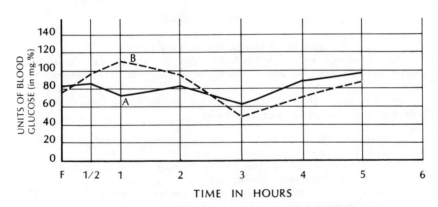

Figure 152. Glucose tolerance test curves

A 9-year-old boy in a private school for learning disorders because of short attention span and hyperactivity.

GTT Curve A is the initial test done in February. Curve B is the followup test done in September of the same year after dietary restriction of sugar. The increased level of blood sugar at 1 hour in Curve B is a characteristic response to the dietary elimination of sugar when the initial curve has been flat. It probably represents a shift in body sensitivity to insulin, an improvement in absorption of glucose, or a decrease in insulin secretion. The boy was clinically about 50% improved when Curve B was obtained. Further treatment is underway. It is expected that the GTT curve will continue to show further change.

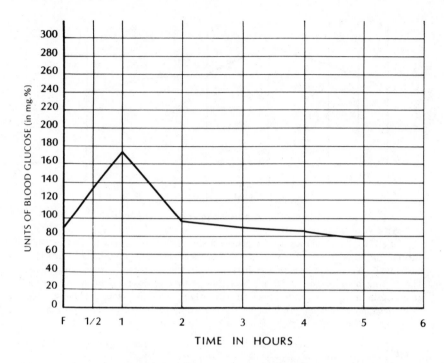

Figure 153. A glucose tolerance test curve

An 11-year-old girl with headaches and nosebleeds. The headaches characteristically came on several hours after a meal and were alleviated by eating.

At first glance the GTT curve appears not to be unusual. On closer inspection, the following features are notable:

1. The peak value of 173 mg.% exceeds the commonly accepted norm of 160 mg.% to 170 mg.%.
2. There is a steady decline from 2 hours to 5 hours. If the test had been continued, the decline may have continued.
3. There is a large drop from the peak to the lowest part of the curve (95 mg.% from 173 to 78).
4. The drop between the 1st and 2nd hour of about 75 mg.% is excessive.

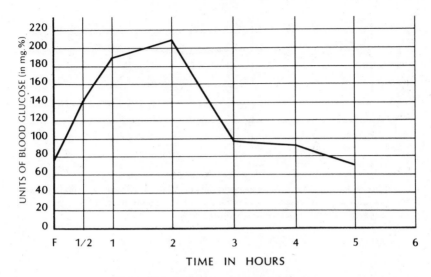

Figure 154. A glucose tolerance test curve

A 10-year-old female with grand mal seizures, overaggressive behavior, and irritable disposition.

This diabeticlike curve has a delayed peak (2 hours) and an excessive rise to 208 mg.%. Had the test been carried on to 6, 7, or 8 hours, a further drop might have been recorded. The child was lost to followup before treatment was instituted.

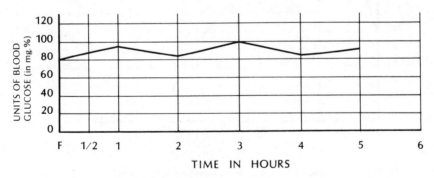

Figure 155. A glucose tolerance test curve

This 13-year-old 8th grade boy was hyperkinetic since birth. He is an under-achiever. He was diagnosed schizophrenic by several psychiatric facilities.

The GTT is essentially a flat curve. This type of curve is frequently seen when the diet is full of candy and other sweets. Such was the case with this boy. Disturbances of carbohydrate metabolism are commonly encountered in schizophrenia.

The family refused to alter his dietary habits and did not wish to try other treatment for his disorder.

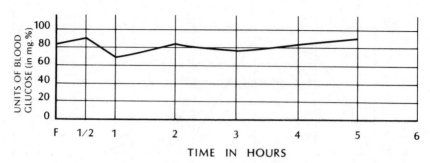

Figure 156. A glucose tolerance test curve

This 6-year-old hyperkinetic child has a "flat" GTT curve which actually declines at 1 hour in response to ingestion of the sugar load. This is consistent with hyperinsulinism, relative or absolute.

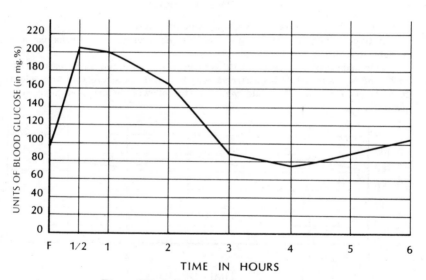

Figure 157. A glucose tolerance test curve

An 11-year-old boy with short attention span, nervous tics, chronic allergic cough, and underachievement in school.

This GTT is a dysinsulinism type of curve — that is, a diabetic type of curve in the early phase and a curve which drops significantly (22 mg.%) below the fasting level in the late phase. This boy had sweating, pallor, dizziness, and dry mouth at the 4th and 5th hours of the test, coinciding with the low points of the curve. Although the lowest blood-sugar level is 77 mg.%, not a particularly low level, the diagnosis of relative hypoglycemia could be entertained. The low point of the curve is 128 mg.% lower than the peak.

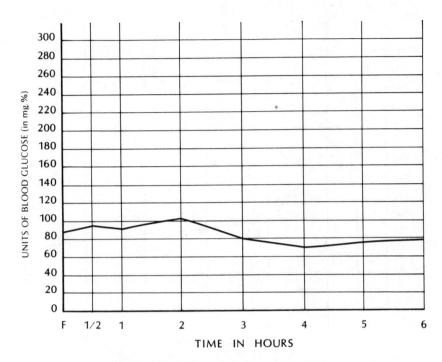

Figure 158. A glucose tolerance test curve

A 6-year-old hyperactive child with hormone tests indicative of an abnormally low function of the adrenal cortex.

The GTT curve shows an essentially flat curve.

Treatment of low and low-normal adrenocortical function is often effective when nutritional measures are used. Severe states such as Addison's disease require hormonal replacement. Treatment with adrenal-cortical extract is used by some physicians for hypoglycemia and lowered function of the adrenal cortex. See Chapter 8 for a discussion of the use of adrenocortical extract.

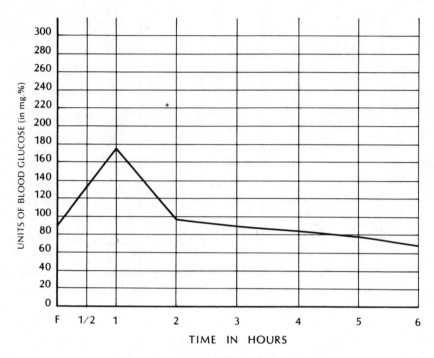

Figure 159. A glucose tolerance test curve

An 11-year-old girl with episodes of headache, nosebleeds, pallor, and malaise.

This GTT appears to be a fairly normal curve in the first few hours. The following symptoms appeared, however, at the 3rd hour and persisted through the 6th hour: tremor, weakness, sleepiness, stomachache, pallor, and mental confusion. These symptoms were relieved by food.

This is a good example of late developing relative hypoglycemia.

One could choose not to recognize this as hypoglycemia, because the blood-sugar levels do not fall below 40 to 45 mg.%. If one did this, he then might not choose to treat the child with dietary alteration.

Elimination of sugar and other refined carbohydrates from the diet is, nearly always, a harmless therapeutic act. Frequent eating of nonjunk food is also a nondangerous procedure. The majority of individuals so treated lose their symptoms.

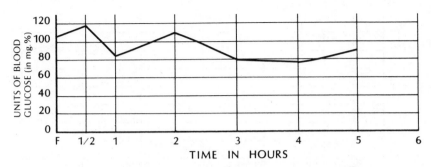

Figure 160. A glucose tolerance test curve

An 11-year-old boy with a severe behavior problem, chronic rhinitis and sinusitis, and hyperactivity. He was described as very irritable since birth. His father was an alcoholic.

The GTT curve is somewhat flat, with a saw-tooth configuration.

The offspring of alcoholics frequently have disturbances of carbohydrate metabolism, especially low blood sugar, according to Dr. Ross Cameron.

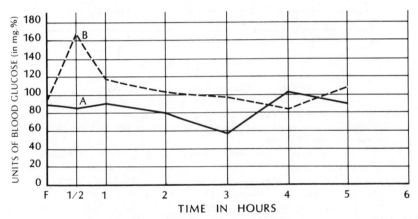

Figure 161. Glucose tolerance test curves

A 10-year-old boy in a private school because of a learning disability.

Curve A was obtained in January 1973.

Curve B was obtained in August 1973.

Between January and August, dietary sugar was restricted and vitamin supplements were given.

Curve A is flat in the early phase, with a hypoglycemic dip at 3 hours. Curve B is much more normal. This type of change is often seen when improved nutrition occurs.

This boy was showing accelerated academic progress at the time Curve B was taken. Improvement of the diet of the child can assist the classroom teacher in educating the child. Too often this aspect of child growth and development is overlooked in the rush to remediate the child who has learning disorder.

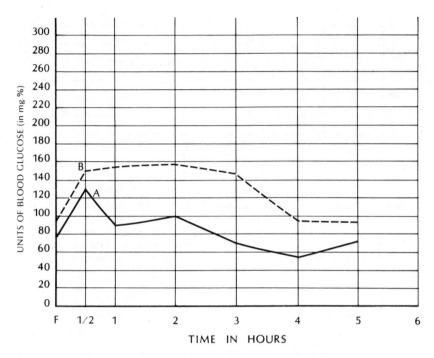

Figure 162. Glucose tolerance test curves

A 7-year-old girl with short attention span, overactivity, difficulty in learning, episodic moodiness, fatigue, and easy crying. The girl's diet was "loaded" with foods containing sugar. There were 10% eosinophils in the peripheral white blood cell count.

GTT Curve A is the initial test done in January. Curve B is a followup test done in August of the same year after sugar consumption was markedly reduced. There is considerable change in these curves. The child had improved in all symptom areas and was no longer considered to be a child with learning problems. The fact that the curve changed from one with a hypoglycemic dip at 4 hours to one whose 2- and 3-hour values are considerably above the fasting level (a diabetic trait) makes one wonder if low blood sugar and diabetes may be different manifestations of a common underlying problem (carbohydropathy or dysglycemia).

Further improvement toward normal is anticipated as the child's needs are more accurately met.

Blood eosinophils dropped to 1% after dietary improvement. An allergy to sugar appears to be a common problem in our society.

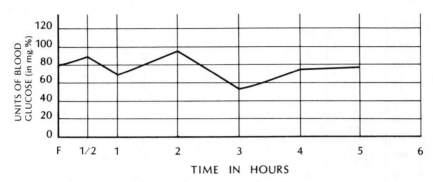

Figure 163. A glucose tolerance test curve

An 8-year-old boy with reading and math problems, nightmares, reckless behavior, and excessive talking.

Somewhat saw-tooth curve. There is a question whether this curve is too low for optimal physiological function in this individual. The drop to 54 at 3 hours is probably not desirable, although it is not low enough to be classified as absolute hypoglycemia. If this patient were an airline pilot, I would not wish to fly with him, especially if he were taking off or landing 3 hours after eating.

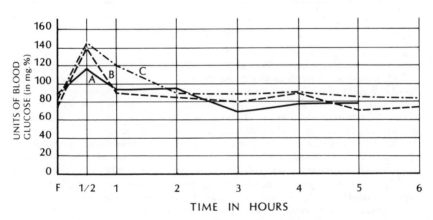

Figure 164. Glucose tolerance test curves

A 4-year-old girl with inadequate nutrition, speech delay, and autistic tendencies.

The GTT of 8/73 (Curve A) was done at a time when her diet was "loaded" with sweets. Elimination of sweets in the diet was followed by the GTT of 2/74 (Curve B) in which there is a higher peak at ½ hour and a reduced dip at 3 hours. These changes are even more pronounced in the curve of 6/74 (Curve C).

The progressive changes, although relatively small, suggest that less insulin is being secreted (decreased hyperinsulinism) in response to elimination of sugar from the diet. Another interpretation would be that adrenocortical function had improved.

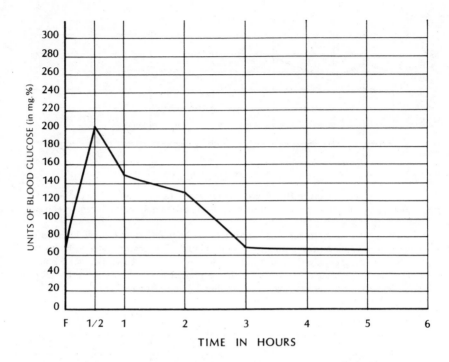

Figure 165. A glucose tolerance test curve

A 28-year-old housewife evaluated because of nervousness, fatigue, and obesity. Weight reduction attempts with Dr. Atkin's diet were unsuccessful.

The GTT shows a diabetic type of curve with elevated peak and 2-hour values. The fasting blood sugar of 70 mg.% is lower than desirable and may indicate that in prior years the GTT may have been hypoglycemic.

Thyroid tests were normal in this patient. When we have refined techniques of assessing the hormones of the hypothalamus, pituitary, gonads, and adrenals, we may find subtle hormonal imbalances in patients such as this.

Diet, exercise, psychological support, behavior shaping, and nutritional supplements are apt to be needed in order to provide an effective program of weight reduction and symptom alleviation. If this woman is not helped, she very well may go on to devleop clinical diabetes.

The availability of fasting, under medical supervision, may be helpful in turning around her chronic overweight problem.

APPENDIX C

HISTORICAL NOTES CONCERNING INSULIN

THE STORY OF DRS. BANTING AND BEST

A LIVING TRIBUTE TO DR. ROSS CAMERON

The person who is *not* diabetic has difficulty in realizing the magnitude of benefit that insulin provides for the diabetic patient. The severe diabetic without insulin is like a scuba diver without oxygen tanks. The availability of insulin has not only saved and prolonged millions of lives, but it has prodded the development of hormonal research leading to many spin-off benefits to humanity.

If one lived before and after the discovery of antibiotics, he is more likely to appreciate the impact that insulin had when it became available for the treatment of diabetes.

To capture some of the flavor of the years of insulin discovery and early use, I have drawn upon the experiences and recollections of Dr. Ross Cameron (M.D.) (see Figure 166). Dr. Cameron is a remarkable individual[1] who, in the later years of his life, carries on advanced medical investigations in the fields of hypertension, hypoglycemia, water chemistry, diabetes, trace minerals, and hu-

[1]See the tribute and the abstract from the curriculum vitae of Dr. Cameron at the end of this appendix.

man ecology. I am privileged to have a continuing professional relationship with this man of quick mind and probing curiosity.

Figure 166. Dr. Ross Cameron (M.D., C.M.) in 1976
Dr. Cameron has been interested in the epidemiology of human illness, with
particular attention to diabetes, hypoglycemia, hypertension, cancer, mental
retardation, and trace elements.

In the early days of insulin use, nearly everyone who was admitted to a hospital with illness was quite seriously ill, and a number were diabetic patients who were in diabetic coma. Dr. Cameron tells the story of a woman who was admitted to Victoria General Hospital in Halifax, Nova Scotia, in those early days. The woman was admitted in diabetic coma; and because she was unidentified, she was placed in the charity ward. The use of insulin at that time was restricted to patients in the charity ward. The woman was given insulin and had a remarkably prompt improvement. The next day, when the woman's identity was established, she was moved to the private section of the hospital. Since the use of insulin was not permitted at that time on private patients, her physician had a difficult time obtaining the life-saving insulin that she needed!

The use of insulin in such "field trials" went on for a number of years after its discovery. Before long, however, it became apparent that this "stuff" was really something! The demand quickened, and the slaughterhouses were hard pressed to supply the pancreas organs of calves for medical use.

Here are the words of Dr. Cameron himself, who tells about his experiences in the very early days of insulin use:

> History is never quite so simple and straightforward as the textbooks suggest. Sometimes historic events become more complex, and perhaps more controversial, as one delves into them more deeply or learns of them from the point of view of a chance observer.
>
> In 1922 - 23 the writer [Dr. Cameron] was a third-year medical student at Dalhousie University, Halifax, Nova Scotia. Physiology became a subject of consuming interest — probably on account of the brilliance of the professor, Dr. David Fraser-Harris (M.D.), a native of Scotland. His eloquence in dealing with a subject that otherwise could be quite prosaic held the students spellbound. It was not unusual to have his lectures end with a round of applause — expressed by the stamping of feet — a method of indicating appreciation that this writer has not experienced elsewhere.
>
> Dr. Harris, as he was known, happened to be a close friend of Dr. John R. R. Macleod (M.D.), Professor of Physiology at the University of Toronto, with whom he frequently communicated. In letters from Dr. Macleod,

Dr. Harris was kept current with development of a new substance, as yet without a name, that resulted from the research of Dr. Frederick Grant Banting and Charles H. Best.[2] It was not unusual to have Dr. Harris share with the advanced class in physiology the contents of the letters that he received from Dr. Macleod.

One morning in late 1922, Dr. Harris told us that he had received "a supply" of this substance from Dr. Macleod for clinical trial. It should be explained that this new "drug" had been demonstrated by Dr. Banting in Toronto and found to be safe and effective.

Because Dr. Harris was "merely" a professor, and not a member of the staff of the teaching institution known as the 350-bed Victoria General Hospital, he prevailed upon Dr. Kenneth MacKenzie, the Chief of the Medical Staff, to administer this new substance to the first diabetic patient admitted to the hospital in coma. Dr. Harris dramatically explained the steps that were taken to be certain that the individual was truly a diabetic, and that the coma was profound. Dr. MacKenzie inserted the needle in the patient's arm and, together with Dr. Harris, the plunger of the syringe was pressed so that each could share the honor of being first in that area to use this new method of treating diabetes — a disease that heretofore had been treated only by a rigid diet. The patient made a remarkable recovery.

A month or two later, Dr. Macleod wrote Dr. Harris, requesting that a name be suggested for this new medication. A student in the back row, Lewis Morrison, put up his hand and said: "Sir, I would like to submit the name 'insulitin' (in-su-li-tin): insula representing the islands of Langerhans in the pancreas from which it is derived, and the suffix tin meaning 'a product of.'"

Dr. Harris considered this a "capital" suggestion. About one month later, he received a reply from Dr. Macleod in which he thanked the student for the suggestion, but a drug manufacturer had taken the liberty of shortening the name by omitting the letters "it," so that the result was INSULIN. Years afterward, we learned that Doctors Banting and Best originally had

[2]At this time, Charles Best was not yet a physician. He received his M.D. degree from the University of Toronto in 1925. (Dr. Best also discovered the vitamin choline and the enzyme histaminase. He was the first to introduce anticoagulants in the treatment of thrombosis.)

used the name "isletin," but Dr. Macleod persuaded them to change the name to *insulin*.

In 1924 - 25 the writer served as intern in a 100-bed children's hospital. Occasionally we transferred a patient in diabetic coma to a nearby general hospital where insulin still was in the clinical trial stage.

In 1925 - 26 this writer first became an anesthetist in the Victoria General Hospital, and later an assistant to the Medical Director, Dr. George MacIntosh (see Figure 167). In that capacity, I was on duty 7:00 a.m. to 7:00 p.m. one week, and 7:00 p.m. to 7:00 a.m. the following week. The salary was $1,000 per year. Among other duties, I was responsible for the administration of insulin to patients in wards 9 and 10 — the "charity" wards.

Figure 167. Dr. Ross Cameron (M.D.) in 1926

Dr. Ross Cameron (first row, seated, far right) in 1926 while a senior medical house officer at Victoria General Hospital in Halifax, Nova Scotia, Canada. Dr. Cameron was in charge of administering the experimental insulin in "field trials" to charity patients. Dr. George MacIntosh (M.D.), the Medical Director, is seated to the left of Dr. Cameron.

One night a patient was admitted in diabetic coma. Insulin was available in two different vials — one containing 20 units per cc. and the other, 40 units. We were told to give one cc. of the 20, or one-half cc. of the 40 units. The ward that contained 20 patients — 10 on each side — was dark. The vial containing insulin bore a handwritten label. After administering the inoculation, it was accidentally discovered that the patient had been given one cc. from the 40 unit vial.

We knew that the Medical Director of the hospital was asleep on the third floor. The elevator was too slow, we thought, so we ran up the stairs and awakened Dr. MacIntosh. He explained, "If the patient lives, she is certain to have severe insulin shock. It might be well to give her orange juice as soon as you can." As I closed the door, the Director said, "Doctor, if you had not given the patient anything, where would she be now?" "Thank you, sir," we replied. "Thank you."

We then were called to the Emergency Room to see a patient who had just been admitted to the hospital, having been brought there by a horse-drawn ambulance!

Upon returning to the ward, we stopped at the nurses' station and asked Miss Hiltz, "How is that patient at the end of the ward doing?" "You mean the one to whom you gave the injection?" she asked. "Yes," I replied. "Why, she is doing fine," Miss Hiltz said; "she is sitting up in bed and has asked for something to eat."

Since we were constantly being asked for suggestions, we recommended that henceforth the vials containing 40 units be labelled in a different color than those containing only 20 units.

In a relatively short time, that recommendation was adopted — and the custom apparently still prevails to this day.

It also was stated that for patients in coma, the dosage might well be increased to 40 units. (It should be explained that since that time the dosage has been standardized so that the values of the new units are not comparable with the old.)

When the announcement was made that this new treatment had been developed, diabetics flocked to

Toronto. This did not occur in Halifax, for the physicians there seemed to be quite skeptical. One of them said to the writer, "I have been advising my diabetics to continue to stick to their diet — for I believe that they would be crazy to start taking an injection three times a day for the rest of their lives. Where is it all going to come from? As you know, it now is being made from calves' pancreases, collected from abbatoirs. How long can that last? I have asked my patients to wait until a better product is at hand."

Well, a better product has become available in the form of synthetic insulin that can be produced in a laboratory. In addition, oral medication is now being manufactured — particularly for the adult onset type of diabetes. In most cases of this variety, insulin is not required. Furthermore, around the corner there is reason to believe that a method of preventing this dread disease may come to light in the not far distant future.

This concludes Dr. Cameron's remarks.

Sir Fredrick Grant Banting, winner of the Nobel Prize in Medicine in 1923 for the discovery of insulin, was born in Alliston, Ontario, Canada, in 1891. Dr. Banting had seen diabetes consume a 15-year-old girl classmate of his into a pathetic child for whom death came quickly.

Dr. Banting had been an average student. In the First World War he served with the Canadian Army. In the 3rd Canadian Division, he was known as a medical officer with a courageous, indomitable heart. Dr. Banting was severely wounded in the battle of Amiens and received the Military Cross for gallantry under fire. (Another Canadian soldier in the Battle of Amiens was Ross Cameron, who describes himself as a lowly signaler.)

In 1921 Dr. Banting was a 29-year-old physician, not experienced in research, whose practice in orthopedic surgery was lagging. There was much interest at that time in finding a satisfactory treatment for diabetes. Dr. Banting read about the pancreatic experiments and the failure in isolating a pancreas extract to treat diabetes. He wondered (correctly) if the protein enzyme (trypsin) of the gland was destroying the desirable substance.

According to Dr. Fraser-Harris, as recounted by Dr. Cameron, Dr. Banting visited the headquarters of several large drug manufacturing companies. He was told, "Young man, we have plenty of young men in our laboratories with bright ideas! If it were so simple as you think, they would have done that sort of thing long ago."

As Dr. Cameron tells it, Dr. Banting was a man with at least one fixed idea. He *knew* he could eventually isolate the product of the isles of Langerhans, and that, when done, it could be successfully used to treat diabetics. Thus, with considerable determination, he set out to find a treatment for diabetes even though the scientists of two continents had previously failed.

In April 1921, Dr. Banting and Charles Best, his co-worker, started work in the "tower room" (attic) of a medical building at the University of Toronto. Without pay, they worked long and hard through a hot, air-condition-less summer. They often wore laboratory coats with little or nothing under them and cooked eggs and sausage over the flame of a Bunsen burner.

Dr. Banting had obtained the use of the lab facility for an 8-week period, with no opportunity for extension of this time period. The search for an active substance from the pancreas of dogs had not succeeded by the end of this time. Since Dr. John J. R. Macleod, head of the Physiology Department, was on vacation, Dr. Banting and Charles Best stayed on in the lab and within the next few weeks succeeded in extracting insulin!

They proved the effectiveness of their substance in dogs who were made diabetic by removing the pancreas. Without insulin, the dogs were doomed to a rapid death. With it, they thrived (see Figure 168)!

After first injecting themselves with insulin to show its lack of toxicity, Dr. Banting and Charles Best were now ready to use the substance in human diabetics. The first patient was a 14-year-old boy at Toronto General Hospital. Diabetes over a two-year period had dropped his weight to 65 pounds, and he was bedridden with weakness. Doomed to die, this boy, like the experimental dogs, "came alive" when he was given insulin! His blood sugar dropped, weight came on, and he was soon strong again.

In 1923, Dr. Banting and Dr. Macleod were awarded the Nobel Prize in Medicine for the discovery of insulin. In typical fashion for this man, Dr. Banting shared his prize money equally with his co-worker, Charles Best.

Figure 168. Dr. Banting and Charles Best in 1922

Dr. Banting (right) and Charles Best (left) in 1922, shown with a pancreatectomized diabetic dog that was given cow pancreas extract. This dog was the first living being caused to survive by the use of insulin. Dr. Banting bought this dog from a boy for $1.00. This expenditure was consistent with the shoestring budget that he had for his research on diabetes.

Note the laboratory in back which is an attic on the roof of a building.

(This photo was provided by the University of Toronto and Dr. Ross Cameron, and is used with their permission.)

Since that time, both Dr. Banting and Dr. Best have had medical research institutes established in their honor. These giants of medical research are pictured in Figures 169 and 170.

Figure 169. Sir Frederick Grant Banting
Dr. Banting's dedication to task paid off in the isolation and use of insulin for the treatment of diabetes mellitus.

In 1941, Dr. Banting was killed in an airplane crash. Before he died, he had dressed the wounds of the pilot who survived.

Figure 170. Dr. Charles H. Best
Dr. Best contributed greatly to the dedicated work of Dr. Banting in isolating insulin and proving its therapeutic effectiveness in diabetes mellitus. He went on to a distinguished career in physiology and medicine.

Dr. Best became a famous physiologist, physician, and author. He is known to medical students as co-author of a well-known physiology textbook. He died in 1978.

Dr. Seale Harris, the "father" of hypoglycemia, worked in Birmingham, Alabama, but was in direct contact with Drs. Banting and Best. One can see how influence spreads.

All of us owe a great debt to Drs. Banting and Best, who spurred on the pursuit of medical knowledge and who provided a means of survival for millions of diabetics. Their accomplishment grew out of determination, sweat, and the prior experiments of French, German, Russian, and American scientists.

Knowledge, goals, hard work, persistence, determination, and self-confidence are powerful traits which paid off handsomely in Toronto in 1921 and 1922.

Dr. W. Ross Cameron (M.D.) has provided me with many fascinating historical vignettes in regard to the practice of medicine. At an advanced age, he continues with his insightful research into the etiology, epidemiology, and management of disease. I am privileged to be able to share his knowledge. I am also fortunate to be able to work with him, presently, on projects pertaining to trace mineral balance, hyperactivity, learning disorder, mental retardation, and degenerative disease.

I am proud to include portions of his curriculum vitae in this book as a well-deserved antemortem tribute to a distinguished medical pioneer.

ABSTRACT FROM CURRICULUM VITAE
OF
WILLIAM ROSS CAMERON, M.D., M.P.H.

1919 - 1925　DALHOUSIE UNIVERSITY, Faculty of Medicine, Halifax, N.S. Degree: M.D., C.M. Intern: Children's Hospital. Resident: Victoria General Hospital.

1926 - 1928　ROCKEFELLER FOUNDATION, Field training in Public Health, Montgomery, Alabama; and Tallapoosa County Health Officer, Dadeville, Alabama.

1928 - 1931　BERKELEY COUNTY HEALTH COMMISSIONER, Martinsburg, West Virginia.

1931 - 1962　MARYLAND STATE DEPARTMENT OF HEALTH, Baltimore, Maryland.

State Assignments

1931 - 1939　Deputy State and Washington County

Health Director, Hagerstown, Maryland.

1939 - 1941   Director, Survey of Medical Care in Counties of Maryland, Baltimore, Maryland. Dr. Cameron developed the program of Medical Care in the Counties of Maryland. (See "Medical Care in the Counties of Maryland," a report of the Committee on Medical Care of the Maryland State Planning Comission, Publication Number 40, April 1944. Maryland State Planning Commission, 100 Equitable Building, Baltimore, Maryland.) This is the medicare program for the state and upon which the national medicare program is based. Dr. Cameron comments in jest about his involvement in the program: "Perhaps that's what's wrong with it."

1946 - 1950   Chief, Division of Cancer and Heart Disease Control, Baltimore, Maryland.

1953 - 1962   Deputy State and Washington County Health Director, and Associate Chief, National Cancer Institutes Field Research Project, Hagerstown, Maryland.

Educational and Military Leaves of Absence

1936 - 1937   Johns Hopkins University, School of Hygiene and Public Health, Baltimore, Maryland. Degree: Master of Public Health.

1941 - 1946   World War II. Served with U.S. Public Health Service in Baltimore, Md., Washington, D.C., Portsmouth, Va., and Charlotte, N.C.

1951 - 1953   Korean War. Served with U.S. Public Health Service as Chief Medical Advisor, Special U.S. Mission to Nationalist Government of Free China, Taipei, Formosa.

1962 - 1974   PINELLAS COUNTY HEALTH DEPARTMENT, St. Petersburg, Fla. Associate Physician, Research Division; Director, Division of Chronic Illness and Adult Health.

SIGNIFICANT POSTGRADUATE TRAINING
    1955 — New York University — Atomic Medicine
    1957 — Yale University — Alcohol Studies
    1958 — University of Maryland — Mental Health

COLLATERAL APPOINTMENTS
  Consultant, U.S. Public Health Service, Division of Public Health Methods; and the National Cancer Institute.
  Consultant, Community Research Associates, New York, N.Y.
  Consultant, American Cancer Society, Maryland Division.
SCIENTIFIC SOCIETIES
  Fellow, American Public Health Association; and American Association for Advancement of Science.
  Member, Delta Omega Honorary Public Health Fraternity; and New York Academy of Sciences.
APPOINTMENTS TO STATE BOARDS
  Atomic Energy Commission of Maryland: Advisory Council on Public Health of Maryland; and American Cancer Society, Maryland Division.
PROFESSIONAL PUBLICATIONS
  Author: "Medical Care in Counties of Maryland — State Planning Commission, 1944."
  Author and/or coauthor of a minimum of 35 papers published in the *Southern Medical Journal, American Journal of Public Health, Journal of the National Cancer Institute, Journal of the American Medical Association, Public Health Reports, Formosan Medical Journal, Public Health Economics,* and other professional periodicals.
  Coauthor with Ray C. Wunderlich Jr., M.D.: "Trace Element Toxicity Associated with a Public Water Supply," a paper presented at International Conference on Trace Elements, University of Missouri, Columbia, Missouri, June 1976.

Dr. Cameron was the first physician in the United States to organize and direct an Emergency Mental Health Service in a County Health Department — first in Maryland and later in Florida.

Much as an ambulance is dispatched to the scene of an accident, one physician and one nurse respond in a crisis to a call from a distraught family, a physician, a law enforcement officer, or a patient. A research team is on duty 24 hours a day and seven days a week.

In five years of service, almost every known type of mental illness was encountered. It often was necessary to deal with agitated persons who had attacked or threatened to attack others, or deeply depressed individuals who had attempted or threatened to attempt suicide.

Originally, the service was designed simply to prevent jail detention of the mentally ill while waiting behind bars to be transferred to a state hospital — quite often a long-delayed event. In less than a decade, however, the major results were as follows:

1. the virtual elimination of jail detention of the mentally ill,
2. a substantial reduction in the number of commitments to a state hospital,
3. significant findings in epidemiological studies conducted in the home, and
4. development of local facilities to serve the mentally ill.

Dr. Cameron wrote up his pioneering mental health project in a report for his department. Details of this mental health service as outlined in his paper were widely discussed in medical journals, magazines, and newspapers. In the November 1965 issue of the *Southern Medical Journal*, for example, this service was characterized as "An important contribution to preventive medicine. This paper points so well to the increasing need for team work in solving a community's health problems."

APPENDIX D

MORE ON THE DANGERS OF SUCROSE[1]

"A great deal of evidence is now at hand implicating dietary sugar (sucrose) in a number of biochemical and other disturbances. The findings come from experiments on human subjects as well as from studies of several species of animals utilizing amounts of sucrose comparable to those taken by many people in the United States and other Western countries.

"Some of the animal work, carried out in several countries, has demonstrated that sucrose may produce enlargement of the liver and kidney; cause considerable alteration in enzyme activity in liver, kidney, and adipose tissue; decrease utilization of dietary protein; and cause histological changes in the retina and in the testes. These experiments suggest involvement of dietary sucrose in human disease.

"One school of medical opinion holds that sucrose is more likely to be the dietary cause of coronary heart disease than is saturated fat. Sucrose, they say, leads not only to an increase in the concentration of blood lipids but also to increased blood concentration of uric acid, diminished glucose tolerance, increased adhesiveness of blood platelets, and increased blood concentration of insulin and cortisol. Furthermore, they believe dietary su-

[1]This material was originally an editorial in the *Medical Tribune*, Wednesday, August 24, 1977. Used with their permission.

crose may be one of the causes of diabetes, especially maturity onset diabetes, not only through its effect on glucose tolerance and blood lipids but also because it may induce insulin resistance at tissue level.

"Recent work has concentrated on the kidney, in order both to follow the suggestion that sucrose may lead to hypertension, and to examine histological changes that may accompany the renal hypertrophy that sucrose produces. Some of these results have now been published, and they add more information to that which we already have about the hazards in respect to cardiovascular and other disorders of the current high consumption of sucrose in affluent societies. The clues they offer to better understanding of the angiopathy [blood vessel disorder] plaguing diabetic patients may be of major significance."

POSTSCRIPT

"What is happening in the world is a projection of what is happening inside each one of us; what we are, the world is."

J. Krishnamurti

GLOSSARY

When writing a definition, it sometimes is necessary to use a word or phrase that also needs to be defined. The author has tried to foresee and provide for such problems. So if you have trouble understanding a word or phrase in a definition, you probably can find in this glossary the meaning of that word or phrase.

Absolute hypoglycemia: A phrase used in this book to indicate blood-sugar levels below 45 mg.% - 50 mg.%.

Academic medicine: The practice of medicine carried on in and around medical schools and research institutions in contradistinction to that variety carried on in clinical practice. Academic medicine tends to be more exact, didactic, categoried, and wed to established precedent as published in medical literature.

"Acetate": A short term used to refer to acetyl-coenzyme A.

Acetone: C_3H_6O. A sweet-smelling substance of the type known as a *ketone*. Acetone is the simplest ketone. It is formed in the body by decarboxylation (removal of a carboxyl group) of acetoacetic acid. Acetone is closely related to two other ketones, *acetoacetic acid* and *D(—)-B-hydroxybutyric acid*. All three of these substances are known as *ketone bodies* and are formed by the liver under certain metabolic conditions in which there is a high rate of fatty acid degradation (oxidation). The elimination of ketones by the kidney involves the loss of alkali and thus leads to a state of acidosis known as *ketoacidosis*.

Acetyl-coenzyme A: "Active acetate" or "acetate." A chemical substance produced in the body by the metabolism of carbohydrate or fat. The final metabolite formed in the glycolytic breakdown of glucose, the beta oxidation of fats (*see* Oxidation), and the metabolism of some amino acids. (It is derived primarily from the first two sources.) Acetyl-coenzyme A combines with oxaloacetate to form citrate in the first reaction of the Kreb's Cycle. Thus, acetyl-coenzyme A is a key chemical intermediate, necessary for the production of energy in the Kreb's Cycle.

Acidosis: A condition of the blood in which metabolic acids accumulate to a degree greater than normal. A shift in metabolic equilibrium to the acid side. An excessively acid blood reaction. Acidosis may develop when large amounts of ketones (acids) are formed in the body.

Acromegaly: A glandular (pituitary) disorder characterized by enlargement of the extremities, thorax, and face.

ACTH: *A*drenocorticotropic *h*ormone. The hormone of the pituitary gland which stimulates the adrenal cortex to produce sugar-regulating hormones (corticosteroids).

Acyl group: An acid radical.

Adrenal-cortical hormones: Also known as *adrenocortical hormones, adrenocorticoids,* or *adrenocortical hormones.* Those 40 to 50 hormones made by the outer part (cortex) of the adrenal gland. They include glucocorticoids, mineralo-corticoids, and sex hormones.

Adrenalin: The word *adrenalin* (or *adrenaline*) is the name of a hormone secreted by the inner part (medulla) of the adrenal glands. Also known as *epinephrine*. Adrenalin, the "emergency hormone," speeds up the heartbeat and increases bodily energy and resistance to fatigue. The word *Adrenalin* (a proper noun) is the name of a brand of adrenalin (epinephrine) that is produced commercially.

Adrenocorticoids: *See* Adrenal-cortical hormones.

Aerobic: With oxygen. Oxygen-consuming. The Kreb's Cycle is an aerobic chemical process.

Aldehyde group: Chemical designation for oxygen connected to carbon by a double bond with a single alkyl group attached to the carbon.

Aldehyde-transfer reactions: Chemical changes involving the movement of an aldehyde group from one substance to another.

Alkaline phosphatase: An enzyme important in the physiology of bone. Alkaline phosphatase is excreted by the liver. Elevated blood levels of alkaline phosphatase may be found in bone and liver disorders.

Alkyl group: A univalent radical occurring in aliphatic (open chain) hydrocarbon derivatives from which a hydrogen atom has been removed.

Alloxan: A chemical used to produce diabetes in experimental animals by selective destruction of the isles of Langerhans.

Alpha cholesterol: High-density lipoproteins in the blood. Currently, these are believed to be positively correlated with cardiovascular health by keeping blood fat circulating within blood vessels.

Amine group: Chemical designation for a combination of nitrogen with univalent hydrocarbon radicals (*see* Radical).

Amino: Of or containing an amine group.

Amino acid: Complex organic compounds of nitrogen that combine in various ways to form proteins.

Amino group or radical: The monovalent radical: $-NH_2$.

Amylase: A starch-digesting enzyme contained in saliva and pancreatic juice. Under the influence of amylase, starch is changed to sugar.

Amylopectin: One of the 2 main forms of starch. Starch that has molecules of branched structure and containing some phosphate; distinguished from the other major group, amylose, which has molecules of unbranched and probably spiral structure. Amylopectin accounts for 80% to 85% of starch.

Amylose: One of the 2 main forms of starch. Starch having molecules of unbranched and probably spiral structure; distinguished from the other major group, amylopectin, which has molecules of branched structure. Amylose accounts for 15% to 20% of starch.

Anabolism: The process by which food substances are changed into living tissues. Constructive or build-up metabolism. Insulin is the primary anabolic hormone in the body.

Anerobic: Without oxygen. Glycolysis is an anerobic chemical process.

Angina pectoris: A heart condition that causes sudden and severe pains in the chest and a feeling of suffocation. It is associated with a diminished supply of blood to the heart muscle.

Anterior pituitary hormones: The hormones secreted by the anterior (front) lobe of the pituitary gland: *somatotropin* (growth hormone) and *corticotropin* (adrenocorticotropic hormone).

Apoenzyme: The protein portion of an enzyme system.

Arteriosclerosis: A progressive thickening and hardening of the walls of the arteries, often associated with high blood pressure.

Artifactual: Caused by an action outside the body.

Ataxia: Loss of normal coordination, especially inability to coordinate voluntary movements of the muscles. Medically, poor balance due to disturbance of the nervous system.

Atherosclerosis: A form of arteriosclerosis characterized by atheroma (fatty degeneration of the walls of arteries) and subsequent narrowing of the caliber of the arteries. The functional consequence of atherosclerosis is deprivation of blood supply to tissues "downstream" of the narrowing. When the blood supply is sufficiently scant, death of tissue occurs (for example, in a heart attack when the process is termed *myocardial infarction*). Lesser degrees of ischemia can create disordered function of tissues without actual death of cells.

ATP: Adenosine triphosphate. A compound consisting of adenosine and 3 phosphate groups. A biologic storehouse of high energy used to power reactions in the body. The removal of phosphate from ATP releases this stored energy that is used for biological reactions such as muscle contraction and the metabolism of sugars.

Autoimmune reaction: Allergy to self. Immune system disturbance in which one's own tissues are attacked (injured) by immune elements such as antibodies or lymphocytes.

Beta cholesterol: Low-density lipoproteins. Currently, these are lieved to be associated with a shortened life span due to a high risk of coronary artery disease. Beta cholesterol is associated with the deposition of blood cholesterol within the walls of blood vessels in the form of atherosclerotic placques.

Bilirubin: The major pigment in human bile. Yellow in color. Bilirubin is the degradation product of hemoglobin from red blood cells. Also, a blood test to determine the level of this pigment in the blood. Bilirubin is elevated in liver disease, blockage of bile, and excessive blood destruction.

Bronchopneumonia: Inflammation of the bronchial tubes and lungs commonly due to infectious agents.

Carbohydrate: Organic compounds containing carbon, hydrogen, and nitrogen manufactured by plants and animals. An energy source for living organisms.

Carboxyl group: A univalent radical, $-COOH$, existing in many organic acids, the hydrogen being replaced by a basic element or radical, thus forming a salt.

Cardiac output: The volume of blood pumped by the heart per unit of time. The cardiac output is the heart rate multiplied by the stroke volume.

Cardiovascular: Pertaining to the heart and blood vessels.

Catabolism: The process of breaking down living tissues into simpler substances with the production of energy. Destructive or tear-down metabolism. Corticosteroids are catabolic in their effects.

Catalyst: A substance that accelerates chemical reactions. A catalyst participates in a reaction and undergoes physical change during the reaction but reverts to its original state when the reaction is complete. Enzymes are protein catalysts for chemical reactions in biologic systems.

Chelating agent: Any organic compound used to attract metallic ions to itself, forming a stable, inert complex that can be eliminated from the body.

Cholesterol: 3-hydroxy-5, 6-cholestene. A fatty substance that serves as the skeleton for the formation of many hormones in the body. It is a product of animal metabolism and occurs, therefore, in foods of animal origin such as meat (skeletal flesh), liver, brain, and egg yolk. The greater part of the body's cholesterol is synthesized within the body. A high level of cholesterol in the blood has been correlated with coronary artery disease and atherosclerosis (*see* Alpha cholesterol and Beta cholesterol).

Circadian rhythm: Any recurrent phenomenon in the body that oscillates with a frequency of once in about 24 hours.

Citrate: The salt form of citric acid. The chemical substance formed in the Kreb's Cycle and subsequently oxidized therein to liberate carbon dioxide and NADH with the release of energy.

Clinical medicine: The practice of medicine carried on in and around private offices or clinics in contradistinction to that variety carried on in medical schools and research institutions.

Coenzyme: A lightweight organic substance, usually containing a vitamin, capable of attaching itself to a specific protein and supplementing it to form an active enzyme system, the holoenzyme. A nonprotein molecule that may be necessary before an enzyme can exert its catalytic action on a body chemical reaction. Coenzymes frequently contain B vitamins as part of their structure.

Coenzyme Q: 2, 3-dimethoxyl-5-methylbenzoquinone. An electron carrier in the respiratory chain linking flavoprotein with the cytochrome enzymes. Ubiquitous in nature. Also known as *ubiquiquinone.* Very similar in structure to vitamin K and vitamin E. Found in large amounts within mitochondria.

Compound sugar: *See* Polysaccharide.

Confabulation: Medical term used to indicate garrulous chatting with the telling of "tall tales."

Congenital: Present at birth.

Corticosteroids: Hormones that are manufactured in the body by the cortex (outer portion) of the adrenal glands. Also known as *corticoids.* Cortiscosteroids exert their principal effects on electrolytes (mineralocorticoids), carbohydrates (glucocorticoids), and the sexual organs (sex hormones).

C-peptide: A chemical fragment connecting the 2 chains of the insulin molecule.

Cytochrome enzymes: Iron-containing enzymes of the respiratory chain within mitochondria. The cytochromes carry electrons from flavoproteins to the cytochrome oxidase enzyme. The cytochrome enzymes are all anerobic dehydrogenases except for cytochrome oxidase. The latter transports electrons from the anerobic cytochromes to the final electron acceptor, oxygen. Cytochrome oxidase contains copper as well as iron. Of the 5 known cytochromes, cytochrome c is best known and most studied. Perhaps this is so because it is the only soluble cytochrome.

Cytoplasm: That part of a living cell contained within the cell membrane but outside the nucleus. It contains important localizations of function such as the mitochondria ("power packs") and the ribosomes (active in protein manufacture).

Decarboxylation: Removal of a carboxyl group from a compound.

Degradation: The process of breaking down into component parts.

Dehydrogenase: An enzyme that enhances the removal of hydrogen.

Delirium tremens: A mental and nervous disorder accompanied by violent trembling and terrifying hallucinations, with associated malnutrition. Usually caused by prolonged and excessive drinking of alcoholic liquor. Also known as *the D.T.'s.*

Dextrose: *See* Glucose.

Diabetes insipidus: A clinical condition of excessive urination and thirst due to insufficient production of posterior pituitary hormone. This disorder involves no abnormality of sugar metabolism and has no spillage of sugar in the urine. The symptoms of excessive urination and thirst are similar to those seen in diabetes mellitus.

Diabetes mellitus: A disorder of sugar metabolism in which high levels of blood sugar are often associated with sugar in the urine, dry mouth, thirst, hunger, excessive eating, excessive urination, weight loss, susceptibility to infection, and other upsets in body metabolism. It is associated with a lack of insulin effect in the body.

Diastolic: Pertaining to *diastole* (relaxation of the heart). The phase of blood pressure related to the relaxation of the heart (diastole) as opposed to *systole* (*see* Systolic). The second or lower number when a blood pressure value is reported. "The patient's blood pressure was 120 over 80." The figure 80 refers to the diastolic blood pressure.

Differential diagnosis: Medical sorting out to establish diagnosis. Consideration of multiple possible causes of a disease or disorder and appropriate investigation to establish the correct diagnosis.

Disaccharide: A double sugar. That is, one containing 2 simple sugars (*see* Monosaccharide). Sucrose and lactose are disaccharides.

Diuresis: Excessive discharge of urine. Urinary outflow of fluid from the body that is greater than usual.

Diuretic: An agent that promotes diuresis. A substance that increases the discharge of urine (water and salts) from the body.

DNA: Deoxyribonucleic acid, a nucleic acid. The hereditary material that makes up the chromosomes of the cell and which contains the genes.

Double sugar: *See* Disaccharide.

Dys-: Prefix implying impairment, abnormality, or undesirable state. Leading away from health.

Dysglycemia: Disordered or dysfunctional blood-sugar levels. Significant deviation from optimal levels of blood sugar. This term includes states of high, low, and combined high and low blood sugar.

Dysmetabolism: Metabolism that leads away from health. Suboptimal physiological function. Deviation from physiological homeostasis. Dysordered body chemistry.

Edema: Swelling caused by abnormal accumulation of watery fluid in the body.

Electrolytes: A term referring to the levels of sodium, potassium, chloride, and carbon dioxide in the blood.

Embden-Meyerhof pathway: Glycolysis.

Endocrine: Producing secretions that pass directly into the blood or lymph instead of into a duct. *Endocrine* is in contrast to *exocrine.*

Endocrine gland: A ductless gland (such as the isles of Langerhans in the pancreas) which secretes a hormone directly into the blood.

Enzymatic: Pertaining to an enzyme or to an enzyme action.

Enzyme: A complex protein produced in living cells that is able to cause changes in other substances within the body without being changed itself. An organic catalyst, an enzyme speeds up the particular chemical reaction in which it participates. An enzyme is specific: It catalyzes one reaction and essentially no others (*see* Apoenzyme, Coenzyme, and Holoenzyme).

Eosinophil: (Also spelled "eosinophile.") A particular variety of white blood cell characterized by its uptake of eosin dye on routine staining of a blood smear. Prominent, large, red granules within the cytoplasm of the cell make the eosinophils distinctive and readily identified. Eosinophils are "normally" present in the blood and may amount to 0% to 4% of the total white-blood-cell count. Elevated levels of eosinophils are found in allergic states and when intestinal parasites are present in the body. Local collections of eosinophils (for example, in the nasal mucus) denote an ongoing allergic process in that local area.

Epinephrine: *See* Adrenalin.

Ester: Any of a class of organic compounds formed by the reaction of an acid with an alcohol, with the elimination of water.

Exocrine: Secreting outwardly, into, or through a duct. *Exocrine* is in contrast to *endocrine*.

Exocrine gland: A gland that discharges its secretion(s) through ducts. The sweat glands and the digestive portion of the pancreas are examples of exocrine glands.

Fast: To go without food.

Fasting: The state of being in a fast. Abstinence from eating food. A fasting blood sugar is a blood-sugar test obtained when the individual has not consumed food for at least 8 hours prior to the test.

Fatty acid degradation: The chemical breakdown of fatty acids into smaller component parts. This complex process is accomplished by a special form of oxidation known as beta oxidation.

Flavoprotein: An enzyme of the respiratory chain found within mitochondria. Flavoprotein is an aerobic dehydrogenase.

Folin Wu test: A laboratory method of blood-sugar determination that results in a value as much as 20 mg.% higher than the true blood-sugar laboratory method. (All blood-sugar values in this book have been determined by the true blood-sugar method.)

Frank diabetes: Clinically evident diabetes. Diabetes accompanied by readily detectable signs and symptoms of the condition.

Fructose: A monosaccharide (simple sugar) found in many sweet fruits. "Fruit sugar." A 6-carbon sugar similar to glucose but not requiring insulin for metabolism in the body. One of the 2 constitutent simple sugars of sucrose (the other being glucose).

Functional: In medical usage, this word is used in contrast to *organic*. The word *functional* implies a nonphysical cause of a disorder. A disturbance of function rather than structure. For example: "No organic cause was found and the origin of the patient's problem was presumed to be functional."

Functional hypoglycemia: A phrase used in this book to indicate blood-sugar levels that lie between 45 mg.% and 78 mg.% in the person who does not have an organic cause for low blood sugar.

Galactose: A 6-carbon simple sugar (monosaccharide). synthesized in the mammary gland to make the lactose of mother's milk. Lactose is composed of 2 galactose units.

Gastrin: A polypeptide gastrointestinal hormone that activates gastric (stomach) acid secretion. Certain pancreatic tumors may produce gastrin in pathologic amounts.

Gastrointestinal: Of or having to do with the stomach and the intestines.

Genes: Packets of hereditary material (DNA) located on chromosomes that determine the synthesis of specific protein enzymes in cells. Genes influence the inheritance and development of hereditary characteristics.

Glucagon: A protein hormone made in the alpha cells of the pancreas that stimulates the release of glucose from glycogen that is stored in the liver.

Glucocorticoids: The hormones of the adrenal cortex that have their predominant effects on carbohydrate metabolism. The glucocorticoids have an anti-insulin action — that is, they are catabolic (*see* Catabolism) and tend to elevate the levels of blood sugar, fats, and amino acids. Glucocorticoids are also anti-inflammatory substances. Hydrocortisone and corticosterone are the predominate glucocorticoids in man.

Gluconeogenesis: The formation of glucose from protein (amino acids) and, to some extent, from fat (glycerol). This process occurs predominantly in the liver.

Glucose: The type of sugar (simple sugar) carried in the bloodstream and ordinarily used by the brain for energy. When large amounts occur in the blood, glucose appears ("spills") in the urine. Glucose is also known as *dextrose.*

Glucose load: *See* Oral glucose load.

Glucose tolerance factor: Mertz's factor. A complex organic molecule containing chromium and niacin. When included in the diet, it is associated with improved carbohydrate function. Some cases of diabetes are associated with insufficient glucose tolerance factor in the diet.

Gluten enteropathy: A form of celiac syndrome (malnutrition due to malabsorption of food) due to the ingestion of gluten (an ingredient of wheat and certain other cereal grains).

Glycogen: A compound (complex) form of carbohydrate (polysaccharide). The storage form of carbohydrate in animals, equivalent to starch in plants.

Glycogenolysis: The breakdown of glycogen to glucose in the liver. This reaction takes place under the influence of a specific enzyme (glucose-6-phosphatase) present in the liver. The absence of this enzyme in the muscle makes it impossible for muscle glycogen to contribute directly to blood sugar.

Glycolysis: The breakdown of glucose within the cell to smaller chemical fragments for metabolism. The process of anerobic metabolism in which the 6-carbon glucose molecule is changed within the cytoplasm of the cell to 2 three-carbon (pyruvate) fragments and lactate. Also known as the *Embden-Meyerhoff pathway*. About 20% of the body's energy supply (ATP) is formed as the result of glycolysis. From the standpoint of evolution, glycolysis is thought to be a primitive form of metabolism, since it proceeds in the absence of oxygen. There are 12 separate chemical steps in the glycolytic pathway. It is necessary for glycolysis to proceed (carbohydrate utilization) in order for the Kreb's Cycle to proceed (fat and protein utilization).

Growth hormone: Somatotropin. A pituitary hormone that encourages growth but that also tends to elevate blood sugar.

GTT: Glucose tolerance test.

Heat labile: Easily changed by heat.

High-density lipoproteins: *See* Alpha cholesterol.

High-energy phosphate: ATP

Holoenzyme: A complete or whole enzyme. For example, a coenzyme in association with an apoenzyme (*see* Apoenzyme, Coenzyme, and Enzyme). The holoenzyme has a protein portion (the apoenzyme) and a nonprotein portion (the coenzyme).

Homeostasis: Maintenance of an internal equilibrium or steady state. Homeostasis is the overall result of a large number of interrelated biochemical actions and reactions in the body that involve negative feedback mechanisms. For example, within the body, temperature is maintained in homeostatic equilibrium — at a relative constant — because of balanced heat production and heat loss. A person works at maximal productive function in homeostasis.

Hormone: A substance secreted by certain body cells directly into the blood. A hormone affects organs, cells, or tissues elsewhere in the body. A primary messenger substance.

Hyaline membrane disease: A breathing disorder (respiratory distress) of the newly born infant, characterized by the deposition of a pink-staining membrane in the air sacs of the lungs.

Hydrogenation: The addition of hydrogen to a substance.

Hyperactive: Inappropriate motor activity. Overactive. Relative inability or unwillingness to control one's motor acts. The term usually applies to a cluster of characteristics in children, including low attention span, distractibility, poor impulse control, and easy crying.

Hyperkinetic: This term is used somewhat interchangeably with *hyperactivity. Hyperkinetic,* however, implies a more hard-core medical disorder than the somewhat more general term *hyperactivity.* The terms *hyperkinetic impulse disorder* and *hyperkinetic behavior syndrome* imply a medical syndrome due to neurological dysfunction.

Hypertensive: The state of having high blood pressure. One who has been found to have chronically elevated blood pressure (hypertension).

Hypoglycemia: Not enough sugar in the blood to meet one's energy needs.

Infarction: A portion of tissue that is dead or dying due to inadequate blood supply.

Insulin: A protein hormone secreted by the beta cells of the isles of Langerhans in the pancreas. Its function is to increase carbohydrate metabolism, glycogen storage, fatty acid synthesis, amino acid uptake, and protein synthesis. Its action in promoting the transfer of glucose across cell membranes and in forming glycogen account for the blood-sugar-lowering effect of insulin. Insulin is the principal anabolic (tissue-building) hormone in the body. The world *insulin* is derived from *insula,* the Latin word for *island.*

Insulinoma: A rare tumor of the pancreas that secretes excessive amounts of insulin with resultant hypoglycemia. An insulinoma is curable by surgical removal of the tumor.

Insult: In the medical sense, the word is used to mean injury or damage to the body or to any portion thereof.

Invert sugar: Equal parts of glucose and fructose prepared by hydrolysis (breakdown) of sucrose. The parent substance, sucrose, is dextrorotatory (it turns the plane of polarized light to the right). The product of sucrose hydrolysis (invert sugar) changes (inverts) the previous dextrorotation because of fructose which is strongly levorotatory (it turns the plane of polarized light to the left).

Ischemia: A relative lack of blood in an area. Local anemia.

Isles of Langerhans: Clusters of cells scattered throughout the pancreas which together make up the endocrine portion of the gland and which secrete the two hormones: insulin and glucagon.

"Junk" food: Food that may be filling or temporarily satisfying but which does not promote health and which may undermine it. Devitalized food. Refined or fabricated foods that contain "empty" calories. Food from which beneficial elements (vitamins, enzymes, nutritive minerals, and fiber) have been removed. Food containing additives such as refined sugar to make it sweet, preservatives to prolong shelf life, colors for eye appeal, and flavors to promote attractiveness to the buying public. Injurious food.

Keto acid: An organic acid, analagous to an amino acid, but containing a keto group instead of an amine group. Keto acids and amino acids are interchanged in the process of transamination.

Keto group: Chemical designation for oxygen connected to carbon by a double bond (a carbonyl radical) and attached to 2 univalent hydrocarbon radicals or to derivatives of these.

Ketones: Acid substances that are formed when fats are burned for energy. When the body cannot get sugar into cells to burn for energy because of a lack of insulin effect, fats are broken down and ketones accumulate in the blood and spill into the urine. One of the ketones is acetone. Acidosis often results.

Ketosis: A blood condition in which ketones, the breakdown products of fat metabolism, are more abundant than normal. Chemical substances known as ketone bodies may be detectable in the urine in ketosis. They may also be detected as a sweet, fruity odor on the breath. Acetone is a ketone.

Kinesiology: The study of movement and the muscles responsible for body tone and movement. Applied kinesiology is the clinical research and practice that investigates the function and coordination of muscles with organ function, health, and disease.

Kreb's Cycle: The principal biochemical pathway for metabolism of the breakdown products of glucose, fat, and protein. The final common metabolic pathway for the oxidation ("burning") of all major foodstuffs in which acetyl-coenzyme A is completely oxidized to carbon dioxide and, ultimately, water. The Kreb's Cycle is aerobic (oxygen-consuming) and results in the production and storage of energy in the form of ATP. Approximately 80% of the body's ATP supply is generated in the Kreb's Cycle. The pathway is also known as the *tricarboxylic acid cycle* or the *citric acid cycle*. The 8 major steps in the cycle take place within the mitochondria of cells. From an evolutionary standpoint, the Kreb's Cycle is believed to be a more recent form of metabolism than glycolysis, because it is oxygen consuming. The Kreb's Cycle cannot proceed without the prior utilization of carbohydrate in glycolysis. The cycle can be viewed as the pathway for aerobic metabolism of fats and proteins, dependent upon prior anerobic metabolism of carbohydrate (glucose).

Kwashiorkor: A severe form of malnutrition involving marked protein deficiency with swelling of the body (edema).

Labile: Easily changeable or altered (for example, by heat).

Lactate: The salt form of lactic acid.

Lactation: The formation or secretion of milk.

Lactic Acid: $C_3H_6O_3$. A principal byproduct of the muscle and red-cell metabolism of glucose. In excess, lactic acid can be responsible for pain, fatigue, and anxiety. Lactic acid is formed as an end product of glycolysis when oxygen is in short supply. Lactic acid (as lactate) is recycled in the liver (the Cori cycle) to glucose. Whenever glycolysis does not proceed because of a deficiency of vitamins and minerals, the more inefficient metabolism of lactic acid is invoked.

Lactic dehydrogenase: *See* LDH.

LDH: Lactic dehydrogenase. A zinc-dependent enzyme that acts as a catalyst for the reduction of pyruvate (*see* Pyruvic acid) to lactate under anerobic conditions. Because of this enzyme, lactic acid is formed in the body. The reverse reaction, the formation of pyruvate from lactic acid (with loss of electrons and a hydrogen ion) is an oxidation reaction and also is dependent upon lactate dehydrogenase. Elevated blood levels of LDH may be seen in liver disease, heart disease, muscle injury, or vigorous exercise.

Lean body mass: That part of the body that is not fat. It includes the vital organs such as heart, brain, lungs, gut, glands, bones, joints, nerves, skin, etc.

Lipid: Any of a group of compounds including fats, oils, waxes, and sterols. Most commonly, the word *lipid* is used to indicate fat.

Lipoproteins: Proteins that have as one of their components a lipid.

Low-density lipoproteins: *See* Beta cholesterol.

Low-level hypoglycemia: Low blood sugar that does not fall into the range of absolute low blood sugar (*see* Absolute hypoglycemia). The actual values vary somewhat according to the belief systems of physicians who interpret laboratory values. Generally, however, the term implies levels of true blood sugar between 50 mg.% and 75 mg.%.

Maltose: Malt sugar. A double sugar (disaccharide) composed of 2 glucose molecules. Sources include starch, cereals, and malt.

Metabolism: The process of building living matter from food and then using living matter for the production of energy by breaking it down into simpler substances. The synthesis or building up of living matter is known as *anabolism*. The degradation or tearing-down aspect is known as *catabolism*.

Mg.%: The number of milligrams of a substance contained within 100 milliliters of a liquid. Usually preceded by a number.

Mitochondria: The "power packs" of a cell. Cytoplasmic (*see* Cytoplasm) structures in which energy is released as the result of the aerobic Kreb's Cycle and the energy-forming respiratory chain (oxidative phosphorylation). Mitochondria contain many enzymes (*see* Enzyme) important for cell metabolism. All the useful energy formed during the oxidation of fatty acids and amino acids, and virtually all of that from the oxidation of carbohydrate is made available within mitochondria.

Monosaccharide: Simple sugar. The fundamental unit or building block of disaccharides (double sugars) and polysaccharides (compound sugars). Glucose and fructose are monosaccharides, the simplest form of carbohydrates (*see* Carbohydrate).

Morphologic: Pertaining to morphology (structure, form).

Myocardium: The heart muscle.

NAD: Nicotinamide adenine dinucleotide. A coenzyme that plays a vital role in metabolism. It is an electron and hydrogen transfer agent. NAD is the oxidized (hydrogen-lacking) form of the coenzyme. In changing to the reduced form (NADH), NAD accepts electrons and hydrogen. NAD is made up of niacinamide, 2 molecules of the pentose sugar D-ribose, 2 molecules of phosphoric acid, and a molecule of the purine base adenine.

NADH: Nicotinamide adenine dinucleotide hydrogen. A coenzyme. The reduced or hydrogen-containing form of NAD. This reduced form indicates that NAD has accepted hydrogen (H) and electrons (*see* NAD).

NADP: Nicotinamide adenine dinucleotide phosphate. This coenzyme is similar to NAD except for the presence of one additional phosphate group (a combination of phosphorous and oxygen). It interchanges with NAD. It transfers hydrogen and electrons from one substance to another.

NADPH: Nicotinamide adenine dinucleotide phosphate hydrogen. A coenzyme. The reduced or hydrogen-containing form of NADP. Reduction occurs when NADP accepts hydrogen (H) and electrons (*see* NAD).

Neonatal period: The first month of life, especially the first few days and weeks after birth.

Nesidioblastosis: Overgrowth of immature beta cells in the pancreas. Excessive insulin release with hypoglycemia may result from nesidioblastosis.

Neuroglycopenia: An insufficient supply of glucose in the nervous system.

NTT: Natural tolerance test.

Nucleic acid: A polynucleotide. A substance consisting of multiple units of nucleotides (*see* Nucleotide). Nucleic acid occurs chiefly in association with proteins in the nucleus of cells and determines genetic specificity. Includes DNA and RNA.

Nucleotide: The structural unit of nucleic acids. Determines the structure of genes. A compound of sugar, phosphoric acid, and a nitrogen base (a purine or a pyrimidine).

Oral: Pertaining to the mouth.

Oral glucose load: The administration of glucose in a single large dose, by mouth, to an individual. This is done to assess a person's glucose absorption and his metabolism of the substance. The amount of glucose given is usually based upon the person's weight (see the glucose tolerance test parts of the book).

Organelle: A minute, specialized part of a cell similar in function to an organ of higher animals. Mitochondria and ribosomes (*see* Ribosome) are organelles.

Organic: In the medical sense, this word is used in contrast to *functional*. It has to do with the underlying origin of a disorder and means that there is a tangible, physical, structural cause. For example: "A tumor was found to be the organic origin of the patient's difficulty, although before surgery her complaints were believed to be functional."

Orthokeratology: A specialized ocular science in which the cornea of the eye is straightened by the use of contact lenses.

Orthomolecular psychiatrist: A psychiatric physician who practices mental and bodily healing primarily through the use of nutritional concepts rather than the use of medications.

Oxalacetate: Also spelled "oxaloacetate." A major chemical substance of the Kreb's Cycle. Oxalacetate combines with acetyl coenzyme A to form citrate in the first reaction of the cycle.

Oxaloacetate: *See* Oxalacetate.

Oxidation: The combination of oxygen with another element to form one or more new substances. It is accompanied by the release of energy in the form of heat. In the living cell, this energy is stored or carried in the compound ATP. In older, less comprehensive terms, oxidation is the addition of oxygen or the removal of hydrogen in chemical reactions (the opposite of "reduction," which is the removal of oxygen or the addition of hydrogen). In modern chemical terms, oxidation is removal or loss of electrons (the opposite of "reduction," which is the gain of electrons). Beta oxidation is a special form of oxidation in which fatty acids are broken down to acetyl-coenzyme A by successive removal of 2-carbon fragments.

Oxidative phosphorylation: Tissue respiration (use of oxygen) accomplished in mitochondria, accompanied by the formation of high-energy phosphate bonds. Oxidation coupled with phosphorylation (the addition of a phosphate group — a combination of phosphorous and oxygen — to a substance). This is a stepwise, efficient, and controlled way to allow the energy from food to become available in the form of heat and ATP (energy storage).

Pancreas: A gland at the rear of the abdomen, near the stomach, that has both exocrine and endocrine functions. The former is accomplished by secretion of digestive juice into the small intestine. The latter is accomplished by the secretion of insulin and glucagon to regulate carbohydrate metabolism.

Pancreozymin: Also known as *cholecystokinin-pancreozymin.* A polypeptide gastrointestinal hormone that stimulates the secretion of pancreatic enzymes and gall bladder contraction.

Pathogenic: Producing disease.

Peptide: Any combination of amino acids in which the carboxyl group of one acid is joined with the amino group of another.

pH: A symbol used (with a number) to express acidity or alkalinity of body liquids. It represents the logarithm of the reciprocal of the hydrogen-ion concentration (in gram atoms per liter) in a given solution, usually determined by the use of an indicator substance known to change color at a certain concentration. The pH scale in common use ranges from 0 to 14, pH 7 being taken as neutral; below 7, increasingly acid; above 7, increasingly alkaline. The pH of the blood is normally 7.4, the saliva about 6.8, the urine ranges from 5 to 8.

Phagocyte: A white blood cell capable of absorbing and destroying waste or harmful material, such as disease-producing bacteria or small foreign particles in the body.

Phenylketonuria: PKU. An inborn error of protein metabolism. An enzyme defect that results in the accumulation of phenylalanine, a product toxic to the nervous system of the young child when present in excessive amounts.

Phosphorylation: The addition of a phosphate group (a combination of phosphorous and oxygen) to a substance.

Photosynthesis: The process by which plant cells make carbohydrate from carbon dioxide and water in the presence of chlorophyll and light.

PKU: *See* Phenylketonuria.

Placenta: The afterbirth. The organ by which the fetus is attached to the inner wall of the uterus (womb) and nourished.

Plasma: The liquid part of the blood in which the blood cells float. The fluid part of the blood before clotting.

Polyneuritis: Multiple neuritis.

Polypeptide: A compound containing 2 or more molecules of amino acids and one or more peptide groups.

Polysaccharide: A compound (complex) sugar made up of many repeating units of monosaccharides (*see* Monosaccharide) linked together in long chains. Starch in plants and glycogen in animals are polysaccharides. They are storage forms of carbohydrate.

Proinsulin: The first form of insulin manufactured by the beta cells of the pancreas. The 80 amino acid precursor (forerunner) of insulin, containing the C-peptide fragment.

Protein: A complex compound containing nitrogen that is a necessary part of the cells of animals and plants. Protein is composed of amino acids (*see* Amino acid) which unite, with a loss of water, to form peptide chains. The peptide chains are linked to form protein.

Purine base: A chemical substance involved in the structure of nucleotides. Adenine and guanine are purine bases.

Pyrimidine base: A chemical substance involved in the structure of nucleotides. Cytosine, thymine, and uracil are pyrimidine bases.

Pyruvate: *See* Pyruvic acid.

Pyruvic acid: A 3-carbon fragment, an end product of glucose breakdown in glycolysis. One 6-carbon glucose molecule gives rise to two 3-carbon pyruvate molecules. (For purposes of this book, the acid pyruvic and the salt pyruvate are used interchangeably.)

Radical: An atom or group of atoms acting as a unit in chemical reactions.

Reducing agent: A chemical substance that reduces or removes the oxygen in a compound.

Reduction: A chemical reaction in which each of the atoms or groups of atoms affected gains one or more electrons. The atom or group of atoms that loses electrons becomes oxidized.

Respiratory chain: A series of catalysts (*see* Catalyst) within mitochondria that are concerned with the transport of reducing agents (hydrogen and electrons) and with their final reaction with oxygen to form water. Electrons flow through the chain in a stepwise manner from the more electronegative components to the more electropositive substance, oxygen. The main respiratory chain proceeds from NAD-linked dehydrogenases (*see* Dehydrogenase) through flavoproteins to cytochrome enzymes to oxygen.

Respiratory distress syndrome: Difficulty breathing in the newly born infant. Hyaline membrane disease is one form of respiratory distress in newborns.

Ribosome: A structure (organelle) in the cytoplasm of the cell that carries on the synthesis of proteins.

RNA: Ribonucleic acid, a nucleic acid.

Secretin: A polypeptide gastrointestinal hormone that stimulates the secretion of water and bicarbonate by the pancreas. Historically, the first substance to be identified as a hormone.

Serum: The clear, pale-yellow, watery part of the blood that separates from the clot when blood coagulates. The fluid part of the blood that remains after clotting.

SGOT: Serum glutamic-oxaloacetic transaminase (*see* Transamination). An enzyme present in liver and heart. An elevation of SGOT in the blood signifies damage to cells in one or both of these organs.

SGPT: Serum glutamic-pyruvic transaminase (*see* Transamination). An enzyme present in liver. An elevation of SGPT in the blood signifies damage to the cells of the liver.

Simple sugar: *See* Monosaccharide.

Skeletal muscle: The type of muscle that is attached to bone and that is responsible for maintenance of tone and movement of the body, its head, and its limbs. Skeletal muscle is under voluntary control. It is known histologically (under microscopic scrutiny) as striated muscle because of the characteristic presence of striae (furrows or ridges) that distinguish it from nonstriated, nonvoluntary, smooth muscle such as found in the gut.

Sludging: A condition of the blood in which the blood is so altered that it does not flow freely through the blood vessels.

Smooth muscle: The type of muscle found in certain body organs (for example, the gut) that is regulated by the autonomic nervous system. Smooth muscle is not ordinarily under voluntary, conscious control.

Starch: A complex carbohydrate (polysaccharide) found in plants. An important ingredient of food. Reacts with certain enzymes to form sugars. Composed of amylopectin (80% to 85%) and amylose (15% to 20%).

Stroke volume: The amount of blood ejected by the heart in one beat.

Subcutaneous: Under the skin. A route used for the introduction of some substances into the body.

Sublingual: Under the tongue. A route used for the introduction of some substances into the body. Large blood vessels close to the surface of the oral mucous membrane make the sublingual route effective for absorption of some medications.

Substrate: The material upon which an enzyme acts.

Sucrose: A double sugar (disaccharide) whose component simple sugars (monosaccharides) are glucose and fructose. Sucrose is found in nature in certain foods such as sugar cane, beets, carrots, and fruits where it is accompanied by other nutrients needed to metabolize it in the body. Sucrose is the whole ingredient of refined white sugar, a substance that lacks the necessary accompanying nutrients for metabolism in the body. Sucrose as table sugar is the ultimate in "naked" or "empty" calories.

Supine: Lying flat on the back.

Surfactant: A substance that exerts its effect on a surface to lower surface tension. A detergent is a surfactant.

Synergistic: Cooperative action such that the total combined effect of 2 substances is greater than the sum of the 2 effects taken independently.

Synthesis: The combining of parts or elements into a whole.

Systemic: Affecting the body as a whole. Involving the general body system. Generalized within the body as opposed to a local effect or condition.

Systolic: The phase of blood pressure related to the contraction of the heart (systole) as opposed to *diastole* (*see* Diastolic). The first or upper number when a blood-pressure value is reported: "The patient's blood pressure was 120 over 80." The figure 120 represents the systolic blood pressure.

Thyroid gland: An important ductless (endocrine) gland in the neck that affects growth and metabolism through the secretion of thyroid hormone.

Titrate: A method of determining the amount of a substance needed to perform some function by the process of trial and error. The addition of progressively increasing amounts of a substance until a desired effect is obtained.

Transamination: The interconversion (change) of a pair of amino acids (*see* Amino acid) and a pair of keto acids. Pyridoxal phosphate (the coenzyme form of vitamin B_6) forms an essential part of the active site of transamination. The process is utilized in protein synthesis and degradation. It is catalyzed (*see* Catalyst) by enzymes. (*See* SGOT, SGPT.)

Triglycerides: A major form of fat in the body. They are also known as *triacyglycerols* or *neutral fats*. Triglycerides are esters of the alcohol glycerol and fatty acids. The blood level of triglycerides can be measured as an index of metabolic well-being.

Univalent: Having a valence of 1.

Valence: The quality of an atom or radical that determines the number of other atoms or radicals with which it can combine, indicated by the number of hydrogen atoms with which it can combine or which it can displace.

Wernicke-Korsakoff syndrome: A degenerative disease of the nervous system characteristically occurring in malnourished, chronic alcoholics. Symptoms include polyneuritis, confusion, disorientation, memory loss, confabulation, eye movement disorders, ataxia, tremor, coma, or obtunded mental state. The origin of the disorder is closely related to malnutrition. The B vitamins, especially thiamine, are of paramount importance in therapy.

Whiskey fit: Slang expression used to denote a convulsive seizure occurring in habitual alcoholics.

Wok: Chinese cooking pan. The wok's curved bottom facilitates stirring and tumbling of foods so they are easily mixed and exposed to little heat.

Work load: A unit of energy required to perform a certain task.

REFERENCES

Amino Acid Metabolism and Its Disorders by Charles R. Scriver, M.D., C.M., F.R.S. (C.), and Leon E. Rosenberg, M.D. Published in 1973 by W. B. Saunders Company, Philadelphia, Pa.

Biochemistry (Fourth Edition) by Abraham Cantarow, M.D., and Bernard Schepartz, Ph.D. Published in 1967 by W. B. Saunders Company, Philadelphia, Pa.

Body, Mind, and Sugar by E. M. Abrahamson, M.D., and A. W. Pezet. Published in 1951 by Avon Books, New York, N.Y. 10019.

Diabetes: A Scope Monograph on the Nature, Diagnosis, and Treatment of Diabetes Mellitus. Published in 1965 by Upjohn Company, Kalamazoo, Mich.

Diabetes, Coronary Thrombosis, and the Saccharine Disease (Second Edition) by T. L. Cleave, M.R.C.P., and G. D. Campbell, M.D., F.R.C.P., with the assistance of N.S. Painter, M.S., F.R.C.S. Published in 1969 by John Wright & Sons, Ltd., Bristol, England.

Diabetes Mellitus (Seventh Edition). Published in 1967 by Lilly Research Laboratories, Indianapolis, Ind.

Diet and Disease by E. Cheraskin, M.D., D.M.D., W. M. Ringsdorf Jr., D.M.D., M.S., and J. W. Clark, D.D.S. Published in 1970 by Rodale Books, Emmaus, Pa.

Eating Right for You by Carlton Fredericks, Ph.D. Published in 1972 by Grosset and Dunlap Publishers, New York, N.Y.

Implications of Hyperglycemia. Published in 1972 by Pfizer Laboratories Division, Pfizer, Inc., New York, N.Y. 10017.

Insta-Tape Recordings of the International Academy of Metabology, Inc., Fall Seminar 1973.

Low Blood Sugar by Peter J. Steincrohn, M.D. Published in 1972 by Signet, 1301 Avenue of the Americas, New York, N.Y. 10019.

Low Blood Sugar, A Doctor's Guide to Its Effective Control by J. Frank Hurdle, M.D. Published in 1969 by Parker Publishing Company, Inc., West Nyack, N.Y.

Low Blood Sugar and You by Carlton Fredericks, Ph.D., and Herman Goodman, M.D. Published in 1969 by Grosset and Dunlap Publishers, New York, N.Y.

Low Blood Sugar, The Hidden Menace of Hypoglycemia by Clement G. Martin, M.D. Published in 1971 by ARC Books, Inc., 219 Park Avenue South, New York, N.Y. 10003.

Management of Juvenile Diabetes Mellitus by Howard S. Traisman and Alvah L. Newcomb. Published in 1965 by C. V. Mosby Company, St. Louis, Mo.

The Poisons Around Us by Henry A. Schroeder, M.D. Published in 1974 by the Indiana University Press, Bloomington, Ind.

Predictive Medicine, a Study in Strategy by E. Cheraskin, M.D., D.M.D., and W. M. Ringsdorf Jr., D.M.D., M.S. Published in 1973 by Pacific Press Publishing Association, Mountain View, Calif.

Psychodietetics: Food as the Key to Emotional Health by E. Cheraskin, M.D., D.M.D., and W. M. Ringsdorf Jr., D.M.D., M.S., with Arline Brecher. Published in 1974 by Stein and Day, New York, N.Y.

Psycho-Nutrition by Carlton Fredericks, Ph.D. Published in 1976 by Grosset and Dunlap Publishers, New York, N.Y.

Review of Nutrition and Diet Therapy by Sue Rodwell Williams. Published in 1973 by C. V. Mosby Company, St. Louis, Mo.

Review of Physiological Chemistry by Harold A. Harper, Ph.D. Published in 1975 by Lange Medical Publications, Los Altos, Calif.

Scope Manual on Nutrition by Michael C. Latham, M.D., Robert B. McGandy, M.D., and Frederick J. Stare, M.D. Published in 1970 and 1972 by the Upjohn Company, Kalamazoo, Mich.

The Story of Medicine by Arthur L. Murphy, M.D., C.M. Published in 1954 by Airmont Publishing Co., Inc., New York, N.Y. by arrangement with Thomas Bouregy and Company, Inc.

Textbook of Endocrinology (Fifth Edition), edited by Robert H. Williams, M.D. Published in 1974 by W. B. Saunders Company, Philadelphia, Pa.

The Trace Elements and Man by Henry A. Schroeder, M.D. Published in 1973 by the Devin-Adair Company, Old Greenwich, Conn.

THE AUTHOR

Ray C. Wunderlich Jr. was born in St. Petersburg, Florida, August 11, 1929. He attended schools in St. Petersburg, graduating from St. Petersburg High School in 1947. He received his B.S. degree from the University of Florida in 1951 and received his M.D. degree from Columbia University College of Physicians and Surgeons in 1955. He served a 2-year rotating internship at Strong Memorial Hospital in Rochester, New York, from 1955 to 1957, with major fields of study in medicine, surgery, pediatrics, and psychiatry.

From 1957 to 1959, Dr. Wunderlich was in military service. He held the rank of Captain, Medical Corps, and was stationed at Tyndall Air Force Base, Panama City, Florida.

Dr. Wunderlich received pediatric residency training from 1959 to 1961 at Strong Memorial Hospital, and he was the recipient of a Wyeth Fellowship.

Since 1961, Dr. Wunderlich has been in private practice. A pediatrician from 1961 to 1975, he now specializes in preventive medicine for people of all ages. In order to serve his patients better, he often consults with professional colleagues in related fields of learning.

In addition to a large practice and the considerable writing that he does, Dr. Wunderlich is called on often for lectures throughout the United States and in Canada. In 1979, he lectured extensively in Australia.

Dr. Wunderlich has written numerous articles in professional journals.

He also is the author of the books *Kids, Brains, & Learning* and *Allergy, Brains, & Children Coping;* the *Wunder-Forms;* and the booklets *Explanatory Notes to Accompany Wunder-Form #11, Improving Your Diet,* and *Fatigue* (What Causes It, What It Does to You, What You Can Do About It).

Dr. Wunderlich's wife, Elinor, a native of Canada and a registered nurse, is the author of the book, *Easy Whole-Food Recipes.* Dr. and Mrs. Wunderlich and their four children, Mary, Janet, Ray, and David, live in St. Petersburg.

INDEX

Abrahamson, E. M., 287, 290
Absolute low blood sugar, 280
Academy of Orthomolecular Psychiatry, 302, 412
Accelerated maturity and sugar consumption, 41-42
ACE, 121, 123, 127, 134-39, 343
Acetohexamide, 335, 339; use of, in treating diabetes, 209, 210
Acetylcholine, 152
Acetyl-coenzyme A, 98, 100, 102
Acid stools, causes of, 81
Acinar cells, 146
Acne and sugar consumption, 41-42
Acromegaly and development of diabetes, 183
ACTH (adrenocorticotropin hormone), 105, 113, 114, 116, 117; deficiency of, 122-23, 324; and drug-induced diabetes, 203; presence of high levels, 124
ACTH stimulation, 117
Addison's disease, 120, 121, 125, 326
Adenoma, 327
Adenosine monophosphate (AMP), 114
Adenosine triphosphate (ATP), 68, 87, 92
Adrenal cortex, 127; deficiency in, 79-80
Adrenal-cortical extract, 348
Adrenal-cortical hormones, 86
Adrenal crisis, 120
Adrenal glands, 101, 108, 119, 136
Adrenal hemorrhage, 119, 120
Adrenal hormones: effect of, on blood sugar, 113
Adrenalin, 15-16, 42, 101, 108, 231
Adrenalin insufficiency, 327
Adrenal-medullary hormones, 118
Adrenal Metabolic Research Society of the Hypoglycemia Foundation, 120, 124, 138, 302
Adrenocortical deficiency, 326
Adrenocortical extract (ACE), 121, 123, 127, 134-39, 343
Adrenocortical hormones, 108, 117, 118, 179, 183, 347

Adrenocortical insufficiency, 118-34, 406
Adrenocortical response, 117
Adrenocorticoid-insulin imbalance in hypoglycemia, 185-86
Adrenocorticotropin: see ACTH
Adrenogenital syndrome, 120
Adult diabetes, 175
Adverse stress, 300
Aerobic, 91
Aerobic conditioning exercises, 419-20
Aggressive behavior and sugar consumption, 41-42
Airola, Paavo, 347
Alanine, 323
Albumin, 319
Alcoholism: and hypoglycemia, 331-32; and vitamin deficiency, 95
Aldosterone, 124
Alexander, David S., 122
Alimentary hyperinsulinism, 330-31
Alkaline phosphatase test, 333
Allergic adrenalitis, 119
Allergic stress, 285
Allergy, 126-27; and development of diabetes, 192-93; and hyperactivity, 417; and sugar consumption, 41-42, 49-51; treatment of, 121-22
Allergy, Brains, and Children Coping, 21, 123, 417
Alopecia, 119
Alpha cells, 145
American Diabetes Association, 215
American Natural Hygiene Society, 347
Amino acids and hypoglycemia, 323
Amino acid sweeteners, use of, 36
Aminophylline, 42
AMP (adenosine monophosphate), 114
Amplification cascade, 348
Amylase enzyme, 79, 82
Amylopectin, 79
Amylose, 79
Anabolism, 107, 152
Anerobic, 91
Anerobic glycolysis, 91, 102
Antacids, 145